SHEET METAL DRAFTING

Second Edition

MELVIN L. BETTERLEY Professor, Engineering Graphics, and Director, Engineering Publications Office, The State University of Iowa.

GREGG DIVISION
McGRAW-HILL BOOK COMPANY

New York	Düsseldorf	Paris
St. Louis	Johannesburg	São Paulo
Dallas	London	Singapore
San Francisco	Madrid	Sydney
Auckland	Mexico	Tokyo
Bogotá	Montreal	Toronto
	New Delhi	
	Panama	

Library of Congress Cataloging in Publication Data

Betterley, Melvin L
   Sheet metal drafting.

   Includes index.
   1. Sheet-metal work—Pattern-making. 2. Mechanical drawing. I. Title.
TS250.B4   1978   604'.2'671   77-2764
ISBN 0-07-005126-7

SHEET METAL DRAFTING, SECOND EDITION

Copyright © 1978, 1961 by McGraw-Hill, Inc. All rights reserved. Printed in the United States of America. No part of this publication may be reproduced, stored in a retrieval system, or transmitted, in any form or by any means, electronic, mechanical, photocopying, recording, or otherwise, without the prior written permission of the publisher.

1234567890 KPKP 783210987

The editors for this book were Don Hepler and George McCloskey, the designer was Tracy A. Glasner, the art supervisor was George T. Resch, and the production supervisor was Regina Malone. It was set in Times Roman by Monotype Composition Company, Inc.
Printed and bound by Kingsport Press, Inc.

CONTENTS

Preface — v
Preface to the Second Edition — vi
1 Introduction — 1
2 Instruments and Materials — 4
3 Instrument Technique — 19
4 Lettering — 31
5 Geometric Constructions — 41
6 Orthographic Projection — 60
7 Auxiliary Views — 71
8 Rotation — 91
9 Parallel-line Development — 97
10 Radial-line Development — 143
11 Triangulation — 201
12 Combination Problems — 249
13 Dimensioning — 274
Appendix — 290
Index — 310

PREFACE

The draftsman, because of the rapid and highly technical progress of industry, has been required to take over, solve, and supervise many of the operations and problems previously handled by engineers. This demand for craftsmen with additional skills and knowledge has influenced many to seek supplementary training. This book was written with these students in mind and as the result of seeing the need for such a book in the author's classes. It was written to train draftsmen—with special emphasis directed at all times toward the sheet metal draftsman; however, all draftsmen should benefit from its use. Information about metals, machines, tools, processes, seams, and allowances was purposely omitted so that the emphasis could be placed on drawing.

The text is designed for courses in technical institutes, trade and vocational schools, and adult evening classes. It should also be particularly useful in apprentice and industrial training programs. The textbook covers all phases of drawing: techniques, projection principles, and descriptive geometry, integrated and applied whenever possible to situations in the sheet metal field. It is not, however, a mere collection of problems and their solutions; on the contrary, it concentrates upon basic principles which can be applied to various situations as they arise. Draftsmen who are well grounded in these principles will be more valuable and hence better-paid workers because they will be able to analyze and solve problems of varied character.

Care has been taken to train the worker to know how to handle more than certain developments and layouts. That is why we have avoided describing how to do a drawing by merely following step-by-step directions. On the contrary, when procedures have been described, the logic behind them has been discussed. The student should therefore consider the information in this book as basic knowledge to be acquired in the same way as shop skill and applied to ever-changing situations. Part of the student's most valuable training will result from a careful study and full comprehension of words and technical terms.

The first eight chapters concern the basic elements of drafting and will be useful to draftsmen in all industries. They also serve as preparation for the sheet metal applications in the chapters on parallel-line, radial-line, and triangulation development. Each chapter is followed by a series of exercises from which the teacher and the student may select practice material. These practice problems are a vital part of the learning process. They have been selected so that they may be solved on American standard-size drawing paper (11 by 17). Some are model size and can be drawn full scale; others, taken from industry, are full size and will have to be drawn to a reduced scale. It is hoped that the student will gain a feeling of accomplishment and professional assurance from working these problems.

The author wishes to express thanks to the many sheet metal manufacturers and industries whose valuable photographs are reproduced in this book. Special thanks are due the late Professor Emeritus H. C. Spencer of the Illinois Institute of Technology for the initial inspiration and encouragement in the preparation of the material, and to the late Professor Emeritus John M. Russ of the State University of Iowa, for encouragement and advice. The author also wishes to express his appreciation to his colleagues at the State University of Iowa for their valuable contributions.

It is the intention and hope of the author that this textbook will become a valuable reference in the classroom and in the drafting office, and that as it is put to use, the user will become a better and more versatile worker in the sheet metal drafting field.

Melvin L. Betterley

PREFACE TO THE SECOND EDITION

The responses to and the successes of the first edition have been very gratifying to the author. Therefore this second edition follows very closely the format and contents of the first edition. Additional topics have been added, material has been updated, and of course some errors are now corrected. The major area of update logically is in regard to measurement. With the United States adopting the international system (SI) the text and the illustrations have been rewritten or changed to conform. Measurements in the text are given in United States customary units (inches) with the SI units (millimeters) in brackets. Specific measurements in the drawings are given in each drawing in one of three different modes as follows: USC; USC with SI in parentheses; SI only. Thus the student can use the text and illustrations to adjust normally but rapidly to the SI system.

The author has received correspondence, questions, and suggestions from throughout the world regarding the first edition and would welcome continued interest and comments regarding the second edition.

Melvin L. Betterley

chapter 1

INTRODUCTION

Sheet metal drafting is an activity associated with many different industries, and the efforts of sheet metal draftsmen are registered in all branches of human endeavor. For example, consider the large and ever-growing heating, ventilating, and air-conditioning industry, of which sheet metal drafting is a vital part. The efforts of this industry are applied to literally hundreds of diverse fields, such as architecture; the aircraft, automotive, railroad, and shipbuilding industries; and farming (see Figure 1-1). "Sheet metal" is generally defined as any metal under ¼ inch [6 millimeters] thick (metal greater than ¼ in [6 mm] thick is commonly called "plate"), but sheet metal drafting principles are also applied to many heavier constructions, for example, water towers, radio telescopes, reactors, band shells, etc. (see Figure 1-2). It is apparent, therefore, that sheet metal draftsmen occupy a commanding position in the general industrial picture, and the demands for their services are increasing.

The exact type of drawing performed by a person in one industry may be quite different from that done by someone in another; likewise, the activities of a person within a certain company may be quite varied and, hence,

Fig. 1-1 Blowpipe systems and fittings. (*Quinn Bros. Inc., Philadelphia, Pa.*)

Fig. 1-2 Type 302 20-gauge [1-mm] stainless steel dome. (*Pittsburgh Auditorium Authority, Pittsburgh, Pa.*)

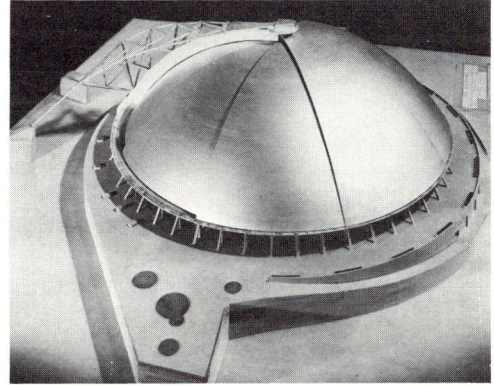

highly demanding of knowledge and skill. While the activities in the sheet metal trades are varied, three main areas of interest may be established for concentration of study by the student or trainee.

First, the area of *design and application*. While a great amount of design is highly theoretical (involving heat transfer, fluid mechanics, and other engineering sciences), some design and application can and should be acquired through practical experience and problem study, using basic mathematics and established procedures. The student draftsman should seek a broad general knowledge. Thus, design situations, wherever and whenever applicable, have been indicated in the following chapters.

Second, the area of *layout and development*. It is in this area that the sheet metal draftsman should excel. Again, the situations from industry to industry may vary: the drafting may take place in the engineering offices, or the layout and development may be performed in the shop. If the layout is done in the drafting room and is to be used for a number of constructions, the patterns are likely to be made of paper, or are made full size on paper for transfer to the metal: if only one or two assemblies are to be constructed, the layout and development would probably take place in the shop, directly on the metal. Development on paper has two advantages: (1) the patterns may be tested for accuracy before cutting the metal and (2) the pieces may be planned for cutting location on the sheet, both advantages resulting in the waste of less metal (see Figure 1-3). Regardless of the place of work, however, the *function of the sheet metal draftsman in all industries is to lay out and develop patterns, templates, and dimensional working drawings which will accurately result in a finished product in the shop.* The above-stated purpose cannot be overemphasized because there is a tendency for students and teachers to consider the drawing as an end in itself. Actually, the *drawing is concerned not only with the beginning of an idea but also must be used in its application and construction.*

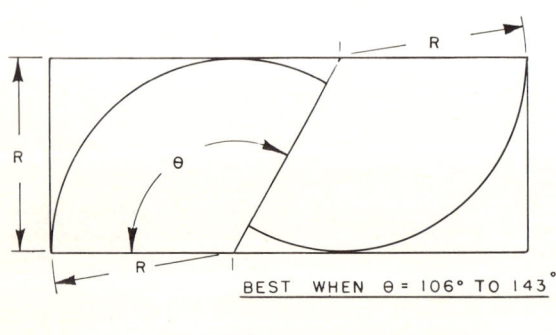

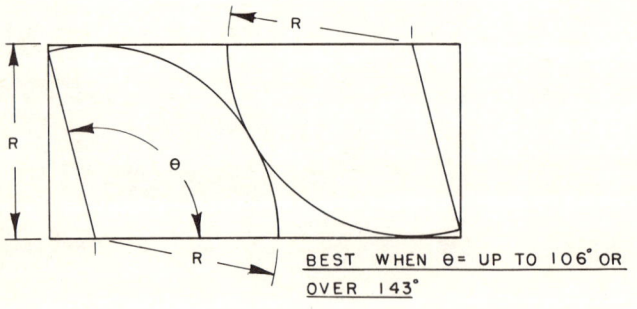

Fig. 1-3 Placement planning for cutting.

INTRODUCTION

Third, the area of *construction* (see Figure 1-4). Obviously, a knowledge of construction materials, tools, and processes is of value to the sheet metal draftsman and should also be acquired through study and practice. The chapters of this book, however, are concerned primarily with design, layout, and development. Thus they may also be useful to the worker in the shop, who is often confronted with design or layout situations for which a handbook is helpful.

Fig. 1-4 Lock seaming. (*Bethlehem Steel Co., Bethlehem, Pa.*)

The international system of units (SI) has begun to replace the United States customary system (USC). Therefore the SI measurements are given along with the USC in the written portion of this text. The base SI units used in measurement of length are the meter (m) and the millimeter (mm), which is one-thousandth of a meter.

The problem illustrations and the specific dimensions of the drawings are given in both the customary (USC) or foot-inch system and the equivalent SI dimensions (SI is tabulated on drawing). If only one or two dimensions are involved, the SI equivalent is parenthesized.

Some problems have only the SI system shown. This is to encourage the student to "think metric" or more correctly to think SI, since the international system does differ from the old European metric system regarding some base units.

It is recommended that SI be used at all times possible and the U.S. customary system not be used unless practical problems or construction constraints require its continuance.

chapter 2

INSTRUMENTS AND MATERIALS

The drafting equipment shown in Figure 2-1 is used by draftsmen in all branches of drawing and therefore is suitable for the student of sheet metal drafting. Small problems or problems drawn to a reduced scale offer the draftsman excellent opportunities to become skillful with this equipment. In shops where large layouts are made directly on the metal and on template material, the worker can apply the skill and techniques learned to larger equipment used later.

The drafting equipment in Figure 2-1 consists of the following items:

1. Drawing board
2. T square
3. 45-degree triangle
4. 30-60-degree triangle
5. Drawing pencils
6. Pencil sharpeners and pointers
7. Erasers
8. Scales
9. Drawing paper
10. Paper fasteners
11. Drawing instruments
12. Irregular curve
13. Protractor

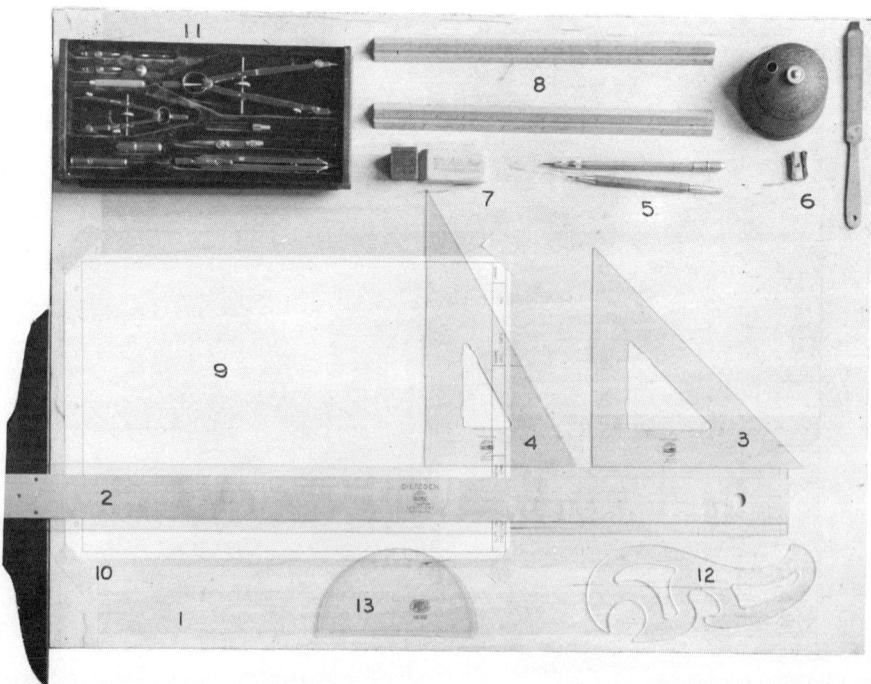

Fig. 2-1 Required drafting equipment.

2-1 Drawing boards

Drawing boards can be made of any smooth close-grained wood, masonite, metal, or other hard, smooth composition surface. Before the extensive use of Scotch drafting tape, boards had to be of soft wood so that the paper could be fastened with thumbtacks. Wood drawing boards should be constructed from several narrow boards glued together so that the annular-ring curvature of each board is reversed to minimize warping. Tongue-and-groove cleats should be used at each end to increase sturdiness and provide a smooth, *straight* working edge. Most draftsmen prefer to cover the board with a backing paper, oilcloth, or linoleum. The backing provides an excellent working surface, gives increased uniformity in pencil touch, and improves line quality. The most suitable drawing board for classroom use is 18 by 24 inches [457 by 610 millimeters] in size. In commercial practice and sheet metal layout, larger boards and tables are usually needed (see section 2-3).

2-2 T squares

T squares, as shown in Figure 2-2, are usually made of hardwood and consist of a head and a blade. The higher-quality T squares have celluloid edges dovetailed to the wooden blade to prevent warping, maintain a straight edge, and afford visibility. The head and blade of the T square should be fastened together with small screws and glue and should be *absolutely rigid*. Some T squares have been manufactured entirely of plastic. These afford increased visibility but their blades tend to be more flexible than the wooden ones; however, they are not as easily broken. The specified length of a T square is the distance from the working edge of the head to the tip of the blade. The most common length used by students is 24 in [610 mm], to match the board. All T squares should be tested occasionally to ensure a *straight* working edge on the blade. To make the test, connect two points with a line, using the working edge as a guide. Now turn the blade bottom side up and connect the points, again using the working edge. The lines should coincide. If any inaccuracy exists, the amount of error is equal to one-half the space between the lines. Inaccuracies may be corrected on T squares and triangles by carefully filing the edges.

2-3 Parallel devices

Parallel-rule drawing boards, tables, and other devices (with or without drafting machine heads) are available. A typical board with drafting machine head is shown in Fig-

Fig. 2-2

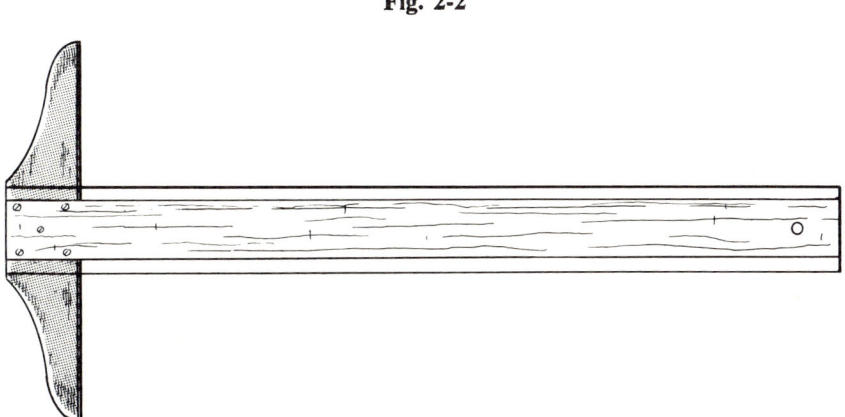

ure 2-3. Adjustable bars, or cables and pulleys make complete coverage of the drawing area possible. The use of the sliding rule or the machine head eliminates the need for T square, triangles, and other tools.

Parallel-rule mechanisms are a great convenience to the sheet metal draftsman because of the large working areas involved and the ease with which a large drawing may be constructed. Most sheet metal drawings involve many parallel and perpendicular lines and these are easily constructed over the whole area.

Fig. 2-3 A parallel device. (*Charles Bruning Co., Mt. Prospect, Ill.*)

2-4 Triangles

The two triangles specified in section 2-1 are the 45- and the 30-60-degree triangles. Both are right-angle triangles; therefore, the specifications refer to the acute angles, as shown in Figure 2-4 (see section 5-1).

High-quality triangles are made of plastics. A 10-in [254-mm] triangle would be considered medium size and is approximately $3/32$ in [2 mm] thick. The specified size of a triangle refers to the longest leg of the right angle. The largest-size triangles available in plastic are 18 in [457 mm]. Some wooden strip triangles are made up to 24 in [610 mm]. If the sheet metal draftsman has occasion to use triangles larger than 18 in [457 mm], they can be made in the shop from sheet stock (preferably aluminum). Large triangles can also be fabricated from heavier-gauge strap.

2-5 Drafting machines

The drafting machine is a device that combines in a single unit the functions of a T square, triangles, protractor, and scales. A typical drafting machine is shown in Figure 2-5. Any of the above drafting features are immediately available by manipulation of the machine with the left hand (left-handed machines are available on special order).

Some of the advantages of drafting by machine are as follows:

Fig. 2-4 Triangles.

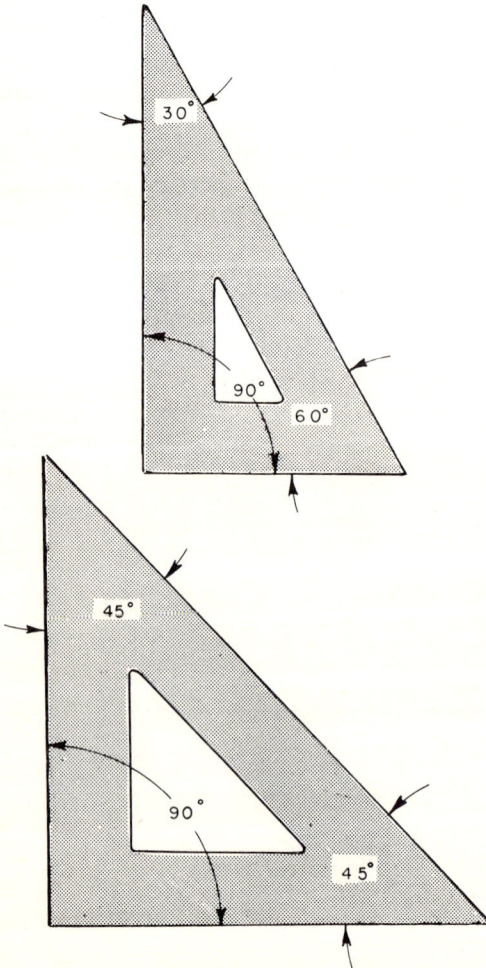

INSTRUMENTS AND MATERIALS

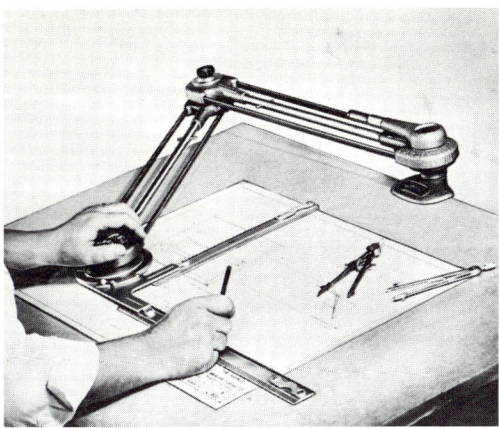

Fig. 2-5 Drafting machine. (*V. and E. Manufacturing Co., Pasadena, Calif.*)

a. Amount of equipment needed is reduced.

b. Multiple function is selected more quickly than by searching for separate items.

c. Work area left free of stray equipment. The one disadvantage of drafting machines is the high initial cost. This is compensated for by the fact that the operator of a machine soon acquires speed and, with the above advantages, saves a great deal of time.

Drafting machines are available in various sizes with arms from 24 to 36 in [610 to 914 mm] long. The larger size may be used on a table up to 4 by 10 feet [1.2 by 3 meters]. For larger drawings, vertical drafting machines are available which are mounted on stainless steel tracks at the top of large vertical drawing boards. The construction of the machine is such that the protractor and T-square blades are accurately available over the entire board area. These machines cover any size of board from 2 to 11 ft [.6 to 3.3 m] in height and any length. They are especially valuable in the automobile and aircraft sheet metal industries.

Fig. 2-6 Drawing pencils.

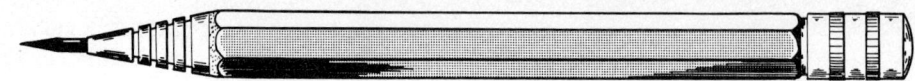

Fig. 2-7   Refill pencil.

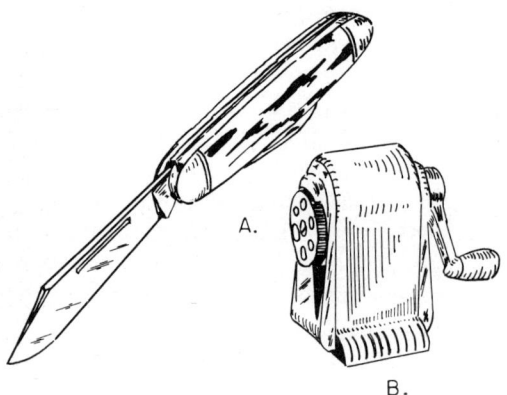

Fig. 2-8   Pencil sharpeners.

2-6   Drawing pencils

Drawing pencils are *special pencils* designed and manufactured to meet the needs of draftsmen. They differ from ordinary writing pencils in that the selection of materials and the manufacturing meet higher standards. This results in higher cost; however, excellent drafting can only be done with top-quality pencils.

Drawing pencils are manufactured in 17 degrees of hardness ranging from 9H (very hard) to 6B (very soft). The complete range of pencil leads is shown in Figure 2-6. The extremely hard pencils are used only in cases of great accuracy and for fine, light lines. The soft pencils are used by artists and architects for pencil sketches, shading, and the rendering of perspective drawings. The selection of the proper pencil for the job at hand is discussed in section 3-3. Drawing pencils are made hexagonal in shape to provide comfortable gripping surfaces and to prevent the pencil from rolling off the drawing table.

Among professional draftsmen and increasingly among students, the refill pencil shown in Figure 2-7 is replacing the wooden pencil. The refill pencil is constructed of a hollow plastic, wood, or metal body and adjustable metal jaws or chuck. The lead is inserted and locked in position with $\frac{3}{8}$ to $\frac{1}{2}$ in [10 to 13 mm] exposed. This is pointed in the conventional manner. (See Figure 3-3.)

The refill pencil is a worthwhile investment in convenience, cleanliness, and time saved.

2-7   Pencil sharpeners and pointers

Typical pencil sharpeners are shown in Figure 2-8 A,B. The older method of sharpening the drawing pencil is with a pocket knife. This technique is described in section 3-4.

Fig. 2-9   Pencil pointers.

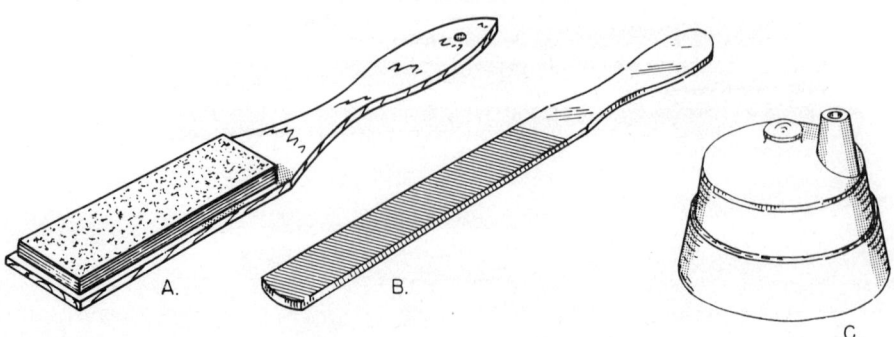

In drafting rooms where wooden pencils are used in great quantity, a special draftsman's sharpener is usually provided. Such a sharpener cuts only the wood at the proper taper and back about 1½ in [40 mm], exposing ⅜ in [10 mm] of lead.

The lead is pointed by using one of the items shown in Figure 2-9 *A, B, C*. The sandpaper pad (*A*) is the most common. The file, shown at (*B*) is used in the same manner as the sandpaper pad. The latter is preferred because it produces a smoother point than the former. The pointer (*C*) is a device that points the pencil lead at the proper angle simply by inserting the pencil and using it to rotate the head of the pointer, the pointer base remaining stationary. Used in conjunction with refill pencils, the draftsman can quickly produce the type of point needed without whittling or messiness. The grindings are collected in a metal cup and the pointer can be opened and cleaned occasionally.

2-8 Erasers

Erasers for draftsmen consist of three basic types, namely, pencil, ink, and Artgum.

The draftsman doing only pencil work is interested primarily in the pencil and Artgum erasers. Pencil erasers are made in several shapes, textures, and colors, as shown in Figure 2-10.

The best way to select a good pencil eraser is to try several different brands, textures, and colors on the type of paper most used. Some factors to keep in mind when testing are (1) the ease with which lines are removed cleanly and completely, (2) the amount of damage to the surface of the paper, and (3) the presence or lack of color smear since certain erasers, especially those containing synthetic rubber, will discolor some papers.

The Artgum eraser is made of a soft, crumbly rubber that is used by the draftsman principally to clean up large areas and to remove smears or messiness caused by dirty equipment.

Most ink erasers are compounded to a firmer texture than pencil erasers and also contain a fine abrasive. In selecting an ink eraser, care must be exercised to avoid getting one that is too harsh for the cloth or paper being used.

2-9 Scales

The sheet metal draftsman is usually concerned with making drawings full, half or quarter size. Layouts made directly on metal are full size. This is done with a carpenter's

Fig. 2-10 Typical erasers. (*Weldon Roberts Rubber Co., Newark, N.J.*)

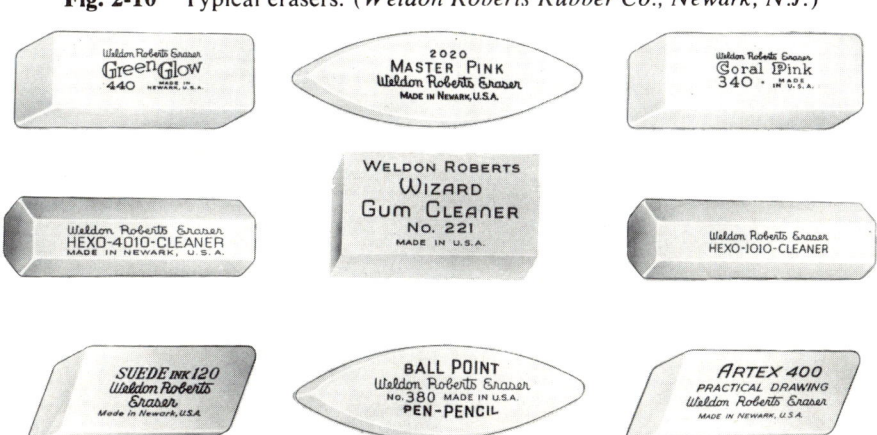

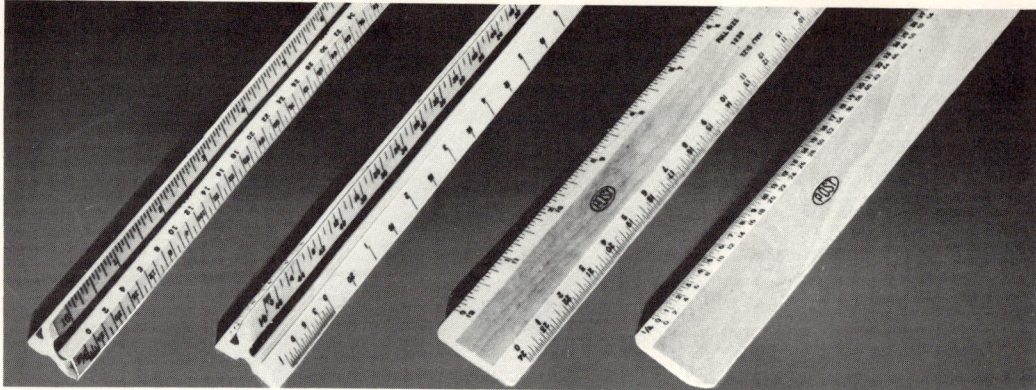

Fig. 2-11 Scales. (*Frederick Post Co., Chicago, Ill.*)

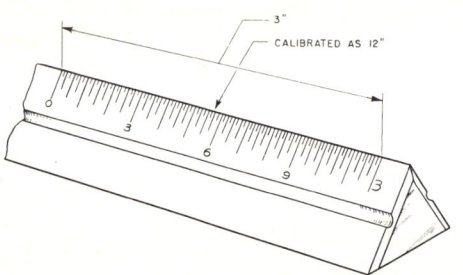

Fig. 2-12 Quarter-size measurement.

square, folding rule, circumference rule, or steel tape. When half, quarter-size, or one-fifth-size drawings are to be made on the drawing board, the most suitable measuring tools are the mechanical engineer's, architect's, or metric scales (Figure 2-11).

On the architect's scale, a convenient measurement is taken as a basic unit. This basic unit is then used to represent one foot on the drawing. It is subdivided into twelve parts, each representing an inch on the draw-

ing. These are divided so as to represent parts of the inch. In this manner, an object of considerable size may be represented on the limited area of a drawing. For example, in Figure 2-12, 3 inches actual size has been selected as the basic unit to represent 1 foot. The basic unit has been divided into twelve parts representing inches and the inches subdivided. Thus the smallest division represents $\frac{1}{8}$ inch on the drawing. In Figure 2-12, the scale selected is, therefore, 3 inches = 1 foot-0 inch. The drawing is said to be one-quarter size. A measurement of 1 foot-7$\frac{5}{8}$ inches is shown quarter size in Figure 2-13.

Using metric measurement, the nearest ratio similar to the one-quarter-size architectural scale would be 1:5 (one-fifth size). A convenient metric measurement is taken as a basic unit. This basic unit is then used to represent 100 mm on the drawing. It is subdivided into tenths and twentieths. Thus

Fig. 2-13

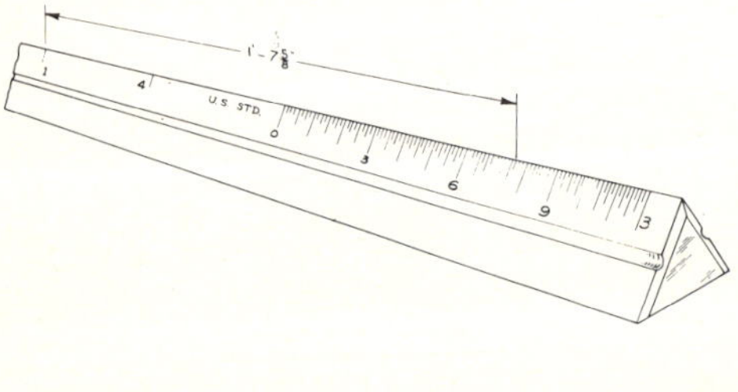

Fig. 2-14

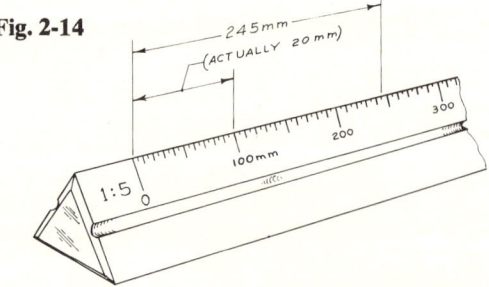

the subdivisions represent 10 and 5 mm, respectively. To use the scale, a larger-sized object can be represented on the limited area of the drawing. For example in Figure 2-14, 2 cm (20 mm) actual size has been selected as the basic unit to represent 100 mm. Thus the smallest division on that scale represents 5 mm.

Most scales are *open divided,* that is to say that the basic units are numbered along the entire length of the scale but only one end unit is subdivided. If the basic units are subdivided for the entire length, then the scale is known as a *full-divided* or *chain scale*.

The standard triangular architect's scale contains six faces calibrated for eleven scales as follows:

12 inches	(each inch subdivided to sixteenths) (full size)
3 inches = 1 foot-0 inch	(quarter size)
1½ inches = 1 foot-0 inch	(eighth size)
1 inch = 1 foot-0 inch	(twelfth size)
¾ inch = 1 foot-0 inch	(sixteenth size)
½ inch = 1 foot-0 inch	(twenty-fourth size)
⅜ inch = 1 foot-0 inch	(thirty-second size)
¼ inch = 1 foot-0 inch	(forty-eighth size)
3/16 inch = 1 foot-0 inch	(sixty-fourth size)
⅛ inch = 1 foot-0 inch	(ninety-sixth size)
3/32 inch = 1 foot-0 inch	(one hundred twenty-eighth size)

For making half-size drawings (6 inches = 1 foot-0 inch), the architect's scale is unsatisfactory because the full-size scale must be used, each dimension divided by two, and the result used in constructing the drawing. For example, in Figure 2-15, 5⅝ inches is shown half size as it would be drawn with the full-size architect's scale. The mechanical engineer's scale is better for making half-size drawings.

On the mechanical engineer's scale, *the inch is represented by one-half or one-fourth inch calibrated to represent parts of the basic unit*. Thus ½ inch = 1 inch means a half-size drawing; ¼ inch = 1 inch means a quarter-size drawing. Engineer's scales are manufactured flat with either two or four beveled edges and are calibrated as follows:

Full size: (12 inches, each unit subdivided to sixteenths)
Half size: (½ inch represents an inch on the drawing) (subdivided to sixteenths)
Quarter size: (¼ inch represents an inch on the drawing) (subdivided to eighths)

As an example, Figure 2-16 shows the measurement 6⅞ inches taken from the half-size scale. From the preceding information, notice that the mechanical engineer's scale is designed for drawing regular sized objects to scale, while the architect's scale is better for reduced scales and large objects.

Architect's scales are usually manufactured with a triangular section, but are also available flat with beveled edges. Engineer's

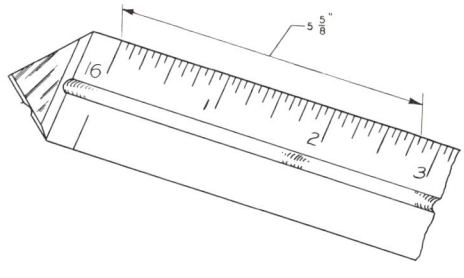

Fig. 2-15 Half-size measurement.

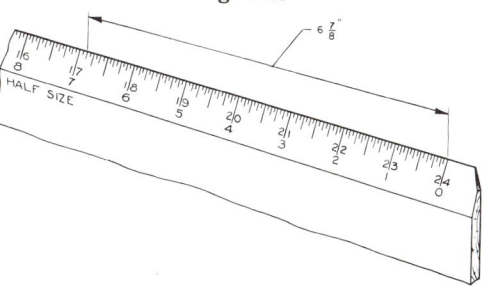

Fig. 2-16

scales are usually flat with two or four beveled edges. The higher-quality scales, both architect's and engineer's, are engine-divided and made of boxwood or plastic.

Metric scales are manufactured in both the flat and triangular form and consist of combinations of the following:

Divided and numbered in millimeters
Scale 1:1; 1:2; 1:5; 1:10; 1:20; 1:50; 1:100; 1:200; 1:500; 1:1000; also 2:1 (double size).

Metric scales do not have the shortcomings mentioned for architect's and engineer's scales. While the scales are divided and identified in millimeters, they can be used to make reduced drawings for small objects (1:2, 1:5) or for large objects (1:50; 1:100) dimensioned in either small or large units.

2-10 Drawing paper

Because of the development of materials more suitable to present drafting room practices, the amount of drawing paper used by industry and the schools is decreasing yearly. It is possible to make excellent reproductions of drawings from penciled work on tracing papers and cloths. This results in a saving in time and number of drawing surfaces needed. The practice of making pencil drawings directly on transparent materials was increased at the beginning of World War II, and had a profound effect upon tooling and increased war production schedules.

Certain draftsmen, however, are very much concerned with drawing papers and their characteristics. For example, patent draftsmen must work on a white high-quality paper; illustrators for technical publications draw upon special papers. In some industries, drawings are made on drawing paper for master filing. These papers are made from a high percentage of rag stock. They are tough, have a hard sized surface with desirable grain, and are usually white, cream, or buff in color. While suitable for sheet metal drawings, they are too expensive in many cases, owing to the large amounts and sizes needed. Detail papers made from manila and other jute stocks are available, which are tough and take pencil lines well. These are satisfactory for sheet metal drawing and are used for stencils, patterns, templates, etc., by patternmakers, modelmakers, stone cutters, ornamental iron and glass workers, and clothing designers as well.

2-11 Tracing papers and cloths

These papers are made of 100 per cent rag stock and may or may not be made more transparent by treatment with wax, oil, or other transparentizing solutions. When treated to become more transparent, they are commonly called vellum. They are used for pencil drawings and the drawings may be reproduced by Ozalid, blueprinting, or other photographic contact printing methods. Tracing papers are used in sheet metal drawing whenever a working drawing of an installation or fabrication must be reproduced for the shop.

Tracing cloth is a fine-thread, closely woven cloth (usually muslin or linen). The cloth is sized with starch to provide body and to give a good drawing and inking surface. Many special tracing cloths are available. Some are intended for pencil only while others may be used for either pencil or ink. Some are highly water-resistant, but most tracing cloths are easily damaged by water. A tracing cloth, woven of glass fibers, is especially transparent. Its most important characteristic, however, is its dimensional stability. It is affected less by moisture or temperature changes than most other suitable materials, including most metals. This feature makes contact printing directly upon the metal practical. The metal is painted or sprayed with a light-sensitizing solution. After exposure with the tracing in contact, the reproduction is developed by dipping or

INSTRUMENTS AND MATERIALS

spraying. This cloth is being used extensively in the aircraft and die tool industries.

2-12 Paper fasteners

There are three types of paper fasteners in general use, namely, thumbtacks, wire staples, and drafting tape.

Thumbtacks are the least desirable for the following reasons: They limit the draftsman to a soft board, they damage the working area of the board if different-sized papers are used, they sometimes obstruct the use of T square and triangle, and they are sometimes difficult to remove.

Wire staples are desirable, especially if the work is to be attached to the board for a considerable length of time. Wire staples are used to advantage in fastening Boardex, oilcloth, or linoleum as a drawing foundation.

Drafting tape is convenient for short-term drawings. The tape has a sticky coating which does not dry out easily; hence it may be re- moved without damaging the paper and it can be used several times. It may be used on any type of surface, which is advantageous when making drawings on hard smooth surfaces. It is available in rolls of standard widths and may be placed in a dispenser. The technique of fastening the paper is discussed in section 3-2.

2-13 Set of instruments

Several American and German manufacturers make high-grade drawing instruments. These are used for engineering drawing and are suitable for many of the operations in sheet metal pattern drafting. The quality and merits of various makes of instruments are difficult for the beginner to judge. It is therefore wise to ask the advice of a teacher, an experienced draftsman, or other engineering acquaintance before investing a considerable amount in a set. The price of drawing instruments, like that of all other products de-

Fig. 2-17 Required drawing instruments. (*Frank Oppenheimer, Denver, Colo.*)

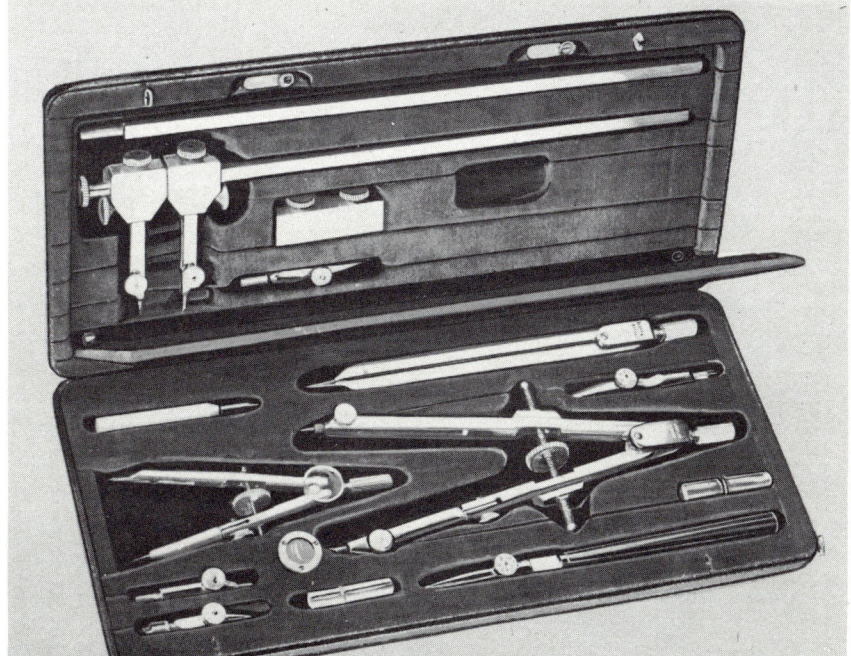

14 SHEET METAL DRAFTING

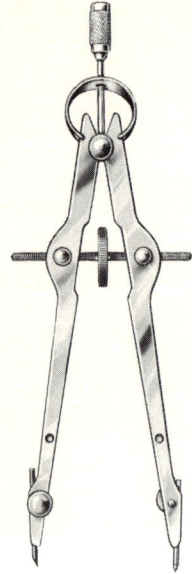

Fig. 2-18 Center-wheel bow. (*V. and E. Manufacturing Co., Pasadena, Calif.*)

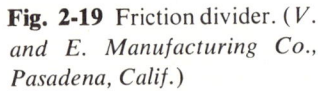

Fig. 2-19 Friction divider. (*V. and E. Manufacturing Co., Pasadena, Calif.*)

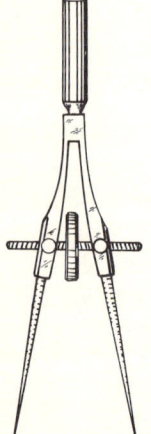

Fig. 2-20 Typical bow divider.

manding craftsmanship in their manufacture, is directly proportional to their quality. Good instruments are a sound investment for the sheet metal draftsman. Draftsmen disagree as to what instruments the ideal set should contain and, of course, the type of drawing being done to a large extent controls the instruments most needed. Students should start with a high-quality basic set and, as experience indicates and pocketbook allows, add items to speed their work and make it more enjoyable. A basic set for sheet metal drafting, which is made by several reputable manufacturers, could consist of the following instruments:

    a. A 6½-in [165-mm] center-wheel bow compass

    b. A 6-in [152-mm] friction divider

    c. A 4-in [102-mm] bow (combination pencil, pen, divider)

    d. A beam compass (pen attachment)

    e. A ruling pen

A typical set of instruments, attachments, and case is shown in Figure 2-17. Each instrument is discussed in detail in the following sections.

2-14 Center-wheel bow compass

Since the making of blueprints directly from pencil drawings has become general practice, the large center-wheel bow compass is the favorite of draftsmen in industry. A typical instrument is shown in Figure 2-18. The instrument may be used to make arcs and circles from $\frac{1}{4}$ to 10 in [6 to 254 mm] in diameter. The center-wheel adjustment provides the rigidity needed in order that pressure be applied to make dense lines without causing a change in adjustment. In addition to rigidity, the instrument selected for purchase should be checked for ease of adjustment. The spring tension, the threads, and the knurled adjustment screw should provide easy one-handed operation. The tension should be sufficient to maintain any setting and a slight difference in feel should be apparent when changing from increased to a decreased setting. The technique of using the instrument is discussed in section 3-8.

2-15 Friction divider

The friction divider is one of the easiest to operate and most useful pieces of equipment in the set of instruments. It is also one of the most abused; so remember that it is a measuring device and should be treated with the respect and care that all fine tools are given. The divider is shown in Figure 2-19. The friction between the legs of the instrument should maintain adjustment and yet be smooth enough to permit one-handed operation (see section 3-9). The points of a high-quality instrument will stay in alignment and can be brought to a zero setting.

2-16 Bow divider

The bow divider is shown in Figure 2-20. A good instrument has all of the characteristics previously mentioned for the large divider and the instrument is especially valuable to the student sheet metal draftsman because it can be used to establish, transfer, and repeat a fixed measurement. Instrumental technique is discussed in section 3-10.

2-17 Beam compass

Because of the large arcs and circles often needed in sheet metal drawing, the beam compass is an important item in the set of instruments (Figure 2-21). Many set manufacturers now include a beam compass in their sets or a space is provided in the case where the instrument can be added when needed.

The beam compass consists of two "heads" and a bar. The heads may be fitted with a needle point, a pencil point, or a pen. They are adjustable by sliding on the bar and, in addition, a vernier adjustment is provided on one head so that accurate settings can be made. Most beam compasses consist of a 6-

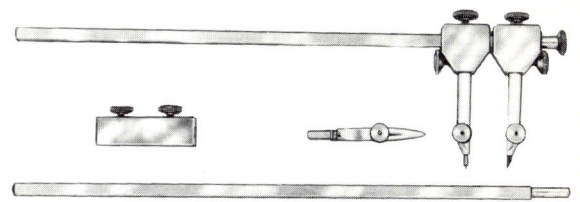

Fig. 2-21 Beam compass. (*Frank Oppenheimer, Denver, Colo.*)

and 10-in [152- and 254-mm] beam, with a coupling to make up a 16-in [406-mm] beam. Thus a diameter of approximately 32 in [810 mm] may be obtained. The beams are usually square, rectangular, or hexagonal, and should be of such a cross section that longer ones can be substituted directly from standard shop bar stock. Some drafting rooms have "trammel points" which are used in the same manner except that they clamp and slide on long wooden bars. The use of the beam compass is discussed in section 3-11.

2-18 Ruling pen

The practice of inking drawings, since just before World War II, has been reduced considerably—yet the ruling pen must remain as a necessary part of the instrument set. Every draftsman will be called on, at some time, to produce quality ink work. A typical ruling pen is shown in Figure 2-22.

The ruling pen is made of hardened steel, ground and polished. It should have a handle that fits the hand, giving the instrument balance. The adjusting screw should be within easy reach of the fingers for one-handed adjustment. The threads should provide easy adjustment yet be rugged enough to avoid stripping. Many pens, as received from the manufacturer, are dull or are ground to a long point. In order to secure a fine line or

Fig. 2-22 Ruling pen. (*V. and E. Manufacturing Co., Pasadena, Calif.*)

before the ink will start to flow readily, these need to be "touched up" on an oilstone. In checking a ruling pen, the following faults may be observed:

a. Dull nibs will make fine lines impossible. Open the nibs and hone the tips on the *outside only*. Maintain a semicircular shape on the tips and sharpen both nibs to the same length.

b. A tip shape that is *too long and sharp* will make the pen difficult to start because the ink is reluctant to flow downward between narrow tips. By means of the stone, change the tips to a more semicircular shape.

c. The pen must be leaned sideways in order to start a line if the nibs are of different lengths. Upon bringing the pen back to vertical after starting the line, one edge of the line becomes ragged. "Joint" the nibs to the same length by holding and stroking the pen perpendicular to the stone surface. This will produce an extensive bright, flat spot on the long nib. Do this until the short nib begins to touch the stone. Now sharpen and shape each nib until the bright spot *just* disappears.

d. If the line floods easily, either the draftsman is attempting to overload the pen with ink or the tip of the pen is shaped too wide and blunt. Try decreasing the pen load or else narrow and elongate the tip shape by stoning.

e. Even a properly shaped and sharpened pen will not start easily or make fine lines if it is dirty, or again if an old bottle of thickened ink is used. Wipe the pen nibs clean and dry after use; do this even when pausing between lines and settings for any length of time. Always use fresh ink.

2-19 Proportional dividers

The proportional divider, as shown in Figure 2-23, is a special drawing instrument not regularly included in the set of instruments. The proportional divider may be of value to sheet metal draftsmen, as with it, patterns at ratios from 1:1 to 1:10 can be made from drawings and developments made to reduced scale. Lines, planes, circles, and solids are easily copied at any of the above ratios. The instrument may be used to increase or decrease, in the prescribed proportion, any length, area, or volume.

The instrument is expensive and should be added to the personal kit only after sufficient experience has indicated need for it. Because of the pointed ends, the instrument is dangerous and should be closed and put away when not in use.

In selecting a proportional divider, the draftsman should examine the instrument for quality workmanship and material. Preferably, the pivot carriage should have a lock screw on one side and, on the other, a knurled knob for moving the carriage by

Fig. 2-23 Proportional divider. (*The Eugene Dietzgen Co., Chicago, Ill.*)

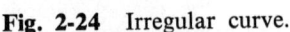

Fig. 2-24 Irregular curve.

Fig. 2-25 Typical adjustable curves.

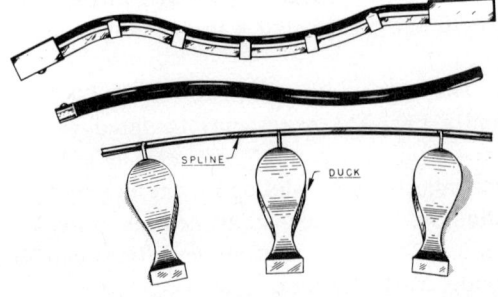

means of a rack and pinion. Proportional dividers are sold with instruction leaflets indicating the proper use for each of the scales.

2-20 Irregular curves

Curved rulers or *French curves* are used to produce curved lines mechanically. They are made of plastic, in various sizes and thicknesses, and are shaped according to portions of ellipses, spirals, hyperbolas, and other mathematically computed curves. A typical curve is shown in Figure 2-24. Sheet metal draftsmen use irregular curves to produce the patterns for elbows, intersections, openings, etc. The technique of using French curves is discussed in section 3-14.

2-21 Adjustable curves

Adjustable curves may be shaped to give the desired line. Typical adjustable curves, as shown in Figure 2-25, consist of a ruling edge attached to a lead strip, steel or lead strips embedded in rubber, or a flexible strip known as a "spline," held in position by means of heavy lead weights called "ducks." Splines and ducks are used for long curves in the aircraft and shipbuilding industries. The technique for adjustable curves is similar to that of French curves, as discussed in section 3-14. The entire curve may be drawn at one time, however, since the curve, being adjustable, fits the entire distance.

2-22 Protractor

The protractor is a metal or plastic device, semicircular or circular in shape, with index lines crossing at a 90-degree vertex. The protractor is calibrated in degrees and is used to establish or measure angular magnitudes (see Figure 2-26). The student should exercise care in selecting a good protractor—one which is large enough and accurately calibrated (preferably 8 in [200 mm]). Avoid the small, thin, inexpensive metal varieties.

2-23 Electronic slide rule calculator

The electronic slide rule, as shown in Figure 2-27, has replaced the mechanical slide rule as the accepted symbol of the engineering profession. Many skilled workers in other professions and trades, however, also find the instrument a valuable time-saver. The sheet metal draftsman may use the device to advantage in solving problems involving diameters, circumferences, surface areas, and container volumes. Problems involving addition, subtraction, multiplication, division, squares and square roots, and cubes and cube roots are easily performed on the electronic calculator. Engineering calculators also handle reciprocals, trigonometric functions, and logarithms which can be of value to sheet metal draftsmen.

The instruments are usually sold with an instruction manual which includes practice problems.

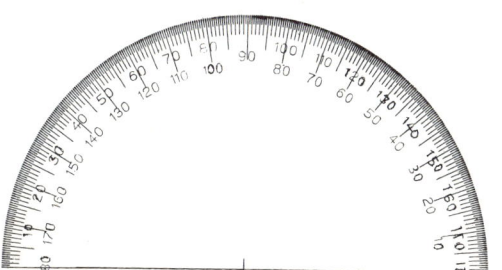

Fig. 2-26 Protractor.

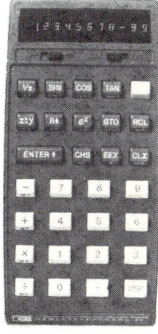

Fig. 2-27 Electronic slide rule. (*Hewlett-Packard, Cupertino, Calif.*)

2-24 Dusting brush

A dusting brush, as shown in Figure 2-28, is a necessary part of the equipment of all draftsmen. The brush is used to sweep away erasings, or any other material that might soil the drawing if allowed to remain or might smear if removed in any other manner.

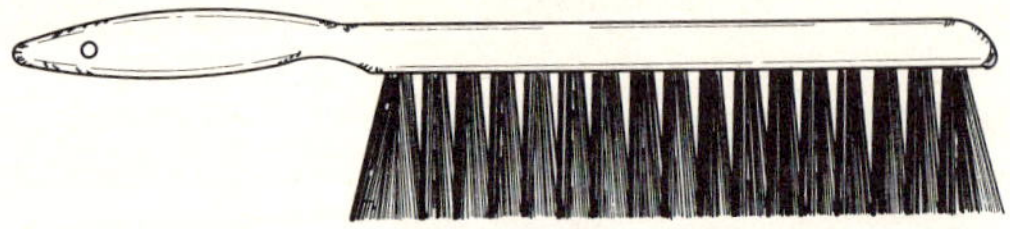

Fig. 2-28

2-25 Sketch pad

A sketch pad of 8½- by 11-in [216- by 279-mm] paper, while not an essential item, is a convenience at the drawing table. The pad offers a ready clean surface for sketching views, jotting down specifications, and making computations in developing problems.

EXERCISES

1. Why do most draftsmen prefer a backing material other than the board surface under the drawing?
2. What equipment items are replaced by the parallel-rule drawing board?
3. What are the advantages of using a drafting machine?
4. How do drawing pencils differ from ordinary pencils?
5. What is a refill pencil?
6. For what purpose is the Artgum eraser intended primarily?
7. What do the scales designated 3, 1½, ¾, etc., on the architect's scale, mean?
8. How many scales does the mechanical engineer's scale contain and what are they?
9. What type of fastener is preferable for fastening paper to a hard surface for a short time?
10. What instruments are considered necessary in a good basic set?
11. What instrument is used to make arcs and circles of 6- to 16-in [152- to 406-mm] radii?
12. For what is a protractor used?
13. If an electronic calculator is available, determine the amount of 16-gauge galvanized iron required to construct a box 1'-0" by 2'-0" by 4'-0" [305 by 610 by 1 220 mm], the box to open opposite the largest surface. What is the approximate weight of the box in pounds? In kilograms?
14. What scale in the metric system would most closely approximate the architectural scale ¾" = 1'-0" (one-sixteenth size)?

chapter 3

INSTRUMENT TECHNIQUE

In Chapter 2, the instruments and materials required were discussed relative to their descriptive characteristics, desirability, cost, and other similar factors. In this chapter, the instruments and materials of Figures 2-1 and 3-1 are considered as they would actually be used.

3-1 Placement of equipment

In order to obtain neat, accurate, and fast drafting skill, one of the first requirements is a definite and logical arrangement of the equipment. The worker who works on a messy, cluttered-up workbench wastes time and makes costly mistakes. The same holds true for drafting. The draftsman should have a place for every item and methodically return each piece to its proper place as soon as he or she finishes using it. The beginner can acquire good housekeeping habits by placing a sheet of wrapping paper or drawing paper on the table, next to the drawing, arranging the tools efficiently, marking the arrangement, and following it in practice. A suggested arrangement is shown in Figure 3-1.

3-2 Fastening the paper

The paper is fastened to the drawing board with thumbtacks, metal staples, or drafting tape. Scotch drafting tape offers the handiest and most practical method because it leaves the drawing board smooth and free of holes, is easily removed, and holds the paper tightly without injury. For the right-handed draftsman, who uses the T square with the head on the left edge of the board, the paper should be placed within $1\frac{1}{2}$ inches [38 millimeters] of that edge and at least 4 in [102 mm] up from the bottom edge of the board. This location permits the T-square blade to be used near the bottom of the sheet and lessens the effects of pencil leverage against it when drawing on the right half of the paper. For left-handers, the arrangement should be reversed, the paper *remaining* well up from the bottom of the board.

3-3 Selection of pencil

Draftsmen generally prefer pencils from 6H to HB. This is a medium-hard range and

Fig. 3-1 Equipment arrangement.

Fig. 3-2 Preferred pencil range.

is described in Figure 3-2. The student in sheet metal drafting, working on average quality drawing paper, may try a 6H for sharp, accurate, extremely light construction lines; an H for finished object and pattern outlines, and an F for lettering. The primary consideration in the selection of the pencil is the type of work being done and the type of line wanted. Not all draftsmen, however, will select the same pencil for a specified line since each will have individual preference, different paper texture, and different brand pencils.

3-4 Sharpening the pencil

The wood is removed from the unlettered end of the pencil and cut away in a tapered manner. The cuts begin about 1½ in [40 mm] from the end and exposes the lead for approximately ⅜ in [10 mm] as shown in Figure 3-3.

Notice the proper way to hold the pencil and knife. Care must be taken to avoid cutting into the lead (see section 2-7). The conical point on the lead is produced by rubbing the point on the sandpaper pad or file; or with the special pencil pointer (Figure 2-9). Most draftsmen prefer the refill pencil since all cutting is avoided and leads can be extended as needed and kept properly pointed at all times (section 2-6).

3-5 Drawing horizontal lines

When drawing horizontal lines, the right-hander holds the pencil as shown in Figure 3-4. The stroke is made from left to right with the pencil perpendicular to the plane of the paper and slanted at about 60 degrees in the direction of the stroke. As the stroke is made, the pencil is rotated in order to maintain a uniform-width line and to keep the

Fig. 3-3 Sharpening a pencil.

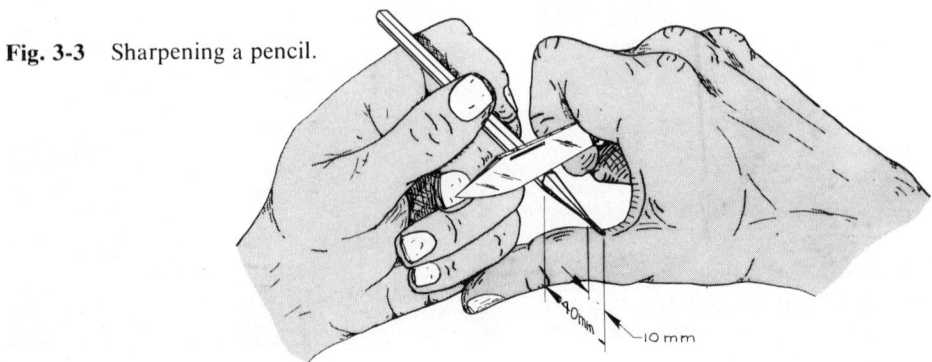

INSTRUMENT TECHNIQUE

pencil tip sharp. The student should practice drawing horizontal lines uniformly spaced and study the results. They should be *uniform, clean cut,* and *black*.

3-6 Drawing vertical lines

Vertical lines are drawn upward, as shown in Figure 3-5. Either triangle, supported by the T square as shown, serves as a vertical ruling edge (the left-handed draftsman reverses the triangle). Again the pencil is held perpendicular to the paper and slanted at about 60 degrees in the direction of the stroke. The pencil is rotated between the fingers as the stroke is made. The student should practice holding the T-square head against the working edge of the board, the triangle on the working edge of the T square, and drawing vertical lines upward, uniformly spaced. They should be of the same quality as those made horizontally.

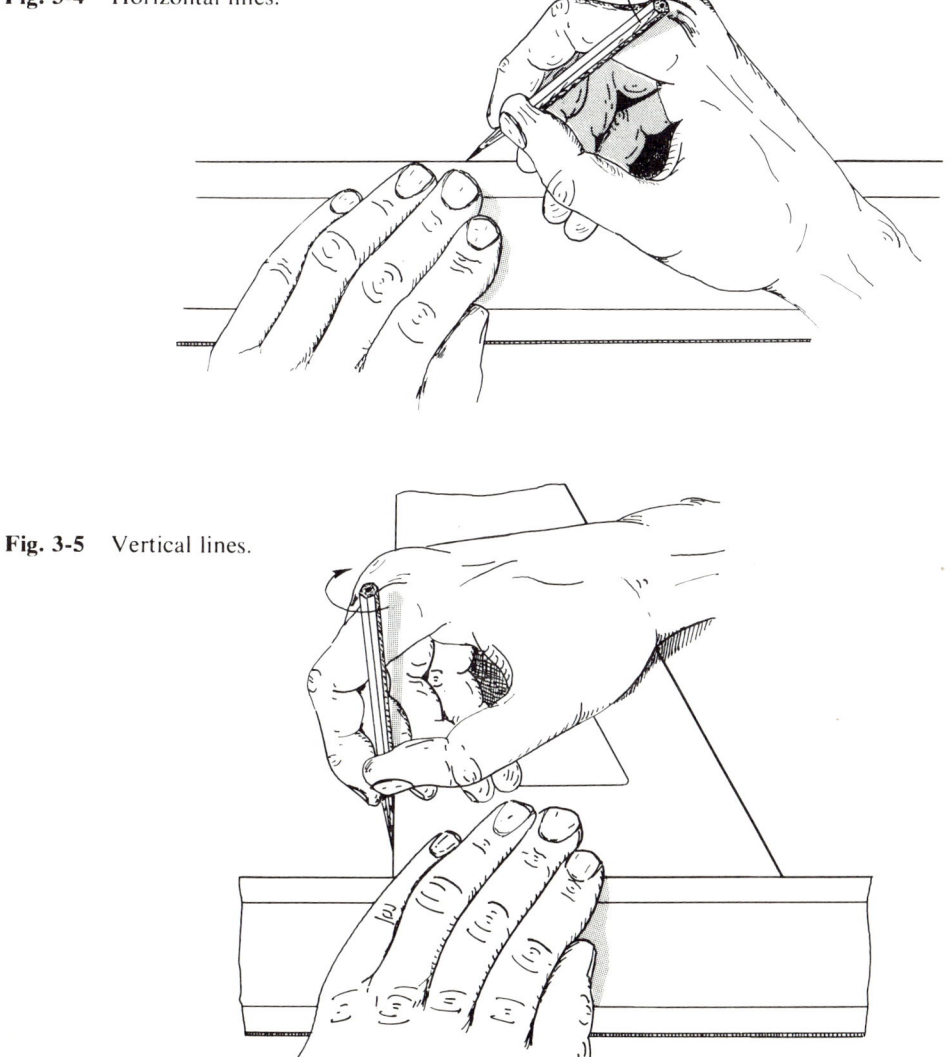

Fig. 3-4 Horizontal lines.

Fig. 3-5 Vertical lines.

3-7 Drawing inclined lines

The hypotenuse of the 45-degree triangle is used, as shown in Figure 3-6A to draw lines at 45 degrees to horizontal or vertical. Likewise, the 30-60-degree triangle, as shown in Figure 3-6B, and C, is used to draw lines at 30 and 60 degrees to horizontal or vertical. Notice, in each case, the proper direction for the stroke.

By combining triangles, additional angles of 75 and 15 degrees may be drawn. Thus twelve lines, spaced at 15-degree intervals, are possible. The lines are drawn in the directions shown in Figure 3-7. Combined triangles are shown in Figure 3-8. The student will find it easier to combine triangles if it is observed that the hypotenuses are in contact, thus the 30- and 45-degree angles add up to 75 degrees. This basic combination can be rotated 90 degrees or "flopped" 180 degrees to obtain the 15- or 75-degree angle to either horizontal or vertical, as shown.

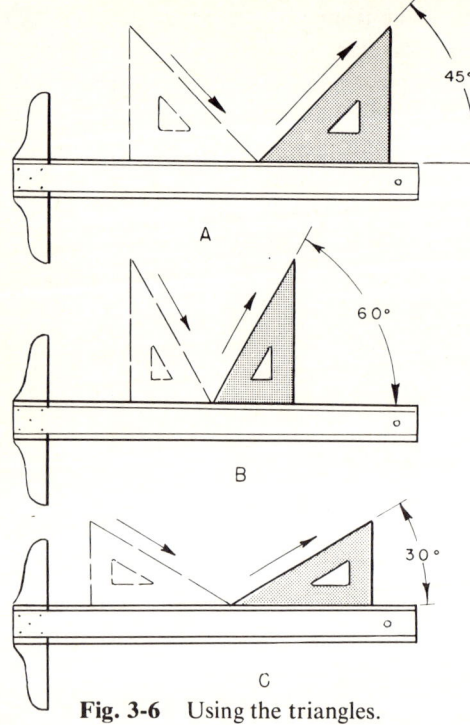

Fig. 3-6 Using the triangles.

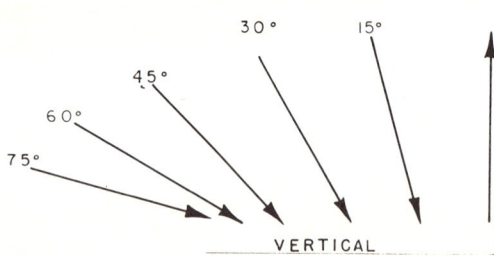
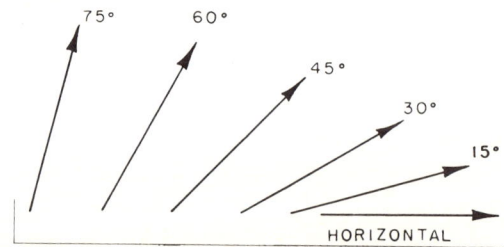

Fig. 3-7 Lines drawn at 15° intervals.

Fig. 3-8 Combining triangles.

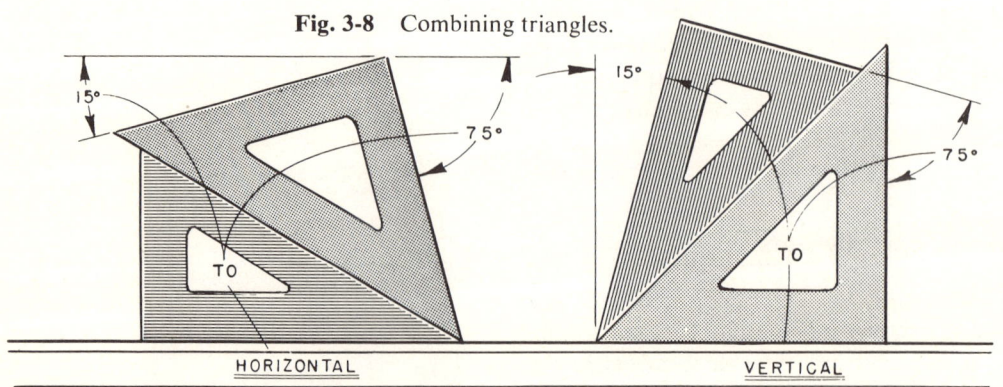

3-8 Use of the compass

The large compass is usually supplied with extra leads, needles, and a pen attachment. Thus the instrument may be used for arcs and circles drawn in either pencil or ink, or it may be converted to be used as an extra divider.

To adjust as a divider each leg of the instrument is fitted with a needle point, conical tip out.

To adjust as a pen compass, place the pen attachment in the proper leg and adjust the needle point of the other leg to the same length as the pen point. Be sure that the shoulder point of the needle is used. It may be necessary to bend the leg containing the pen at the "knee" in order to have both lips of the pen touch the paper properly.

For use as a pencil instrument, the compass is fitted with a lead sharpened as shown in Figure 3-9. A long semiconical point is produced on the outside of the lead. Hold the compass at an angle, as shown, and stroke the lead on file or sandpaper with a rocking motion. After the lead is sharpened, the shouldered needle may need readjustment as to length. Final adjustment should always be shoulder to lead so as to avoid possible turning of the lead.

To make the arc or circle, hold the compass as shown in Figure 3-10 and lean the instrument slightly in the direction of rotation. With a center-wheel instrument, sufficient pressure may be applied to produce a jet black line without danger of springing the compass setting. Features of the large center-wheel bow compass are discussed in section 2-14.

3-9 Using dividers

The friction divider is adjusted with one hand, as shown in Figure 3-11. Notice that the thumb and index finger are outside the legs for closing them, while the second and

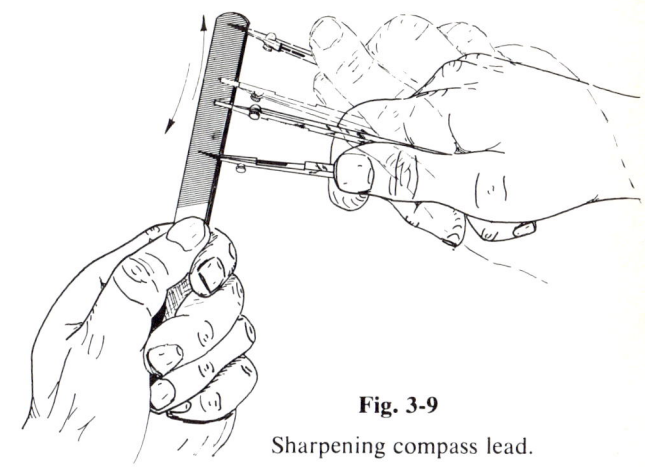

Fig. 3-9 Sharpening compass lead.

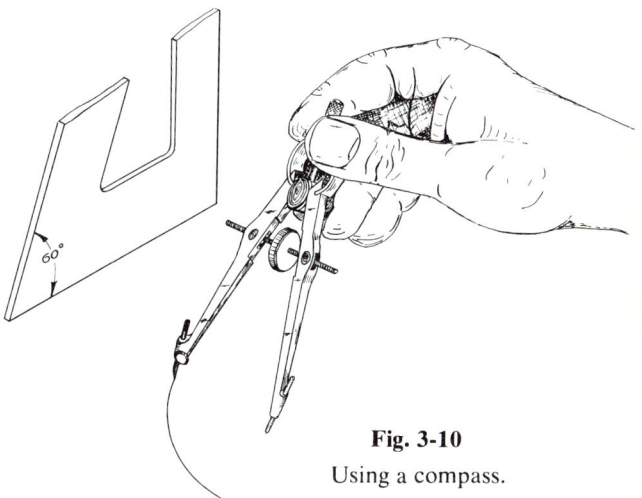

Fig. 3-10 Using a compass.

Fig. 3-11 Adjusting dividers.

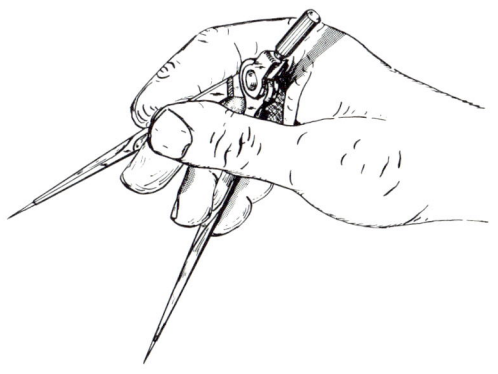

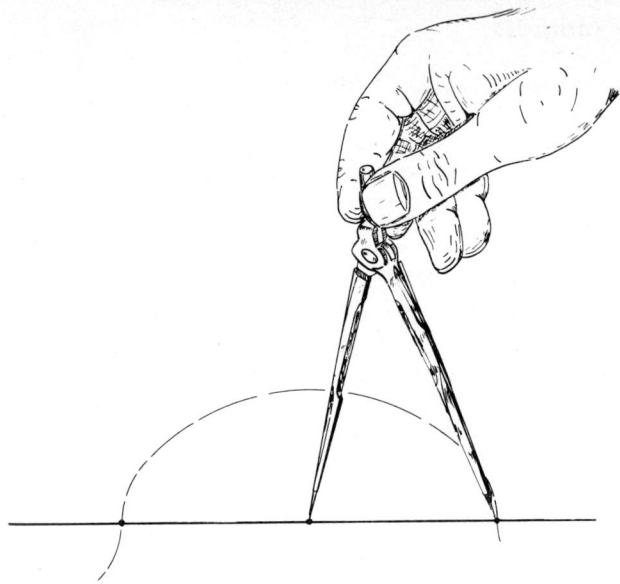

Fig. 3-12 Using dividers.

third fingers are inside and are used to increase the setting. Adjustment of the instrument in this manner should be practiced until it becomes natural and easy.

The divider is used to transfer measurements, and to divide lines, arcs, and circles into a specified number of equal parts. When transferring a measurement, hold the instrument as in Figure 3-11. When dividing a line, manipulate the divider as shown in Figure 3-12. Proportional dividers are discussed in section 2-19.

3-10 The bow instruments

Each of the large instruments have as their counterparts, the bow pencil, bow compass, and the bow spacer. These are referred to as

Fig. 3-13 Bow instruments.

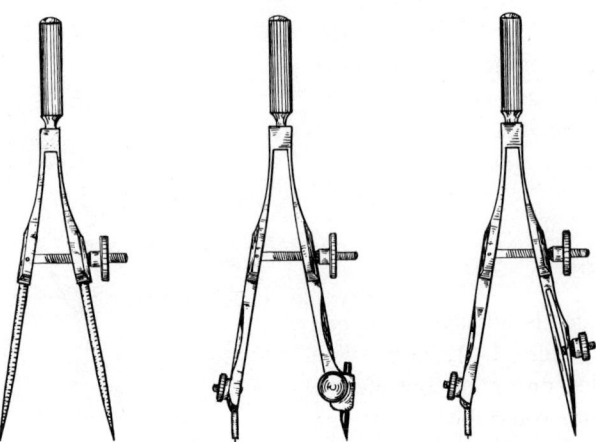

INSTRUMENT TECHNIQUE

the bow instruments. They are used in the same manner as the parent instrument but for smaller settings. Typical bow instruments are shown in Figure 3-13. The bow pen and pencil are used for arcs and circles of 1-in [25-mm] radius or less. The bow spacer is used to pick up and repeatedly transfer a small measurement. The bow instruments are well suited for small settings that remain constant during the entire construction of the drawing.

3-11 Using the beam compass

The beam compass is used to draw curves of a radius of 6 in [150 mm] or more. The instrument is held in both hands and manipulated as shown in Figure 3-14. One head of the beam compass contains a vernier adjustment and, depending upon whether the draftsman is right- or left-handed, the vernier head is held and adjusted, both for approximate and vernier settings, by that hand. If the beam compass heads are to be used as trammel points on long stock for radii greater than 36 in [914 mm], it is preferable to have a helper hold the point while the draftsman scribes the necessary curve.

3-12 Line symbols

Drawings are composed of lines of different types arranged so as to convey information. The usual information conveyed consists of shape and size of objects and the solution of problems involving points, lines, and planes. It is important, therefore, that symbolic lines be established as standard for

Fig. 3-14 Using a beam compass.

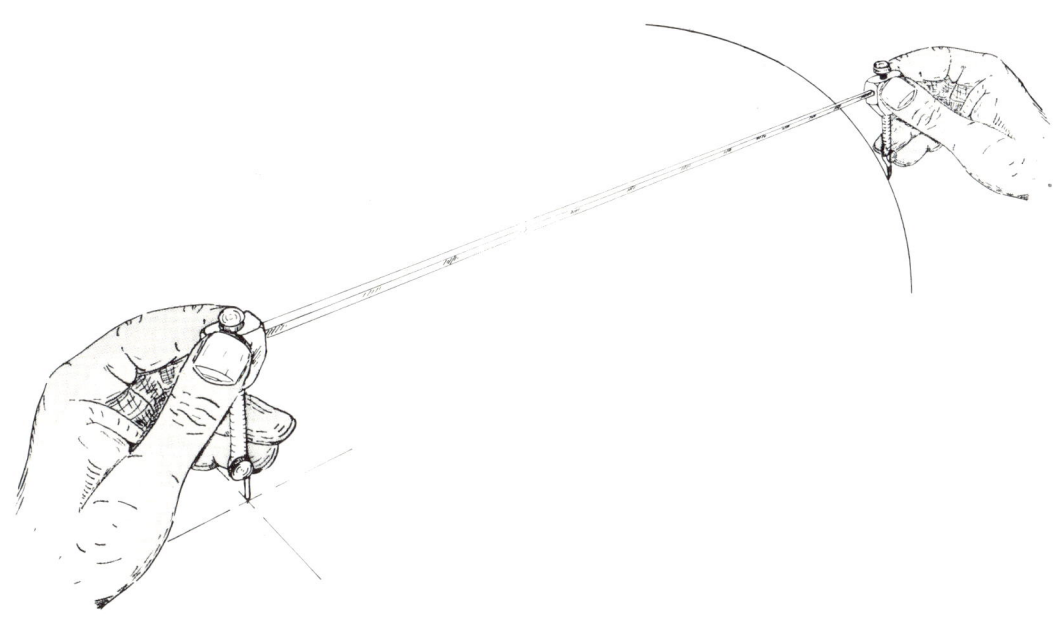

the needs of the drawing profession. An alphabet of eleven lines is shown in Figure 3-15. While the entire alphabet of lines is given, the sheet metal draftsman is concerned primarily with the first six.

Line Characteristics. The pencil lines of any drawing possess two distinguishing characteristics, namely, weight and color. The weight of a line actually refers to its width. For example, a thin line is approximately .003 to .008 in [.5 mm] wide; a medium line .012 to .017 in [1 mm] wide; if the penciled line is greater than 0.020 in [.5 mm] wide, it is generally regarded as a thick line. The proper weight for each line symbol is shown in Figure 3-15.

The *color* of all penciled lines, except construction or layout lines, must be *black* because the lines must be sufficiently opaque to reproduce well upon sensitized paper or metal. Thin, sharp, accurate, gray lines can only be used by the layout draftsman working on template material. Line *color* is dictated by the use to which the drawing is to be put and is controlled by pencil hardness, pencil pressure, paper texture, and humidity.

Types of Lines. "Construction lines" are very thin gray lines just visible to the draftsman at close range. They will not show on a print if properly made on the drawing. They need not be erased when the drawing is completed.

"Object lines" are used to represent, on the drawing, the outline or edge view of surfaces of the object, the sharp intersection of two surfaces, or the contour elements of a curved surface. They are *medium* weight, *black* lines, which take precedence over all other lines of the drawing.

"Hidden lines" are used to represent edges and surfaces which can not be seen because they lie behind visible surfaces of the object. Hidden lines are made of carefully spaced dashes, as shown in Figure 3-15, and have been misnamed and somewhat standardized as "dotted lines." Hidden lines, although classified as medium weight, should be made a little narrower than object lines in order to produce some contrast with the object lines of the drawing. The readability of the drawing is affected materially by the manner in which the hidden lines are executed. Notice in Figure 3-16 that a hidden line always starts as a dash, unless it would thereby become a continuation of an object line. In that case, the object line would incorrectly appear to extend beyond its required limits and a space between object line and hidden line is in order. Notice that dashes intersect at corners; also, that the illusion of depth in a view is helped by spacing the dashes of the hidden line so as to avoid contact with any object lines crossed. Hidden lines are "real" lines and therefore take precedence over "imaginary" lines, such as center, dimension, and extension lines.

"Center lines" are used to indicate the existence and location of circular features on a drawing. Center lines are made *thin* in weight, but *black* in color. Since they are a valuable aid in reading a drawing, they should be included for all circular features in all views where appropriate. Pay particular attention to the length and spacing of the dots and dashes and the amount of center line used for various cylindrical features. As stated in section 3-13, center lines are the first lines drawn and furnish the framework upon which the rest of the drawing is constructed.

"Dimension" and "extension" lines are used to add size description to a drawing. They are the same weight and color as center lines.

All the above line symbols are used and illustrated in the drawing of Figure 3-16.

3-13 Pencil drawing

A pencil drawing is made up of the standard line symbols properly arranged, but

INSTRUMENT TECHNIQUE 27

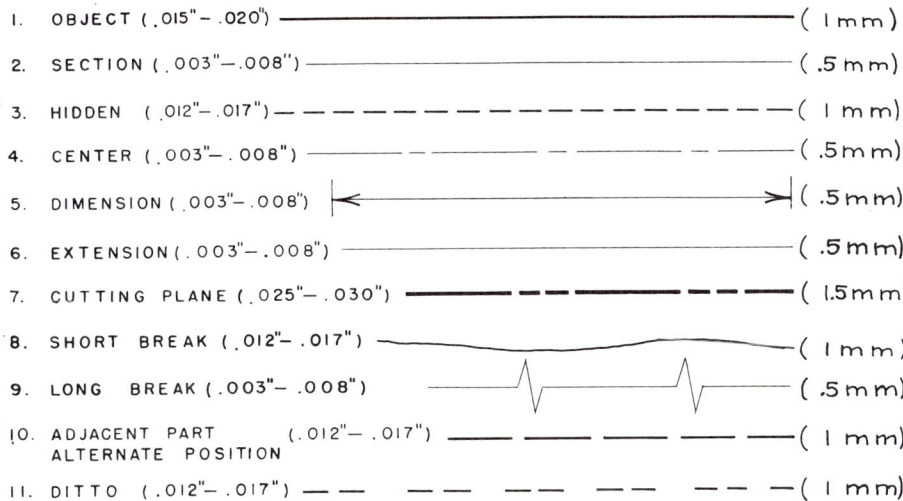

Fig. 3-15 Line symbols.

Fig. 3-16 A sheet metal drawing.

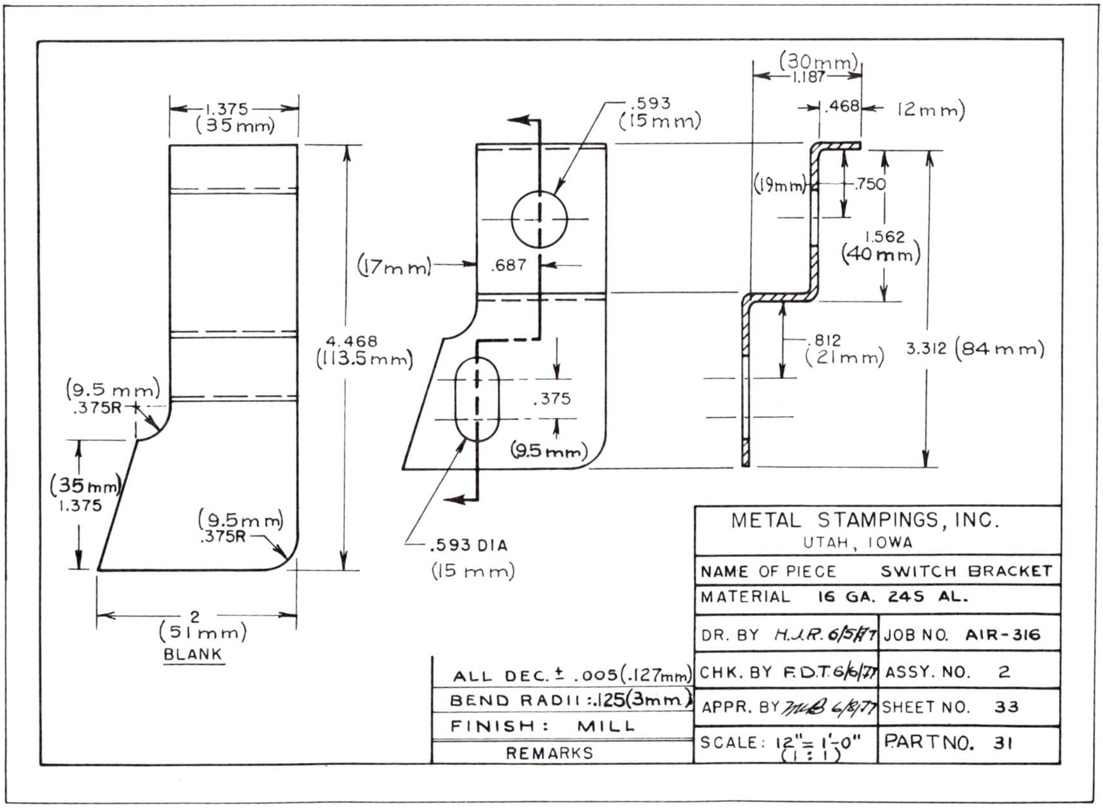

it should be made by following a logical plan of procedure.

a. Relative to a sheet metal drawing, the problem normally starts with *a sketch,* upon which the necessary specifications are shown, the arrangement of views decided, and the procedure for the solution of the problem determined.

b. The scale and size of the drawing is decided, based upon accuracy, drawing space, and the paper available. Sheet metal drawings are made full size if possible. In some cases, half-size, quarter-size, or one-fifth-size scales may be necessary and the final pattern enlarged for working purposes (see section 5-12).

c. Lay out the *center lines.* For measurement purposes this may be, at this stage of the drawing, a continuous line of construction weight.

d. Block in the views with construction lines.

e. Make circles and arcs, locate tangency points.

f. Darken the *object lines* of the required views.

g. A sheet metal drawing may require the location of a *line of intersection* and the *development* or *stretchout* of a pattern. Solve the problem using sharp black lines with emphasis upon accuracy.

h. Add any *dimensions, specifications,* or *notes* needed. This is done with clean, sharp lettering.

The outstanding characteristics of the professional draftsman are *speed* and *accuracy.* The beginner will, necessarily, sacrifice speed in order to obtain skill, accuracy, and knowledge. Gradually, speed will improve and a style and technique will be acquired that will give the drawings personality. This is an important quality because it adds to the ability of the drawing to convey the draftsman's ideas to the person using the drawing. A finished style will result from the thorough study and practice of these fundamentals.

3-14 Using irregular curves

An irregular curve is used to produce a smooth, uniform mechanical line through a series of points which do not lie on a straight line.

After the points have been established, a fine light line is *sketched freehand* through them. This is done in order to secure a continuous, smooth, and uniformly curved line to which the variable curvature of the ruler may be fitted. When a portion of the curves have been matched, the line is drawn solidly for a distance *somewhat less than* the true matching distance. The curves are rematched and an additional portion drawn, again stopping short of the over-all matching distance. In this manner, a smooth mechanical line is produced which is free of humps, bumps, and other sudden changes in direction. It is important that the draftsman change position, when necessary, in order to be on the opposite side of the curve being used. The correct and incorrect methods and their results are shown in Figure 3-17.

For curves which are symmetrical, the irregular curve can be "flopped" about a center line so as to use the same portion for identical but opposite curvatures. The center lines and the exact starting and stopping points should be marked on the curve. To this end, the plastic can be "frosted" by rubbing with an eraser; it will then "take" a pencil mark which can be easily rubbed off after use.

3-15 Cleanliness, neatness, and erasures

Cleanliness applies to the draftsman and to the equipment. The hands should be clean.

INSTRUMENT TECHNIQUE

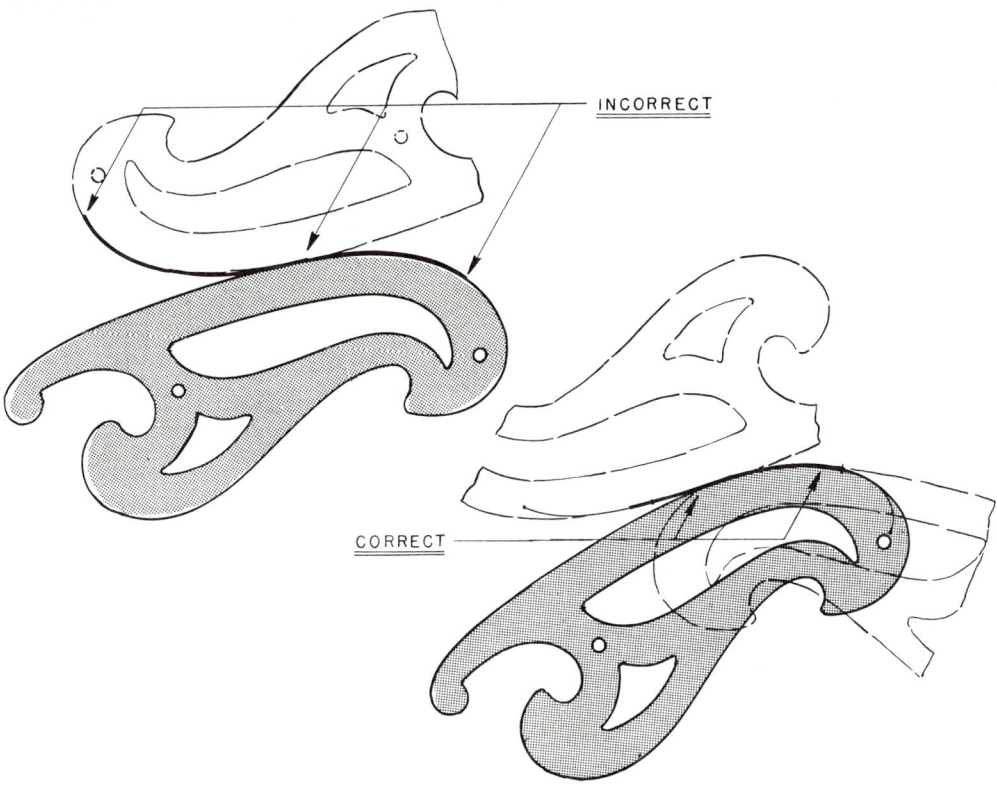

Fig. 3-17 Using an irregular curve.

If necessary, wash them frequently with soap and water. In cases of extreme perspiration or oily skin, talcum may be used to avoid soiling drawings. The hair should be neatly combed and out of the face so that running the hands through the hair, of necessity, or out of habit, is avoided. Most professional draftsmen present a neat appearance which is reflected in the quality of their work.

The equipment should be kept clean. Cover the table and board when not in use. This applies, especially, when drawings are left on the board between work periods. The drawing instruments should periodically be wiped clean and polished with a soft dry cloth. The triangles may be washed with soap in cool water, rinsed and wiped dry. The T-square blade may be cleaned by wiping with a damp cloth and soap; wipe dry immediately to prevent warping. The blade may also be cleaned with the Artgum eraser.

The neatness of a drawing is determined, to a large extent, by preparation and the application of correct methods of procedure. For example, sharpen and point the pencil when away from the drawing. Keep all sharpenings and loose graphite off the hands, drawing, or equipment. The file or sandpaper pad should *hang below* the drawing surface. A cloth should be used to wipe the pencil after sharpening. After *black* object lines are drawn, they should be "flicked" with the dust brush to remove loose graphite or a clean, soft cloth may be passed lightly over them to pick up the loose particles.

Erasing on drawings is a necessary evil,

obviously to be avoided if possible. Care in planning a drawing and following proper procedure will result in fewer erasures. When erasing is necessary, proceed with caution, using a shield, and stop as soon as the mistake has been satisfactorily removed. Object lines may be more easily erased if a hard surface, such as a triangle, is placed under the paper. When erasing on tracing paper or cloth, it is sometimes necessary to erase both sides as a certain amount of "pickup" may still show through. Electric erasing machines are sometimes available. They should be used with little pressure, always moving the eraser and always working through an erasing shield. Use the brush to remove the erasure particles. Artgum may be used for the final cleanup of large areas.

EXERCISES

1. Where should the drawing paper be fastened and located on the drawing board?
2. What pencil range is preferred by draftsmen?
3. Practice drawing some horizontal object lines spaced ¼ in [6 mm] apart. What appearance should they have?
4. Practice drawing some vertical object lines spaced ¼ in [6 mm] apart. What direction are vertical lines drawn?
5. Singly and in combination the two triangles with T square provide all angles at what degree intervals? By using only triangles and T square, divide a semicircle into twelve parts.
6. What are the friction dividers used for?
7. Pencil lines are judged by what two characteristics?
8. Besides neatness, what other work characteristics are important?
9. Why is a line sketched freehand through a series of points before the irregular curve is applied to them.
10. How may a plastic triangle be made to take a pencil marking?
11. To what, other than the drafting equipment, must the principles of cleanliness be applied in order to produce clean drawings?
12. Name some factors which contribute toward neat drawings.

chapter 4

LETTERING

The lettering used in engineering and drafting is known as *single-stroke gothic*. The "gothic" form of the individual letters is based upon Old English and German text. The gothic alphabet of today, however, is a simplified form of the older alphabets and is made with all strokes of uniform thickness. "Single stroke" simply means that a single stroke of the pen or pencil determines the uniform thickness of all the lines necessary for a given letter. For example, a capital E is made using four separate strokes, all of uniform weight, in the proper order, direction, and length (Figure 4-1).

Single-stroke gothic lettering is approved by the American National Standards Institute as acceptable lettering on all industrial drawings and may be executed in either the vertical or the inclined style. Both capital and lower-case alphabets are approved and may be used separately or in combination, depending upon the purpose of the lettering. Speed and legibility are of primary importance. The ability to letter freehand with speed, neatness, beauty of form, and legibility is a skill much desired by all engineers and draftsmen. Skill in lettering can be acquired by any normally intelligent person who is willing to *study* and to *practice*.

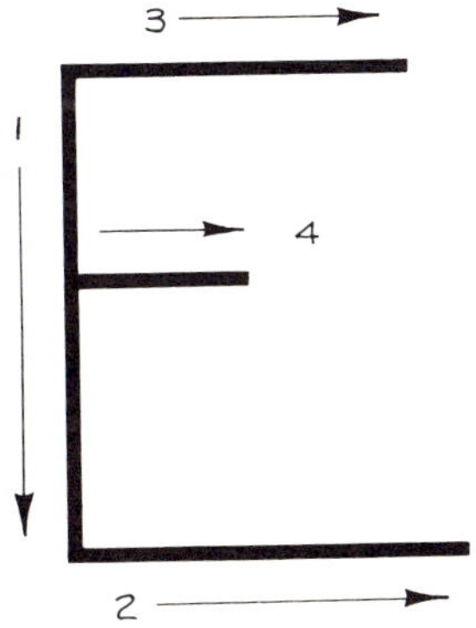

4-1 Selection of the lettering pencil

Since most drawings are made only in pencil (sections 2-10 and 3-12), the pencil used for lettering must produce *black* letters. Because the amount of pressure applied is less than that used in drawing lines, the lettering pencil should generally be one or two degrees softer than the pencil used for the drawing. On detail paper and pencil cloth, an HB or F pencil may be found suitable. On tracing papers, a harder pencil, perhaps an H or 2H, may be preferable because of the "tooth" of these papers. The pencil should be sharpened to a long conical point as described in section 3-4. In order that the pencil *remain* sharp and produce strokes of uniform thickness, it should be rotated slightly from time to time between strokes.

Fig. 4-1

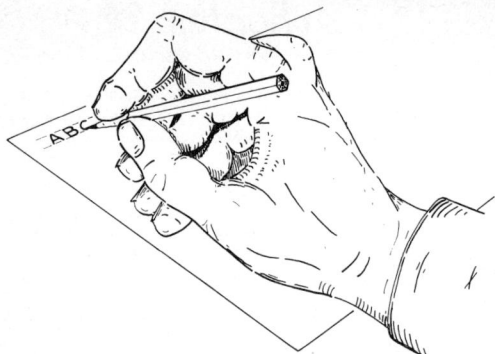

Fig. 4-2 Position of hand when lettering.

Hold the pencil as shown in Figure 4-2 and produce vertical strokes downward with a finger movement. Produce horizontal strokes with a wrist motion. Finger and wrist motions are combined to produce curved strokes. Curved letters, especially the S, may need attention and practice, since they are the most difficult to produce consistently.

4-2 Guide lines

All freehand lettering is made with the aid of horizontal guide lines for the tops and bottoms of the letters. Vertical or inclined guide lines are made at random to help in maintaining uniform slant or verticality. While the professional draftsman realizes the importance of guide lines, the student is apt to regard them as unessential and a nuisance. The results, when using guide lines, however, are more than worth the investment in time and trouble.

Guide lines must be made light in weight and color like construction lines (Figure 3-15) because they cannot be erased when the penciled lettering is finished. A sharp, hard pencil (6H) should be used for these lines. On inked lettering, the guide lines may be erased with Artgum after the ink has *dried thoroughly*.

The standard height of letters for most drawings will vary from $\frac{1}{8}$ to $\frac{1}{4}$ inch [4 to 6 millimeters], depending upon the size of the drawing and the relative importance of the lettering. Letters $\frac{1}{8}$ in [4 mm] high are most common and are a good size to practice with.

The space between lines of lettering may vary from $\frac{1}{2}$ to $1\frac{1}{2}$ times the letter height. A good spacing between rows of $\frac{1}{8}$ in [4 mm] lettering is $\frac{3}{4}$ letter height ($\frac{3}{32}$ in [2 mm]). Avoid using a space equal to letter height.

Vertical or inclined guide lines are spaced every $1\frac{1}{2}$ to 2 in [38 to 51 mm] as an aid in maintaining verticality or slant. Vertical lettering looks well only when all letters are absolutely vertical. Inclined lettering is slanted at approximately 68 degrees to horizontal.

While guide lines may be constructed with regular drawing equipment, they are most easily made, especially for single letters and fractions, by using a special device with holes for selection of size and automatic spacing of the lines. The *Braddock-Rowe Lettering triangle* and the *Ames Lettering instrument* are available (instructions included) at drafting supply stores. They are a convenient device for spacing guide lines. The draftsman may also drill, notch, or otherwise alter areas of regular triangles so as to produce desired guide lines.

4-3 Capital letters

The location and length of each of the strokes used to make any capital letter determine the form or characteristic shape of that letter. The form of each letter must be impressed upon the memory to such an extent that, in the imagination, the letter can be seen on the paper as the hand performs the strokes. The form of each vertical and inclined capital letter is shown and discussed in the paragraphs and illustrations which follow. Suggestions and hints are given relative to the location and length of strokes, and the general over-all appearance of the letter. Pay particular attention to the following:

a. The proportion of width to height. Students have a tendency to make small letters too narrow. Large letters may be condensed, but small letters should always be expanded.

b. The position and length of mid-bars in the A, B, E, F, G, H, P, and R. The position varies for the different letters.

c. The location of the crossover or intersection points in the K, R, T, X, Y.

d. The starting and ending points for the curved strokes.

The strokes in Figure 4-3 and Figure 4-4 are numbered, indicating their proper sequence. The numbered strokes will thereby serve as a basis for developing a consistent manner of making each letter. The order may be varied to suit individual preferences but the important thing is that each letter always be made the same way.

The direction of the strokes is indicated in Figures 4-3 and 4-4 by means of arrows. The directions given are for right-handers; left-handers will vary the direction somewhat as indicated in Figure 4-5.

The capital letters of Figures 4-3 and 4-4 are not in alphabetical order; rather, they are arranged with the easiest straight-stroke letters first and the more difficult curved-stroke letters last.

The capital I has no width and is the easiest letter to make since it consists of a single vertical or inclined stroke (Figure 4-3).

The L is five-sixths as wide as it is high.

The T is as wide as it is high.

The H is five-sixths as wide as it is high. Strokes 1 and 2 are parallel. The cross stroke is three-fifths up from the bottom in order to provide stability and avoid a top-heavy appearance (Figure 4-3).

The F is five-sixths as wide as it is high. The horizontal stroke is slightly above the center.

The E is another letter which is five-sixths as wide as it is high. The first two strokes are the same as for the letter L.

The N is five-sixths as wide as it is high. As in the H, strokes 1 and 2 are parallel. Notice the slant direction of stroke 3 (Figure 4-3).

The X is as wide at the base as it is high. It is narrower at the top, with the crossover above center in order to maintain stability. *Notice that the inclined X consists of a short stroke 1 and a long stroke 2 at about 40 degrees to horizontal.*

The width of the top of the V and its height are equal. The inclined letter consists of a stroke, almost vertical, and a long stroke at 45 degrees (Figure 4-3).

The A is as wide as it is high. The A is an inverted V with the crossbar, stroke 3, added at one-third the height of the letter from the bottom.

The K is five-sixths as wide as it is high. Stroke 2 intersects stroke 1 two-thirds of the way down from the top. Stroke 3 aligns with starting point of stroke 1 and is almost perpendicular with 2 (Figure 4-3).

The Z is five-sixths as wide as it is high. Stroke 1 is shorter than stroke 3 but is centered over 3 for stability.

The Y is as wide as it is high. The mutual intersection point of all three strokes is the center of the appropriate square "box" (Figure 4-3).

The M is as wide as it is high. Strokes 1 and 2 are parallel and strokes 3 and 4 are like the V.

The W is the only letter wider than it is high. Its width is equal to one and one-third times its height. Actually, it is made by lettering two Vs which are two-thirds normal width. For inclined letters, strokes 1 and 3 can be vertical (perpendicular to horizontal guide lines). (See Figure 4-3.)

The curved letters O, C, and Q when vertical are based on the circle; therefore, width equals height. The inclined letters are based on the ellipse, the major axis being the diagonal of the enclosing box, as shown in Figure 4-4. Thus the curvature will tend to be sharp in the upper right and lower left corners and

Fig. 4-3 Order and direction of strokes for straight-line letters.

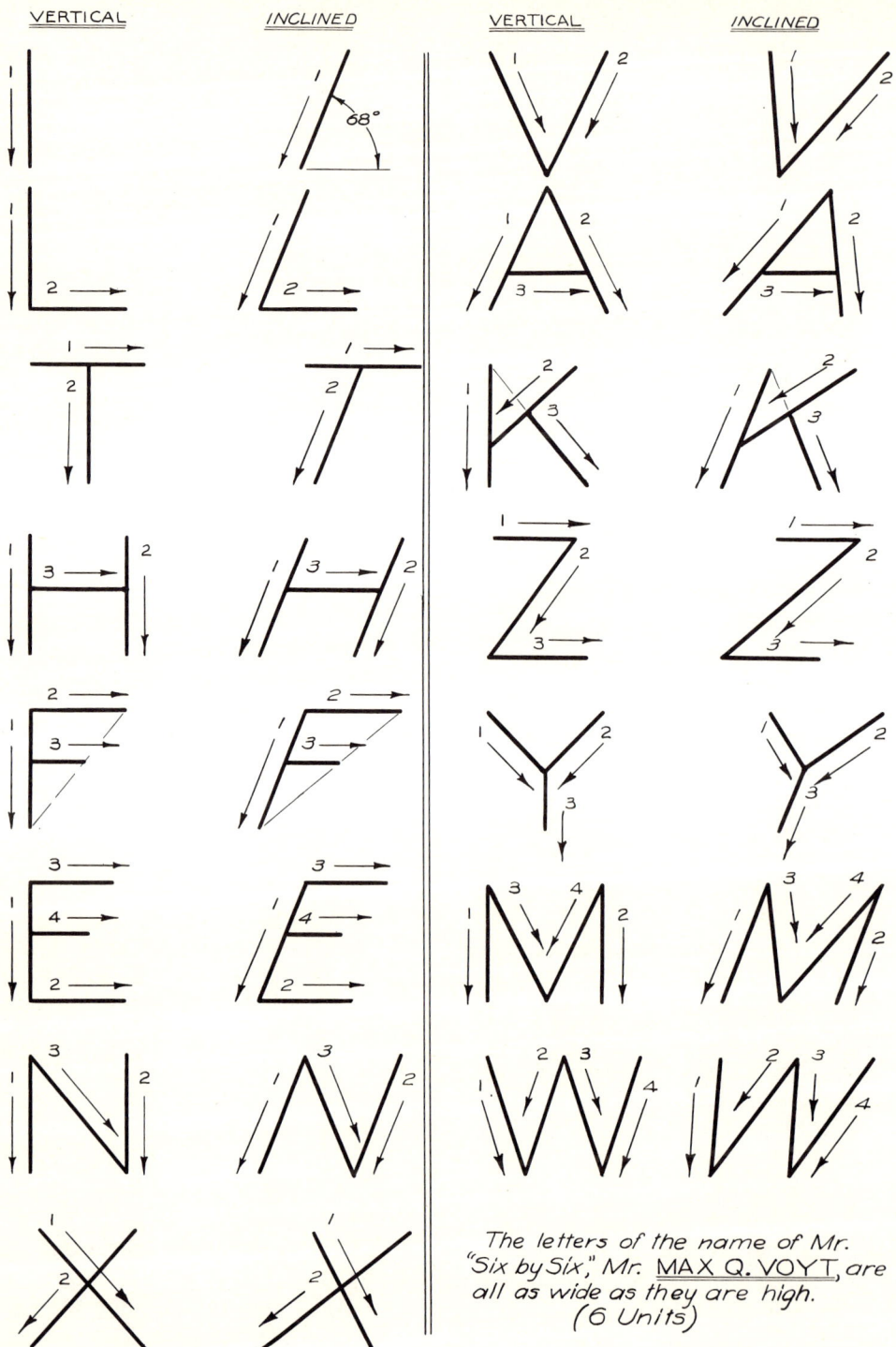

The letters of the name of Mr. "Six by Six," Mr. MAX Q. VOYT, are all as wide as they are high. (6 Units)

Fig. 4-4 Order and direction of strokes for curved-line letters.

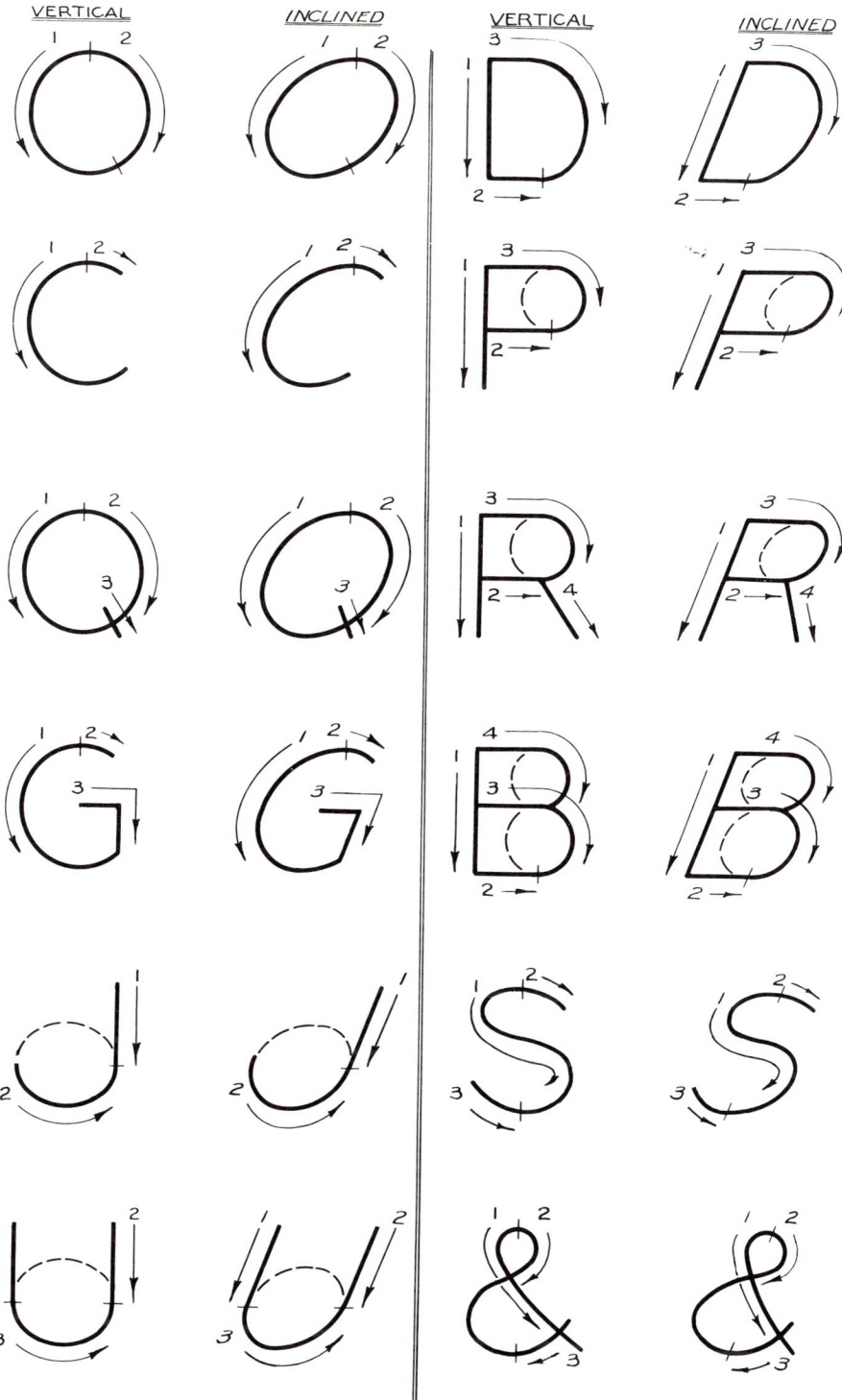

long in the upper left and lower right areas of the letters. The third stroke of the Q is a radial line. The G is the same as the C except the inverted L-shaped third stroke is added, making the letter five-sixths as wide as it is high.

The J and the U are similar and are five-sixths as wide as they are high.

The D is a short L with a third stroke added (Figure 4-4).

The P is five-sixths as wide as it is high. Stroke 2 ends at the center of stroke 1. The P has a circular end.

The R is the same as the P with stroke 4 added.

The form of the B is similar to the P, R, D and short L.

The S is a difficult letter to make look well, and it usually needs more practice than the others. The bottom loop is larger than the top loop. Three strokes are shown in Figure 4-4 but variations may be found easier, especially in the smaller sizes. Left-handers may find the S easier to make in a single stroke.

The ampersand is the symbol for the word *and* (see Figure 4-4).

4-4 Numbers and fractions

With the exception of number 1, which is identical in form and stroke with the letter I, *the width of a number is equal to five-sixths its height*. Thus the curved numbers are ellipses or portions thereof. The form, order, and direction of strokes for each number are indicated in Figure 4-6.

Fractions are always lettered with a horizontal bar. The numerator and denominator of the fraction are of such a size that the entire fraction is *twice* the height of capital letters and integers. Guide lines are required for the top and bottom of the fraction, but are not needed inside the fraction for small lettering. Notice, however, that an open space exists between the numbers of the fraction and the horizontal bar (see Figure 4-7). In a mixed number, such as 2¼, the spacing is important so that it is read as one rather than two separate values. The numerator and

Fig. 4-5 Order and direction of strokes for left-handers.

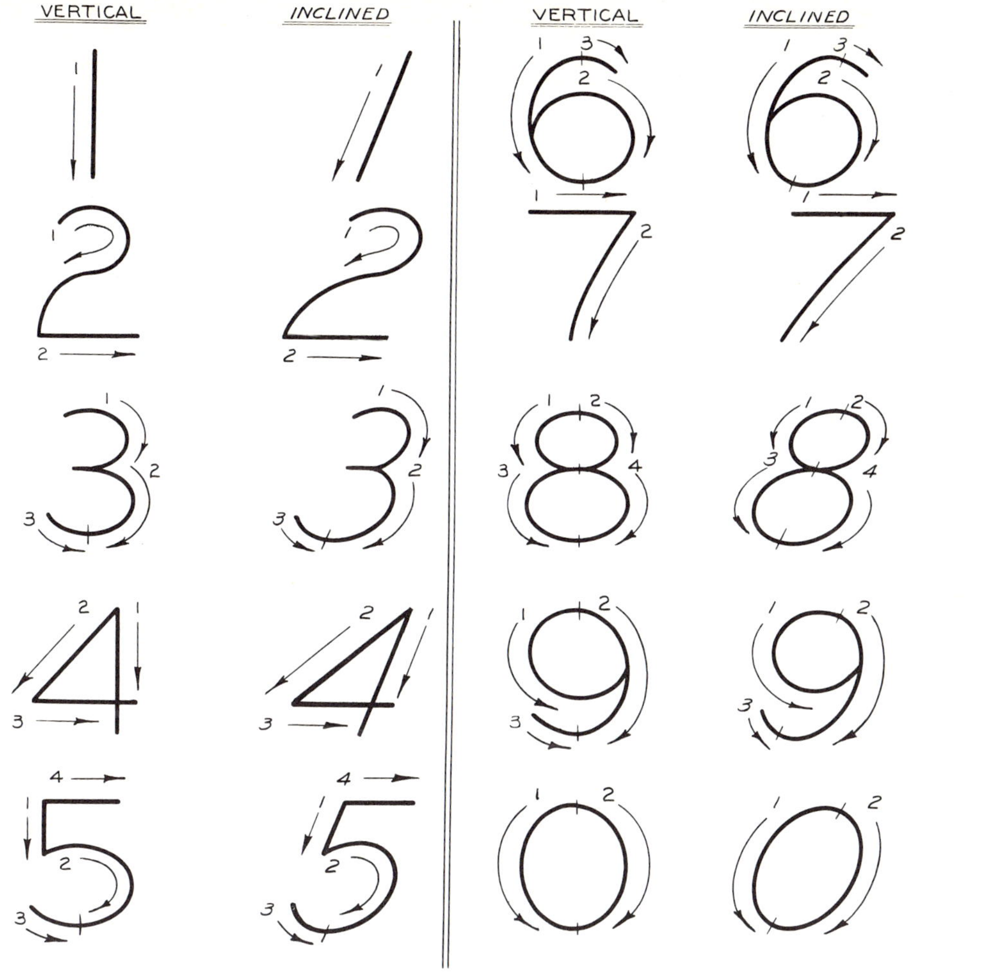

Fig. 4-6 Numbers.

Fig. 4-7 Fractions.

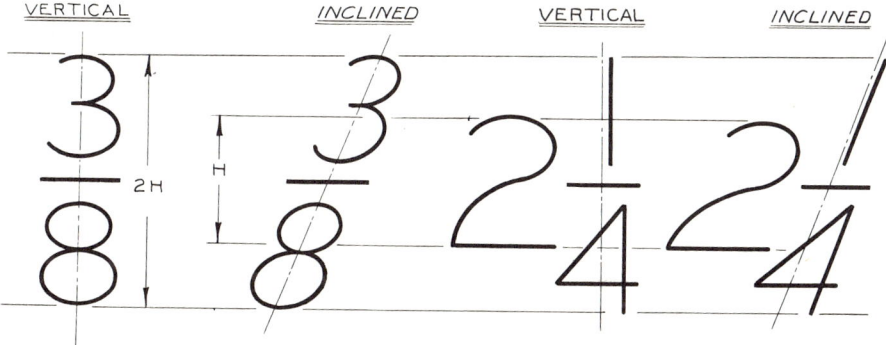

denominator in inclined fractions must be positioned carefully so as to create the proper slant for the fraction (see Figure 4-7).

4-5 Lower-case letters

Lower-case letters are made between the base and the cap guide lines. A waist guide line two-thirds up regulates the height of the main body of the letters. The lower-case b, d, f, h, k, and l ascend to the cap line; g, j, p, q, and y descend to a drop line. The lower-case t ascends almost to the cap line. The guide lines and the basic strokes for all lower-case letters are shown in Figure 4-8. Notice that vertical letters are composed of circles, arcs, and straight lines; inclined letters are composed of ellipses and straight lines. The vertical and inclined lower-case alphabets are illustrated in Figure 4-9.

Fig. 4-8 Fundamental strokes for lower-case letters.

Fig. 4-9a

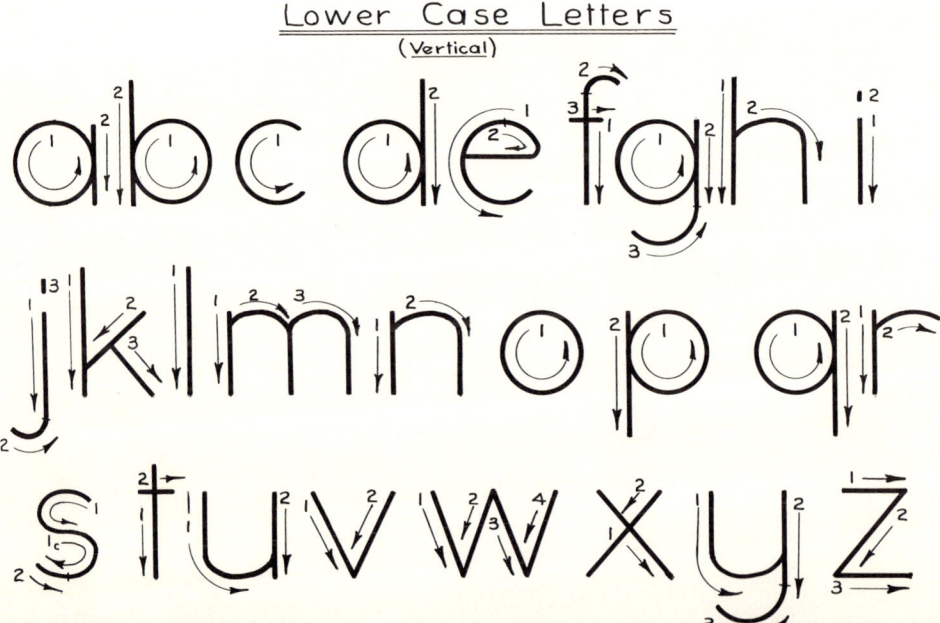

LETTERING

Fig. 4-9b

4-6 Spacing

Excellent composition in lettering depends not only on good letter form but also on correct spacing of letters into words and words into sentences. Best word appearance results when the white spaces between letters are as nearly equal in *area* as possible. Since the spacing then depends on the letters involved and their order, it is obvious that no specific rules for spacing are practical. Good spacing depends largely on the draftsman's judgment, experience, and artistic sense.

Two spacings of the same word are shown in Figure 4-10. Certain letter combinations present more difficulty in spacing than others. Combinations with adjacent straight sides, such as IN, HE, etc. must be opened up more than combinations involving pocketed areas. Other groupings may require overlapping top and bottom in order to reduce the area between letters. Typical ex-

Fig. 4-10

amples of overlapping are AT, PA, FA, LT, LY, etc. The beginner has a tendency to space the letters in words *too far apart*.

The spaces between words in a sentence should be uniform, sufficient to separate the words, and range from a space sufficient to accommodate a capital I up to a space equal to letter height. Sentences should be separated by a space slightly greater than letter height.

Professional draftsmen take pride in their lettering. Good lettering technique is recognized by letter form, adherence to verticality or proper slant, and crisp, sharp strokes. Some draftsmen provide additional "snap" to their lettering by "pointing," i.e., touching up the initial and terminal points of each stroke. Lettering style, like drafting skill, should improve in quality and speed with experience and practice. A certain amount of aptitude and interest is absolutely necessary in order to progress in lettering.

EXERCISES

1. To what does "single stroke" refer?
2. A 2H pencil was found suitable for making black object lines on a certain detail paper. Recommend a suitable lettering pencil for the same paper.
3. What motions are used for producing horizontal strokes in lettering? Vertical strokes?
4. List important specifications regarding the form of each of the following capital letters: A, W, S, N.
5. Which, of all letters, is considered most difficult to make?
6. Make an alphabet of capital letters ⅛ in [4 mm] high. (If right-handed, use inclined letters; if left-handed use vertical letters.)
7. Other than the 1, what are the proportions of all numbers?
8. Make three sets of numbers ⅛ in [4 mm] high (1 ... 9-0).
9. Make an alphabet of inclined lower-case letters (use cap guide line ⅛ in [4 mm] high, waist line ⅔ cap height).
10. Make a series of mixed numbers, such as: 1⅛, 2¾, 4⁵⁄₁₆, 7⅜, etc. (use ⅛-in [4-mm] cap height).
11. Letter the first sentence of section 4-6 using ⅛-in [4-mm] caps.
12. Letter, using ⅛-in [4-mm] lower case and caps where appropriate, the first two sentences of section 4-4.

chapter 5

GEOMETRIC CONSTRUCTIONS

The paragraphs and illustrations which follow define the geometric terms and symbols used in this book. This is necessary in order that ideas may later be expressed clearly, and easily conveyed to fellow workers. The terms and symbols defined are those most used in the text to abbreviate the discussion and clarify the illustrations.

5-1 Terms and symbols

Line. The term "line" is assumed to mean a "straight line" with a constant direction. A line is indefinite in length but the term is usually applied to what is more correctly stated as a "line segment." A line segment has only length, with a beginning and an end respectively known as the initial and terminal points.

Angle. An angle is formed when two lines intersect. The point of intersection is called the "vertex" of the angle. The two lines are called the sides of the angle. Angles are measured in right angles, degrees, or radians. The sheet metal draftsman is interested primarily in degree measurement. The degree is $1/90$ part of a right angle, $1/360$ part of a circumference. Degrees on the drawing are measured with a protractor or by means of the protractor scale on the drafting machine (see sections 2-5, 2-22).

Right Angle. A right angle is defined as two lines meeting so as to form four equal angles, each of which is a right angle. The two lines of a right angle are perpendicular to each other (a 90-degree angle).

Acute Angle. Any angle of a value less than a right angle is called an acute angle.

Obtuse Angle. Any angle of a value greater than a right angle is called an obtuse angle.

Complementary Angle. Two angles whose sum is 90 degrees are said to be complementary. Each angle is the "complement" of the other.

Supplementary Angles. Two angles whose sum is 180 degrees (a straight line) are supplementary angles. Each angle is said to be the "supplement" of the other.

Plane. A plane is a flat surface. The plane may be real or imaginary. It may be of fixed size and shape or it may be unlimited in size. The position of a plane on a drawing is established (1) by locating three points other than in a straight line, (2) by locating two intersecting lines, (3) by locating two parallel lines. Each of these is shown in Figure 5-1.

Plane Figure. When a plane is of fixed size and shape it is called a plane figure. Plane figures are therefore composed of three or more sides. They may be classified as triangles, quadrilaterals, or polygons.

Triangle. A triangle is a plane figure having three sides. The sum of the internal angles of any triangle is always 180 degrees.

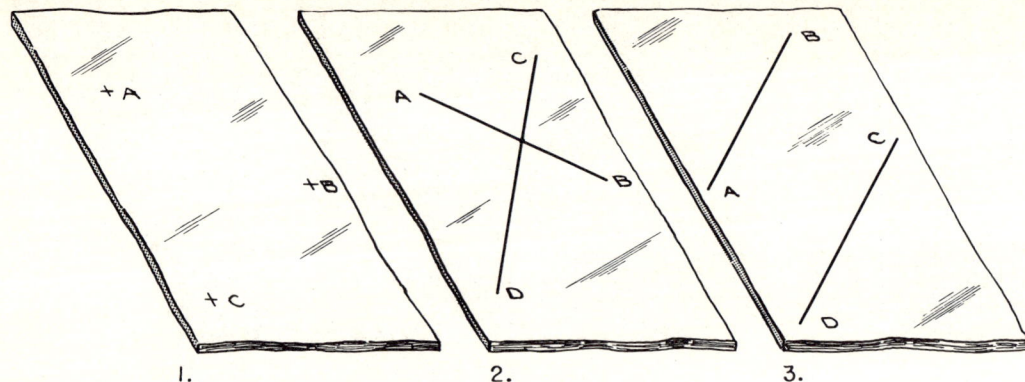

Fig. 5-1 Planes.

Right Triangle. A right triangle contains one angle of 90 degrees and two acute angles whose sum is 90 degrees.

Isosceles Triangle. An isosceles triangle has two equal sides and two acute angles that are equal.

Equilateral Triangle. An equilateral triangle has three equal sides. Since the three angles are 60 degrees each, the triangle is sometimes called equiangular.

Scalene Triangle. Any triangle which is not right, isosceles, or equilateral is called a *scalene triangle*. The three sides are all of different lengths and the three angles are all of different values. If one angle is obtuse the triangle is called an *obtuse scalene triangle*.

Quadrilateral. A quadrilateral is any plane figure having four sides. The quadrilaterals, illustrated in Figure 5-2, are (1) square, (2) rectangle, (3) rhombus, (4) rhomboid, (5) trapezoid, (6) trapezium.

Polygon. A polygon, according to literal translation, is a plane of many angles, hence of many sides. The term is usually applied to plane figures of more than four sides. If all sides are of equal length and all enclosed angles equal, the polygon is called a *regular*

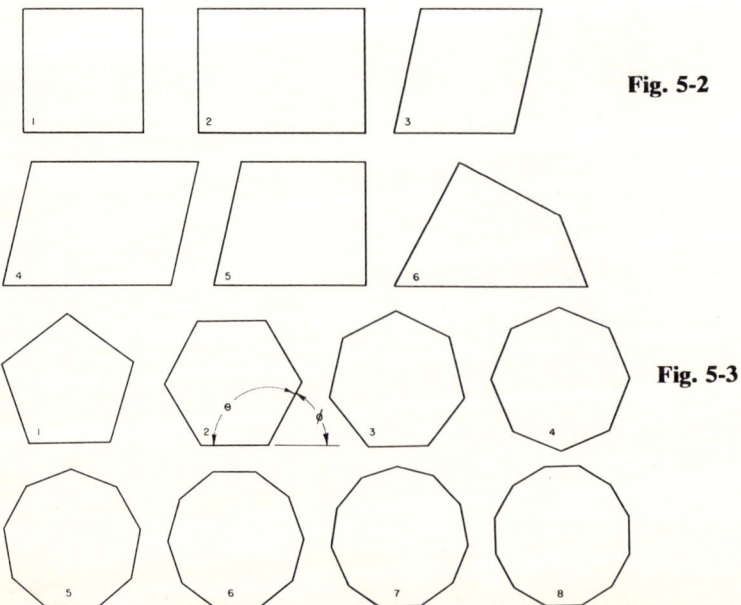

Fig. 5-2

Fig. 5-3

GEOMETRIC CONSTRUCTIONS

polygon. Typical regular polygons, shown in Figure 5-3, are (1) pentagon, (2) hexagon, (3) heptagon, (4) octagon, (5) nonagon, (6) decagon, (7) dodecagon, (8) duodecagon.

For any regular polygon, the enclosed angle θ or the external angle ϕ may be computed by using the following formulas

$$\theta = 180 - \frac{360}{N} \qquad \phi = \frac{360}{N}$$

where N = number of sides or angles

Cylinders, especially in heavy sheet, are sometimes "bent" rather than rolled. "Brakes" and "folders" are sometimes calibrated in degrees. If the brake is calibrated, the above formulas become useful in calculating cylinders and prisms. For example: A cylinder made up from flat stock using 48 brakes for forming would have the brake set to bend 7.5 degrees since $360/48 = 7.5$.

Circle. A circle is a plane figure enclosed by a curved line. All points on the curve are equidistant from a single point within the enclosure. This point is called the *center* of the circle.

The uniform distance from the center to the curve is called the *radius* of the circle.

The distance across the circle through the center is called the *diameter* of the circle, and is equal to twice the radius.

The complete curved line is called the *circumference* of the circle.

Circle Arc. Any part of any circumference is called a *circle arc*.

A *chord* is a straight line connecting the ends of an arc.

Segment. That area of a circle bounded by an arc and its chord is called a *segment* of the circle.

Sector. That area of a circle bounded by two radii and an arc is called a *sector* of the circle.

Ellipse. An ellipse is a plane figure enclosed by a curved line; having a major and a minor diameter perpendicular to each other, and crossing at a center. The elliptical figure is a *true ellipse* only when the sum of the distances from two points, called the *foci*, to any and all points on the curve remains constant. This value is equal to the major diameter of the ellipse (see Figure 5-4). The ellipse is a common and practical curve often encountered by the sheet metal worker. Notice in Figure 5-4 that a foreshortened view of any circle, or the true size and shape view of an angular cut on a cylinder will appear as an ellipse.

Oval. A plane figure with circular ends

Fig. 5-4 Ellipse.

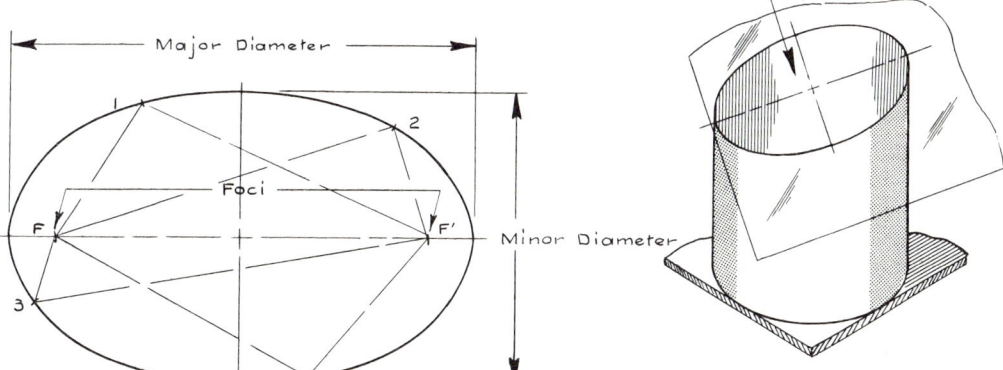

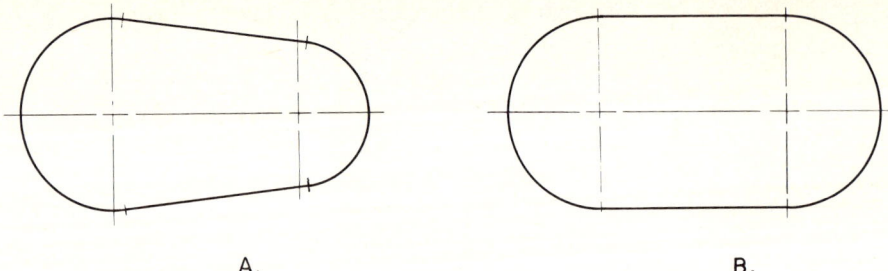

Fig. 5-5

of different radii connected by straight sides tangent to the circle arcs is called an *oval* (see Figure 5-5*A*).

Oblong. A plane figure with circular ends of equal radii connected by straight sides tangent to the circle arcs is called an *oblong* figure (see Figure 5-5*B*).

Pipe Openings and Shapes. The various terms defined thus far, such as triangular, square, rectangular, hexagonal, elliptical, oblong, oval, are used regularly by the sheet metal tradesman to describe pipe and duct openings as well as cross-sectional shapes of sheet metal fabrications.

Geometric Solids. Three-dimensional objects are known as *geometric solids*. The most common are the *prism, cube, cylinder, cone,* and *pyramid*. Typical sheet metal objects, such as pipes, ducts, containers, ventilators, separators, hoods, guards, fittings, cabinets, heaters, are shaped as prisms, cubes, cylinders, cones, pyramids or as combinations of these *basic geometric shapes*. While referred to as geometric solids because of their shape, they are made of sheet metal and are hollow or open.

Prism. A prism is a geometric solid whose ends or *bases* are identical plane figures. The bases are parallel and the sides of the prisms are either rectangles or parallelograms. If the bases are perpendicular to the sides, the prism is a *right prism* and each side will be a rectangle. If the bases are oblique to the sides—some angle other than 90 degrees—the prism is an *oblique prism* and each side will be a parallelogram. The shape of the base determines whether the prism is called triangular, rectangular, square, pentagonal,

Fig. 5-6   Prisms.

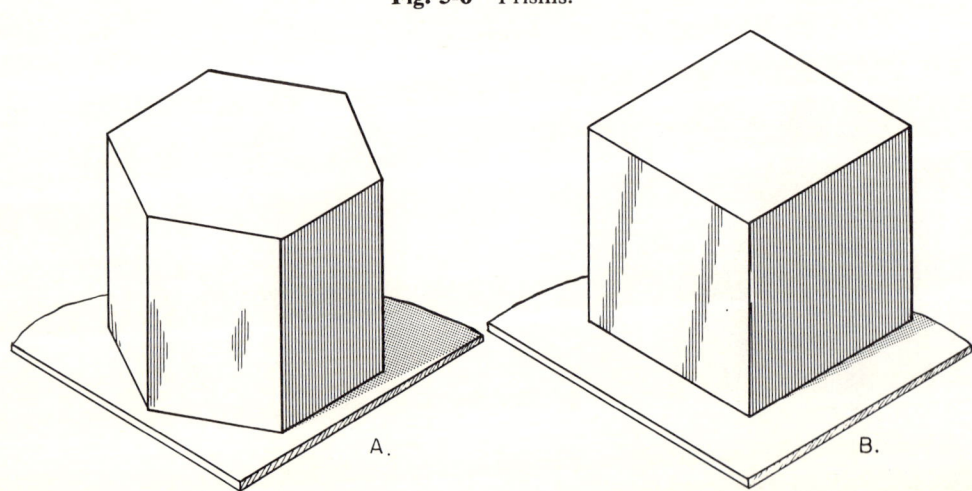

GEOMETRIC CONSTRUCTIONS

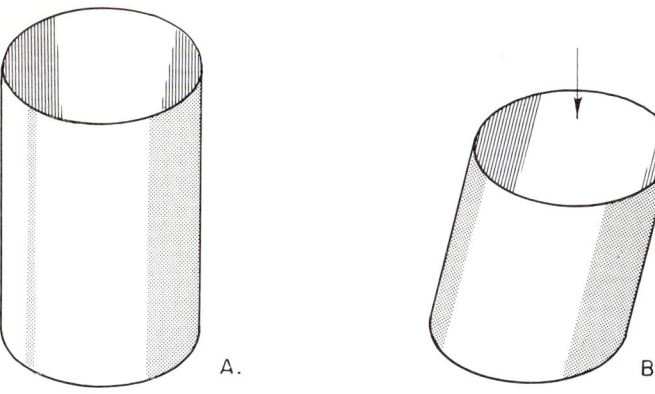

Fig. 5-7 Cylinders.

etc. Hence Figure 5-6A illustrates a *right hexagonal prism*.

Cube. The *cube* is a right square prism whose six faces (bases and sides) are all the same sized squares (Figure 5-6B).

Cylinder. A right, circular cylinder is a prism of infinite sides whose bases are circles. The cylinder is one of the most common geometric shapes used in sheet metal (see Figure 5-7A). A cylinder may be circular, elliptical, oval, or oblong depending upon the true shape of its right section. A cylinder is *right* when the specified bases are perpendicular to the axis and *oblique* when the bases are at some angle other than 90 degrees to the cylinder axis. Figure 5-7B illustrates an *oblique, elliptical cylinder*.

Pyramid. A pyramid is a geometric solid with triangular sides meeting at a common point called the *apex* of the pyramid. The base is a polygon which determines the name of the pyramid as triangular, square, octagonal, etc. A *right* pyramid has a base which is perpendicular to its axis (the axis is a line from center of base to the apex). The base of a regular pyramid is a regular polygon (see Figure 5-3). Hence, the pyramid in Figure 5-8A is a *right regular pentagonal pyramid*. A pyramid may also be *oblique* if the axis is not perpendicular to the base.

Cone. A cone is a *right* cone if the base is perpendicular to the axis, and is *oblique* (or *scalene*) if the base is not perpendicular to the axis. Cone bases are usually circular

Fig. 5-8

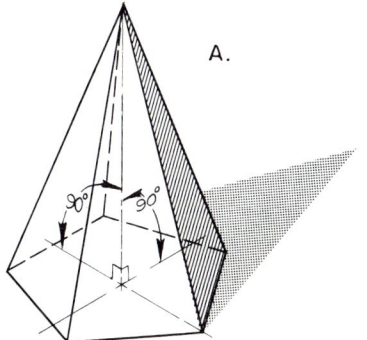
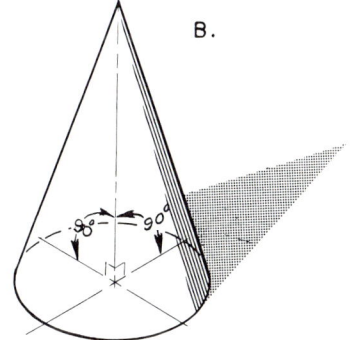

or elliptical. Right sections of cones may then also be either circular or elliptical. A *right circular cone* is shown in Figure 5-8B.

The *warped* cone is used in many sheet metal fittings and is constructed when the shape of the cone must change from base to frustum. Hence in the warped cone, the elements on the surface do not intersect at a single apex as they do for the right or oblique cone. The surface of a warped cone can be approximated by triangulation and will be discussed further in Chapter 11.

Elements. Ruled lines on the surface of geometric shapes are known as elements. On the surface of a prism or cylinder, they are parallel to the axis of the geometric solid. On the surface of a cone, they radiate from the apex to the base.

Frustum. If a cone or pyramid is cut by a plane parallel to its base, the portion between planes is called a *cone,* or *pyramid frustum.*

Truncation. If a cone or pyramid is cut by a plane not parallel to its base, the portion between the planes is called a *cone truncation* or *pyramid truncation.*

Symbols. Occasionally it is convenient to use standardized symbols or abbreviations in the text. Those used and their meaning are as follows:

Symbol	Meaning
$+$	add
$-$	subtract
$\times$	multiply, also separate dimensions
$\div$	divide
$=$	equals
$\therefore$	therefore
$\angle$	angle
$\llcorner$	right angle
$\perp$	perpendicular to
$\parallel$	parallel to
π	Greek letter pi; used to denote ratio of diameter to circumference (approximately 3.14)
$\alpha, \beta, \theta, \phi$	Greek letters alpha, beta, theta, and phi used to denote angles.

5-2 Parallel lines

By using the regular tools of the draftsman, any number of lines can be constructed parallel to a given line. A triangle and T square or two triangles are arranged as shown in Figure 5-9. First, line up the edge of the working triangle with the given line; support the triangle with the T square or another triangle. "Freeze" the supporting T square or triangle to the board; slide the working triangle to the location of the desired lines and draw them. Keep the working triangle in contact with the supporting edge.

5-3 Perpendicular lines

It should be noticed in Figure 5-9 that the hypotenuse of the working triangle is the edge used to slide against the support. The arrangement of Figure 5-9 may then be used to construct any number of lines perpendicular to a given line. Again, first line up the edge of the working triangle with the given line. Support the triangle with the T square or other triangle. Freeze the supporting T square to the board and *slide* the working triangle so that the perpendicular leg is in position to construct the desired lines.

5-4 To divide a line into a specified number of parts

a. To Bisect a Line. See Figure 5-10. Set the compass to any radius greater than one-half the length of the given line. Use each end of the given line as a center and strike intersecting arcs on each side of the given line. Connect the two arc intersections with a straight line, which is the required *perpendicular bisector* of the given line.

b. Three or More Parts. Use the *parallel-line method,* as illustrated in Figure 5-11, to quickly and easily divide a line of any length into any number of parts.

GEOMETRIC CONSTRUCTIONS

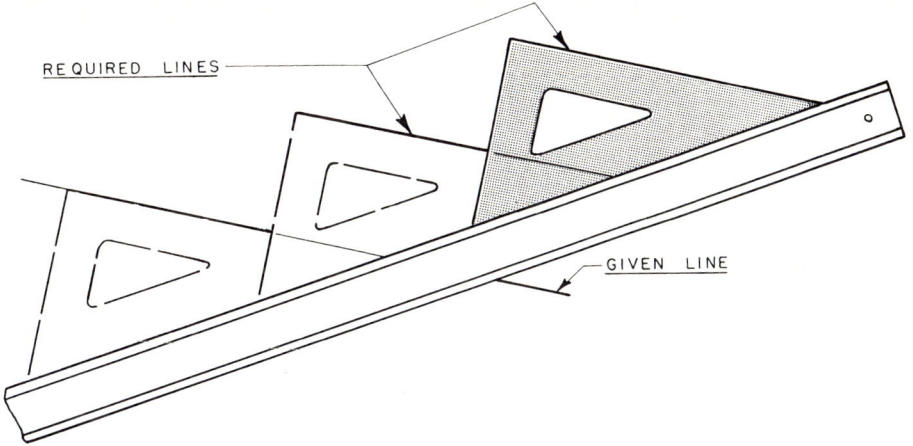

Fig. 5-9 Parallel-line construction.

A "measuring" line is constructed through one end of the given line (see Figure 5-11A). The desired number of parts are spaced along the measuring line using any convenient sized unit to represent one part (see Figure 5-11B). The last point is connected by a straight line with the terminal end of the given line. Lines then drawn parallel to this "connecting line" through the remaining points will divide the given line into the required number of parts (see Figure 5-11C).

5-5 To divide a circle into a specified number of parts

The sheet metal draftsman is often required to divide a circle or circle arc into a convenient number of equal parts. Circles may be divided into 3, 4, 6, 8, 12, or 24 parts using the T square and triangles only (see Figure 5-12). A circle may also be divided

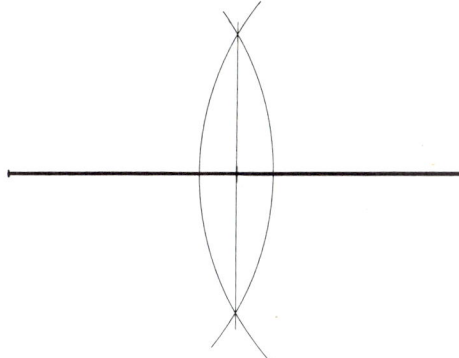

Fig. 5-10

Fig. 5-11

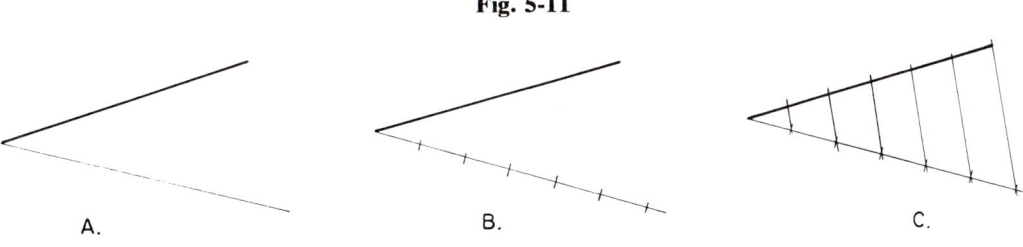

A.   B.   C.

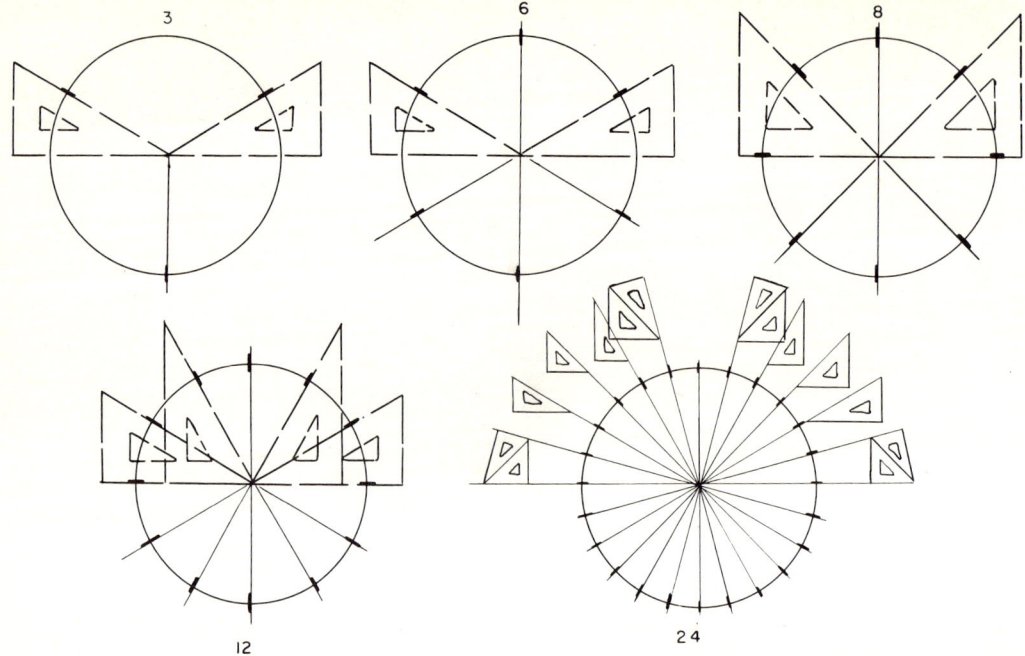

Fig. 5-12

into 6 parts by using the radius as shown in Figure 5-13. Circles may be divided into any number of equal parts by using the dividers and setting them by trial and error.

Fig. 5-13 Dividing a circle into six parts.

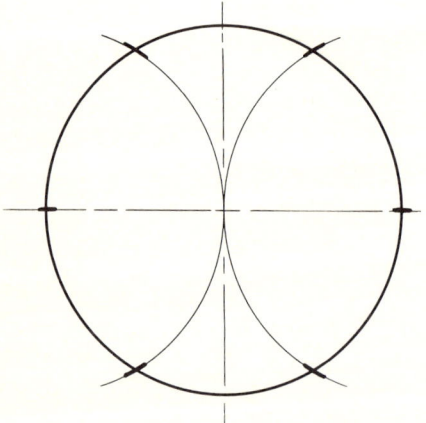

5-6 To divide an angle into a specified number of parts

a. To Bisect an Angle. Using any convenient radius and the vertex of the angle as a center, strike arc points on each leg of the angle. Using these points as centers and the *same radius each time,* strike intersecting arcs. A line through the intersection point and the vertex is the bisector of the angle (see Figure 5-14).

b. Three or More Parts. Strike an arc of any convenient radius with the vertex of the angle as center. Use the dividers and, by trial and error, divide the arc into the required number of equal parts. An angle may also be divided mathematically and the divisions made on the drawing with the aid of a protractor. For example, an angle of 105 degrees divided into seven parts would mean that each sector would contain 15 degrees.

GEOMETRIC CONSTRUCTIONS 49

5-7 To construct a line tangent to a circle

a. Through a Point on the Circle. A straight line is tangent to a circle when it makes contact with the circle arc at one point only. A radial line perpendicular to the tangent line determines the exact point of tangency; therefore, if a line is required tangent to a circle through a given point on the circle, draw the radial line. Use the triangles to establish a line through the point perpendicular to the radial line. (see Figures 5-15 and 5-9).

b. Through a Point outside the Circle. Align the triangle as in Figure 5-16A. Before drawing the tangent line, construct the radial line perpendicular to the tangent line. This determines the exact point of tangency.

5-8 To construct a line tangent to two arcs

The triangle is arranged as for perpendicular constructions (Figure 5-9). The arrangement is *aligned so as to produce the desired straight line tangent to both arcs.* Before drawing the line, however, construct a radial line for each arc perpendicular to the required line. Mark the exact tangency points by means of tics (see Figure 5-16B). Now the line may be drawn the correct length between *tic marks.*

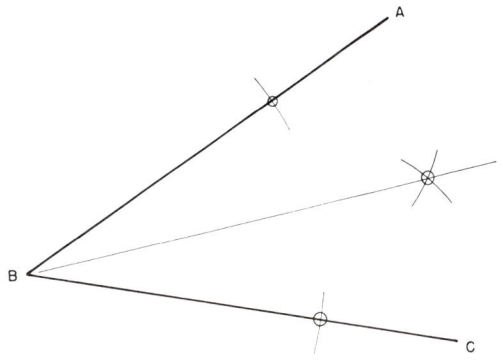

Fig. 5-14   Bisecting an angle.

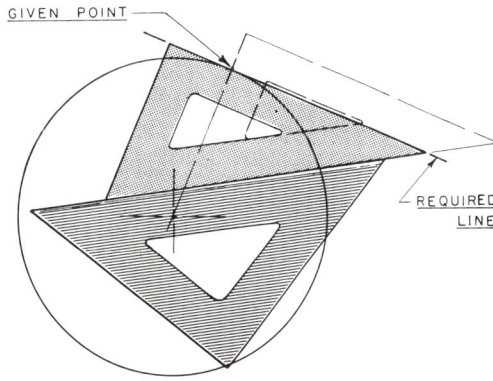

Fig. 5-15

Fig. 5-16

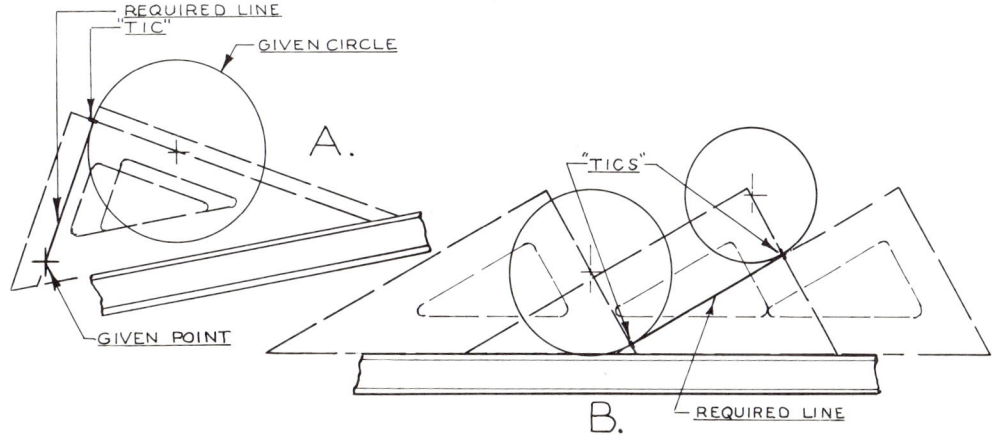

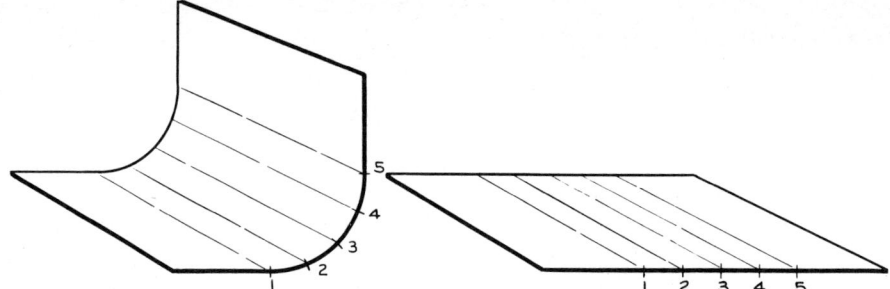

Fig. 5-17 Stepping-off a curved distance.

5-9 To find the length of an arc

The friction dividers provide a satisfactory means of measuring an arc or cylinder so as to provide an equivalent measurement "in the flat." Figure 5-17 illustrates the stretchout of a corner plate, in which the curved portion is divided into chord segments 1-2, 2-3, etc. These are transferred, by dividers, to a straight line.

Mathematics may be used to compute the length of metal needed to form a cylinder:

Circumference $= \pi D = 3.14 D$

(section 9-2)

If the angle of curvature is less than 360 degrees, use the following formula:

$$\frac{\text{Radius} \times \text{angle, degrees}}{57.3} = \text{circular distance}$$

For example, in Figure 5-18, a heel strip for a duct elbow is computed to be $28\frac{1}{4}$

Fig. 5-18 $\dfrac{18 \times 90}{57.3} = \dfrac{1620}{57.3} = 28.27$ or $28\frac{1}{4}''$.

$$\frac{457 \times 90}{57.3} = \frac{41\ 148}{57.3} = 718 \text{ mm}.$$

Fig. 5-19 Equilateral triangle made with T square and 30-60-degree triangle.

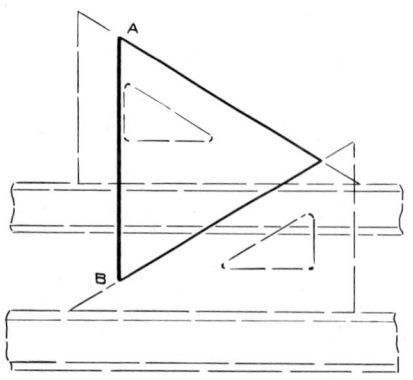
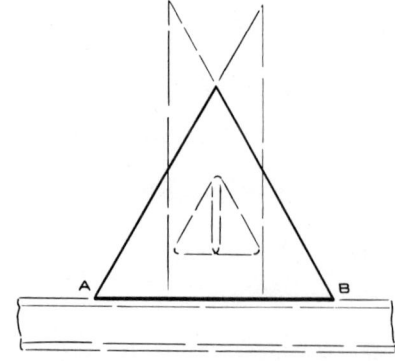

GEOMETRIC CONSTRUCTIONS

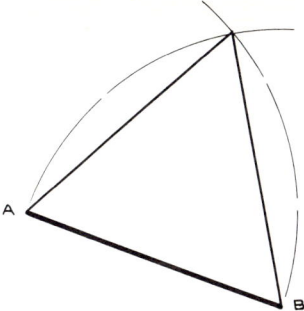

Fig. 5-20 Equilateral triangle constructed with a compass.

in [718 mm] (additional allowance necessary for seaming, and bend allowance necessary in heavy metal).

5-10 To draw an equilateral triangle

An equilateral triangle is also equiangular, and, therefore, may be drawn with T square and 30-60-degree triangle (Figure 5-19). This is a convenient method if one side is to be either horizontal or vertical. If one side is to lie along a given line, the compass may be used to locate the other sides (see Figure 5-20).

5-11 To transfer a figure by triangulation

If the lengths of all three sides of a triangle are known, a triangle of only one shape and size may be constructed. On the other hand, if the lengths of all four sides of a quadrilateral are known, a variety of figures may be constructed by varying the angles between adjacent sides. Therefore, if it is necessary to transfer a quadrilateral or other polygon to a new position, the figure is divided into triangular areas and these are constructed in the new position in their proper relationship to one another. This process is known as *triangulation* and is used extensively in the development of sheet metal patterns (Chapter 11). As an example, the quadrilateral *ABCD* of Figure 5-21 is reconstructed in a new position by dividing the figure into two triangles, *ABC* and *BCD*.

5-12 To enlarge or reduce a pattern

Because of the large areas involved, the original drawings for large sheet metal developments may be made to a reduced scale. Enlarged copies of the original pattern are

Fig. 5-21 Transferring a figure by triangulation.

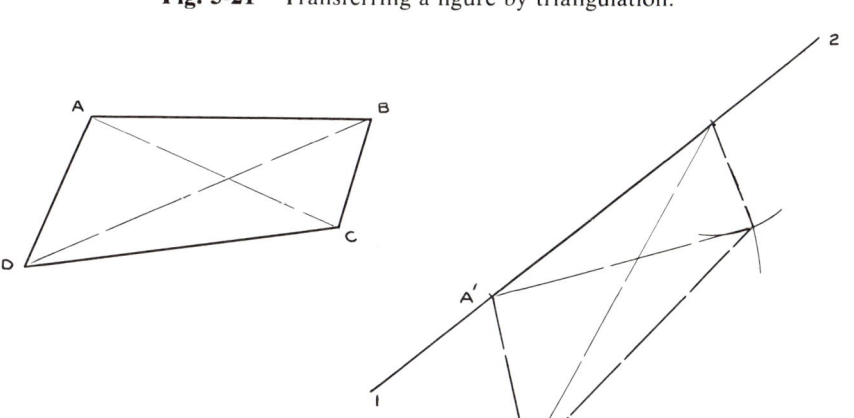

Fig. 5-22 Enlarging or reducing a figure.

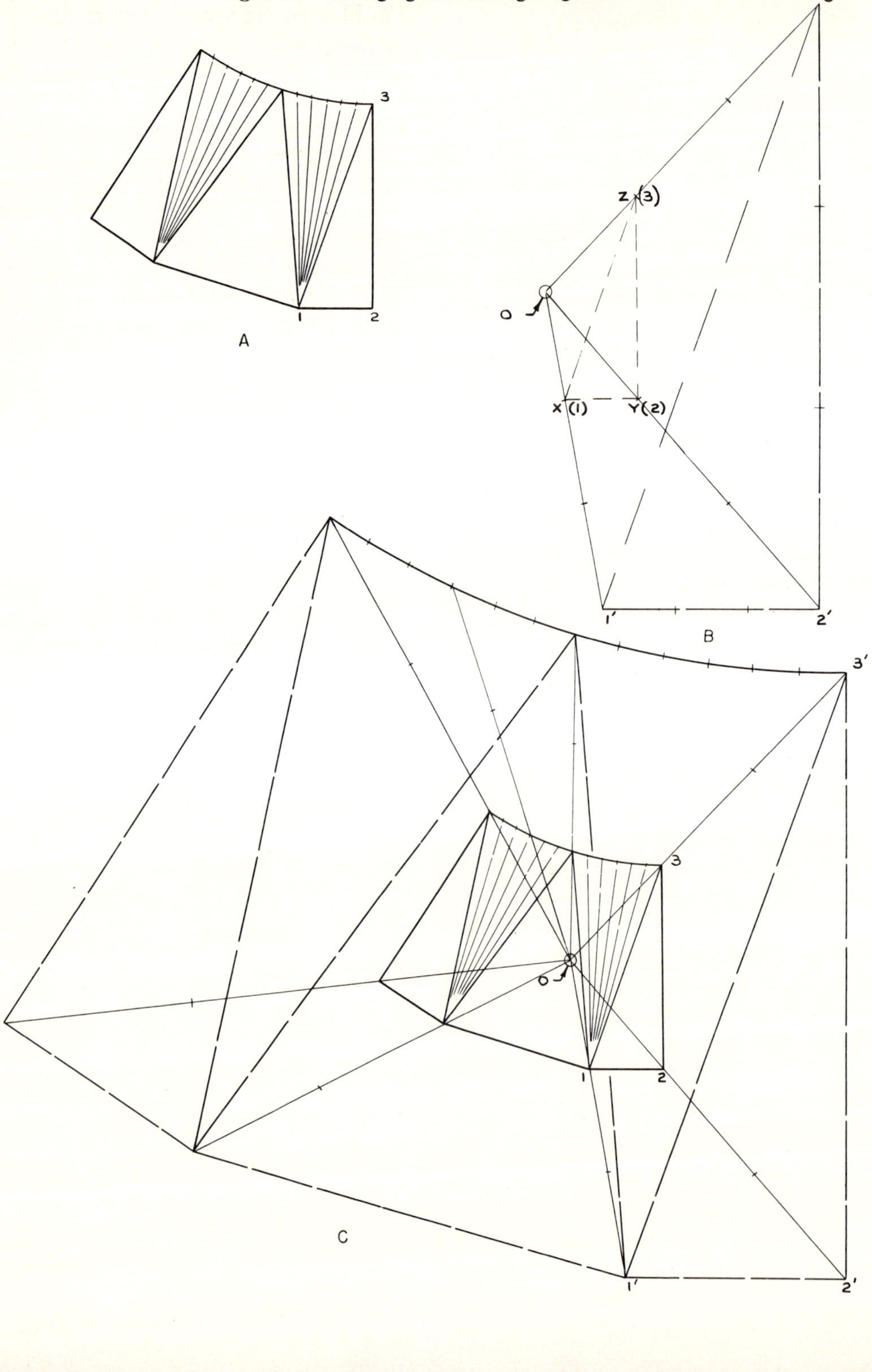

GEOMETRIC CONSTRUCTIONS

then made on, or transferred to, the working metal.

There are several methods of doing this.

a. Dimension to Scale (see section 2-9). The drawing and pattern are made to as large a reduced scale as possible. The pattern is measured to reduced scale and dimensioned full size. The *full-size* pattern is laid out on the metal from the dimensioned pattern.

b. Proportional Dividers (see section 2-19). Proportional dividers are useful only for minor enlargement or reduction because of the small size of the instrument.

c. Photographic Projection. Projection enlargers are used in the aircraft and ship-building industries to project and enlarge a pattern directly upon the working metal. An accurate drawing or slide is used as the original. The metal may be sensitized with an emulsion and the large sheets developed in tanks just as photo negatives are processed. In ship plate lofting, the master pattern is projected downward and the layout is made by workers directly upon the metal while the projection is maintained. Such methods require optical equipment of the highest quality in order to minimize distortion.

d. By Similar Triangles. Two similar triangles are two triangles with corresponding angles equal. Similar triangles may be of any size. The *corresponding sides of any two similar triangles are proportional.* This law of similar triangles may be used to enlarge or reduce, according to a specified ratio, the size of any pattern.

A half-pattern for a transition is shown in Figure 5-22*A*. As an example, increase the size of this pattern to *three* times the original size. The corner 1- 2- 3 on the original is a right angle. The corner 1'- 2'- 3' can therefore be made three times as large, on a corner of pattern material. Draw construction lines from 1', 2', 3' *to any arbitrary point O.* See Figure 5-22*B*. Points *XYZ* located one-third the length of each line from *O* form a triangle similar to 1'- 2'- 3' and the same size as 1- 2- 3 on the original. Use points *XYZ* to position the original pattern, as shown at Figure 5-22*C*. Lines are then drawn from *O* through the remaining critical points of the original pattern and the distance from *O* to the original point is added two more times outside the original. This increases the pattern size to three times the original.

Sometimes, in order to build a model of a large installation, a small pattern of a large fitting is desired. The same method, as used for enlargement, may be used for reduction; except that the lines from the critical points are drawn to converge at any arbitrary point *O*. The specified reduction is then applied to each line to determine points for the model pattern.

5-13 To construct a hexagon

a. Flat Diameter Specified. A regular hexagon has six sides of equal length and six internal angles of 120 degrees. If the diameter across flats is specified, the hexagon may be drawn as shown in Figure 5-23. Begin by constructing a circle of the specified diameter. Two sides of the hexagon are constructed tangent to the circle, parallel to each other on opposite sides of the circle, and in the required direction. The remaining four sides are drawn tangent to the circle and at 120 degrees to the first sides. Use the 30-60-

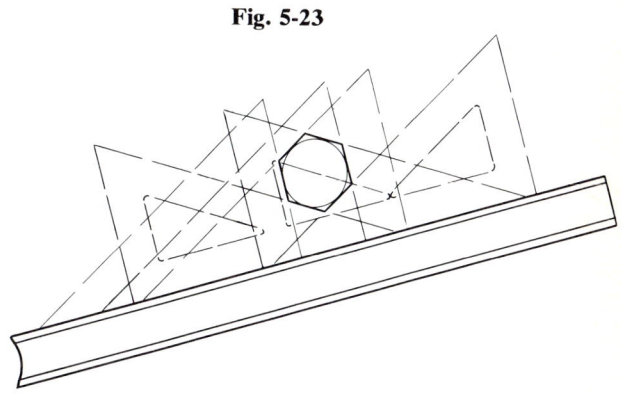

Fig. 5-23

SHEET METAL DRAFTING

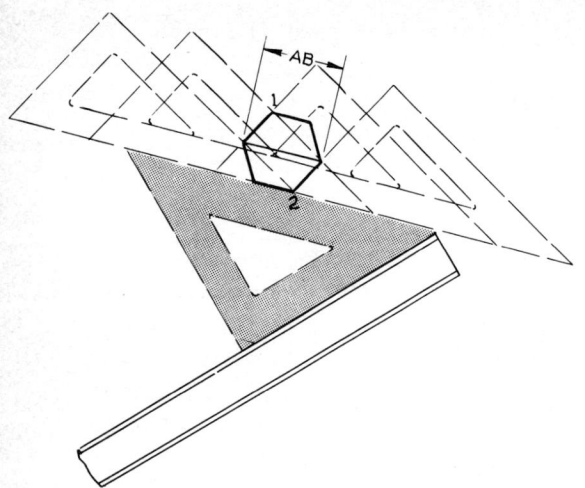

Fig. 5-24

and *B* at both 30 and 60 degrees to *AB*. Thus points 1 and 2 are located and the hexagon is completed by drawing lines through 1 and 2 parallel to *AB*.

Another method is shown in Figure 5-13. Construct a circle of the specified diameter. Use the radius distance, begin at a specified point, and locate the remaining corners of the hexagon.

degree triangle arranged on the T square in two positions and slid along the blade to produce opposite sides parallel (Figure 5-23).

b. Corner Diameter Specified. If the diameter across corners is specified (long diameter) the hexagon may be drawn as shown in Figure 5-24. *AB* is the given diameter. Use the 30-60-degree triangle above the 45-degree triangle to construct lines through *A*

5-14 To construct an ellipse

The ellipse is a common shape encountered in sheet metal fittings (section 5-1; Figure 5-4). There are several methods for constructing true ellipses which range from geometric to mechanical, involving pins, strings, and trammels. Many ingenious devices, called *ellipsographs,* have been patented which mechanically produce ellipses to specification. For all practical purposes however, there are three methods which work well in sheet metal drafting, namely, the *trammel method,* the approximate *four-center method,* and, for large work, *the pin and string method.*

Fig. 5-25 Drawing an ellipse by the trammel method.

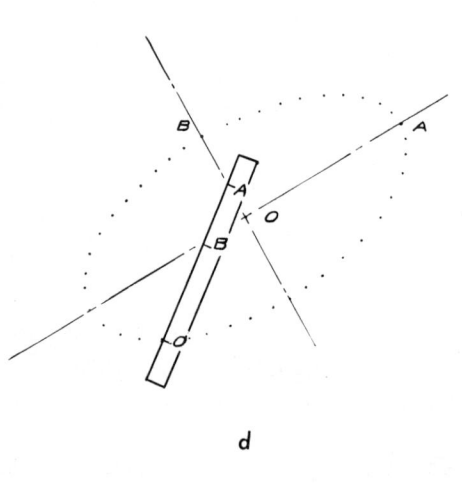

d

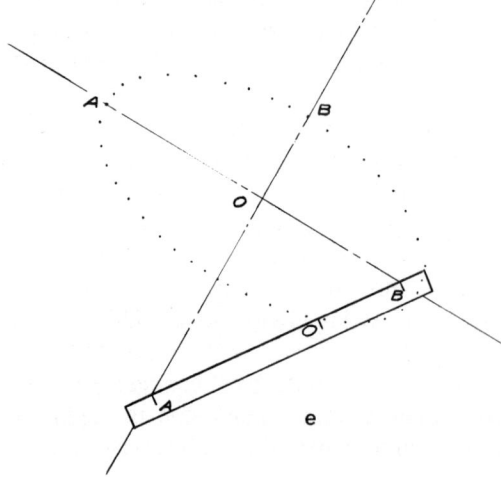

e

GEOMETRIC CONSTRUCTIONS

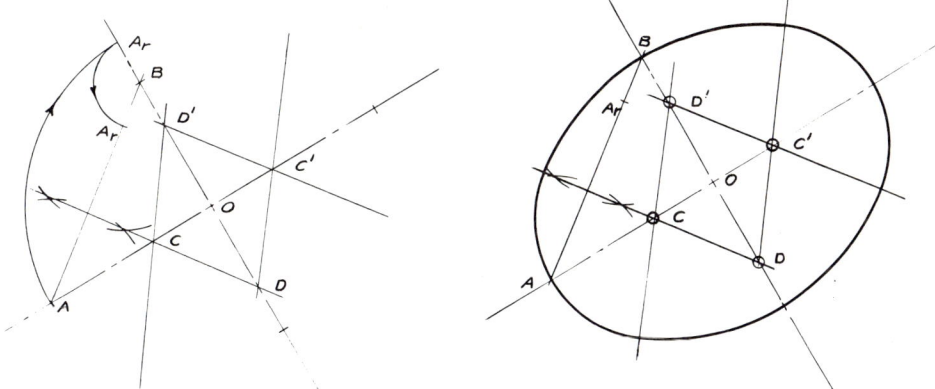

Fig. 5-26 The four-center method.

a. The Trammel Method. The major and minor diameters of the required ellipse, as shown in Figure 5-25, are determined and located on the center lines. A *trammel* is made from heavy paper, wood strip, or a strip of scrap sheet metal. The *semimajor* axis OA, and the semiminor axis OB are marked on the trammel edge. This may be done so that the distances overlap, having a common starting point O as in d, or the distances may "add up," as in e. In either case, the three points on the trammel are used in d and e as follows. The semiminor point B is placed on the major axis; the semimajor point A is placed on the minor axis; the common point O is used to locate one point on the curve. Additional points are located by resetting the trammel slightly. Be extremely careful to keep the points A and B exactly on the *opposite* center lines. When a sufficient number of points have been located, the curve is completed as discussed in section 3-14.

b. The Four-center Method. This method uses four tangent circular arcs to produce a figure which closely *approximates* the true ellipse. The similarity between the approximation and the true curve increases as the difference between major and minor diameters decreases; as obviously, the curvature becomes more circular in character.

The major and minor diameters of the required ellipse, as shown in Figure 5-26, are determined and located on center lines. Subtract the semiminor diameter OB from the semimajor diameter, AO. The remainder A_rB is subtracted from the hypotenuse AB. Construct the perpendicular bisector of the line AA_r thereby locating C and D. Duplicate the distance OC as OC^1 and the distance OD as OD^1. Connect them, as shown in Figure 5-26, to determine the tangency points of the four arcs. Draw the two minor arcs using C and C^1 as centers with AC as radius. Draw the major arcs using D and D^1 as centers with BD as radius.

c. The Pin and String Method. A true ellipse exists when the sum of the distances from the two foci to any and all points on the curve remain at one value—this value being equal to the major diameter of the ellipse (see Figure 5-4). The pin and string method is a mechanical rigging whereby this constant value is maintained while tracing the curve.

The major and minor diameters of the

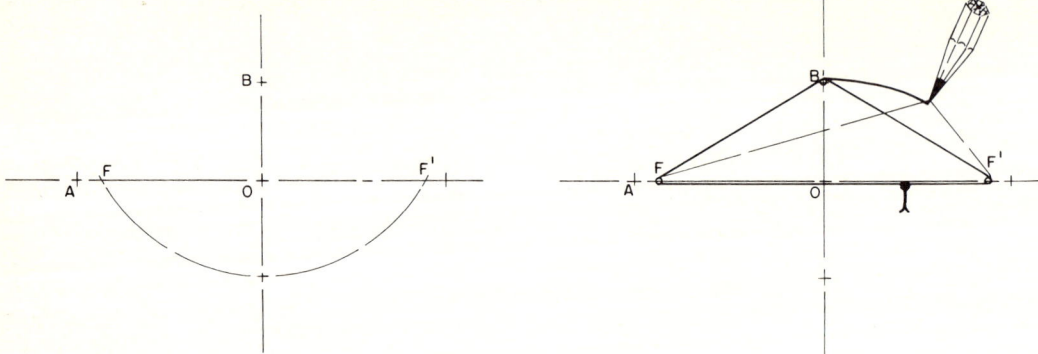

Fig. 5-27 The pin and string method.

required ellipse, as shown in Figure 5-27, are determined and located on center lines. With *B* as center and *OA* as a radius, an arc is struck which establishes the foci *F* and *F'*. Pins are located at *F, F'*, and at *B*. Use a nonstretchable string and tie a tight loop around the three pins, or use a fine wire and splice it securely to form a tight loop. Remove pin *B* and substitute a marking device. A true ellipse, having the required major and minor dimensions, will be generated by moving the marking device within the loop while the loop is held taut. This is a good method for all large ellipses.

5-15 Conic sections

The sheet metal draftsman is often confronted with a development problem which involves the intersection of a plane and a cone. Therefore, the draftsman should be acquainted with, first, the *types* of curves produced when right circular cones are cut at various angles by planes; and, second, the methods for constructing the resulting curves.

The four possible curves, called *conic sections*, are the *circle*, the *ellipse*, the *parabola*, and the *hyperbola*.

a. Circle. A right circular cone cut by a plane perpendicular to its axis results in a conic section which is a circle (Figure 5-28*A*). The radius, or the diameter, may be obtained from a view of the cone which shows the plane as an edge.

b. Ellipse. If the cone is cut by a plane which makes a greater angle with the axis than do the elements of the cone, the resulting conic section is an ellipse (Figure 5-28*B*). The major diameter may be obtained from a view of the cone which shows the plane as

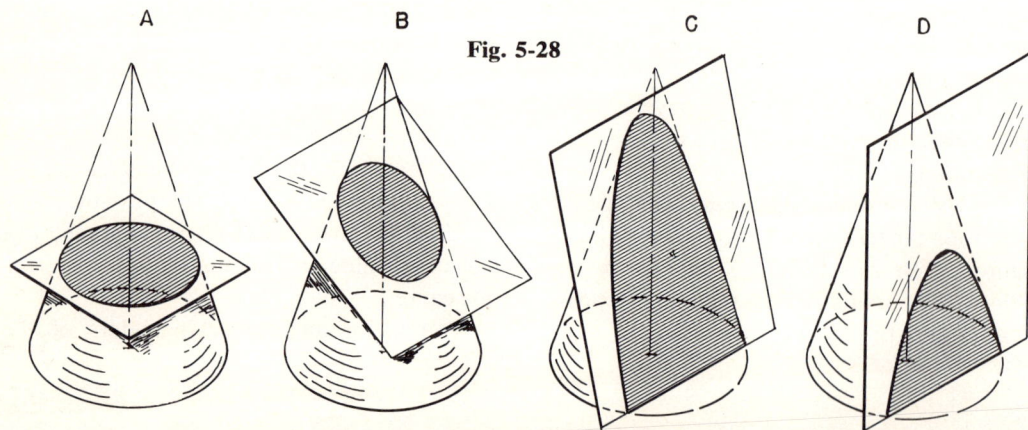

Fig. 5-28

an edge and the minor diameter as a point (see Figure 7-13 and section 7-7). The ellipse may then be constructed, using the methods described in section 5-14.

c. Parabola. This conic section is produced when a right circular cone is cut by a plane which makes the same angle to the axis as the cone elements (Figure 5-28C).

Mathematically speaking, a parabola is defined as a curve produced by moving a point so that its distance from the *focus* (a fixed point on the axis) is always equal to its distance from the *directrix* (a straight line perpendicular to the axis).

Two methods will be given for constructing a parabolic curve. In the *tangent method,* Figure 5-29, the three points *A, B,* and *C* are given. Make *EC* equal to *CD.* Divide the tangent lines *AE* and *EB* into the same number of equal spaces (five in the example). Number the points in opposite directions, omitting the previously lettered end points. Straight lines connecting the correspondingly numbered points represent tangents of the required parabola. The curve is drawn tangent to these lines at the mid-point of their enclosed portion.

In the *parallelogram method* of Figure 5-30, three points *A, B,* and *C* are given. Construct the enclosing rectangle. Divide each half into the same number of spaces vertically and horizontally. Number them as shown. Connect the vertical points by straight lines to *C.* Drop vertical lines from the horizontal points to locate the required points on the curve. For practical applications of the parabola see section 10-8c.

d. Hyperbola. The hyperbola is produced by cutting a right circular cone with a plane which is parallel to the cone axis but does not intersect the axis, or by cutting with a plane which makes a *smaller* angle to the axis than do the elements of the cone (Figure 5-28D).

Mathematicians define a hyperbola as a plane curve produced by moving a point so that the difference in distance from the point to two fixed points, called foci, is a constant value; (the constant is equal to the transverse axis of the hyperbola).

The most practical method for constructing a hyperbolic curve, when required, is to make the necessary drawing and plot the points for the curve as discussed in section 10-8d.

Fig. 5-29 Drawing a parabola by the tangent method.

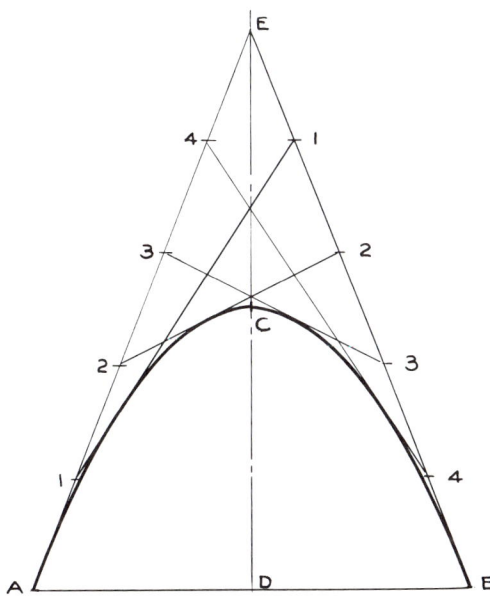

Fig. 5-30 Drawing a parabola by the parallelogram method.

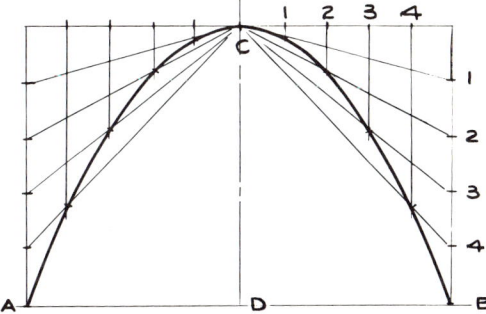

5-16 Ogee curve

Many sheet metal designs involve a reverse or ogee curve. It is often necessary, for example, to have offsets in air-conditioning and heating ductwork. In aircraft, spacecraft, and shipbuilding, offsets are likewise sometimes required for conduit and hydraulic systems. In many product designs, a reverse curve of the proper proportions is desirable from an aesthetic viewpoint.

In Figure 5-31*A*, the parallel lines 1-2 and 3-4 represent a construction which is to be connected using an ogee curve. A straight line is drawn connecting the initial and terminal points of the curve (2-3 of Figure 5-31*B*). Next, *the cross-over or tangency point* (5) of the two curves must be placed. (See Figure 5-31*C*) The cross-over point can be placed arbitrarily or perhaps its location is dictated by the specific design conditions. In all cases, knowing that the curves flowing into the straight lines are *tangent* to the straight lines at 2 and 3, perpendiculars to the straight lines can be erected which will contain the center of the desired arcs (Figure 5-31*C*). The center for one arc is found by constructing a perpendicular bisector of either 2-5 or 5-3. The perpendicular bisector crosses the straight-line perpendicular at one of the required centers (circled in Figure 5-31*D*). A straight line through this center point and the tangency point 5 locates the second center point.

Ogee construction principles are applied in the design of curved duct elbows (section 9-16).

Fig. 5-31

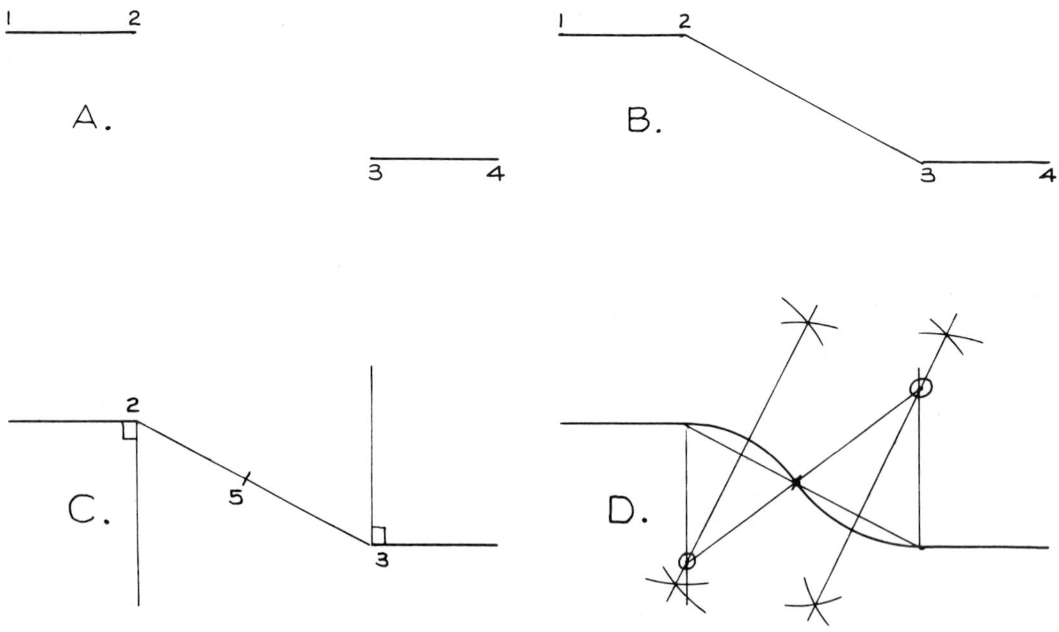

GEOMETRIC CONSTRUCTIONS

EXERCISES

1. What unit of measurement is most commonly used for measuring angles?
2. When are two angles complementary? Supplementary?
3. What is the difference between an oval and an oblong shape?
4. By means of a sketch, show the draftsman's arrangement of T square and triangle to construct parallel and perpendicular lines.
5. By means of a sketch, illustrate the parallel line method of dividing a line into a number of parts.
6. What is the true length of an arc of 5-in [127-mm] radius:
    a. When the angle of curvature is 80.2 degrees?
    b. When the angle of curvature is 57.3 degrees?
7. What is the shape of the conic section, if a cone is cut by a plane which makes a *greater* angle with the axis than the angle between the cone elements and the axis?
8. What are tic marks used for?
9. Name two industries in which photographic enlargements of patterns are used.
10. Name three methods for establishing a plane surface on a drawing.
11. Use the trammel method and construct an ellipse measuring 3¼ by 6 in [83 by 152 mm].
12. Use the four-center method and construct an ellipse measuring 4 × 7 inches.
13. Construct an oval. Use radii of 1½ and 2 in [38 and 51 mm] with centers spaced 4½ in [114 mm] apart.
14. Construct a figure to represent the open end of an oblong pipe. Use 2-inch radii spaced 4 inches center to center.
15. Using triangulation, transfer the true size pattern of Figure 8-5 to a sheet of drawing paper. Use similar triangles to enlarge the pattern to four times its transferred size.
16. Construct the front view of a reverse curve duct elbow as specified in Figure 5-32.
17. Compute the approximate surface amount of metal required for the offset fitting of problem 16.

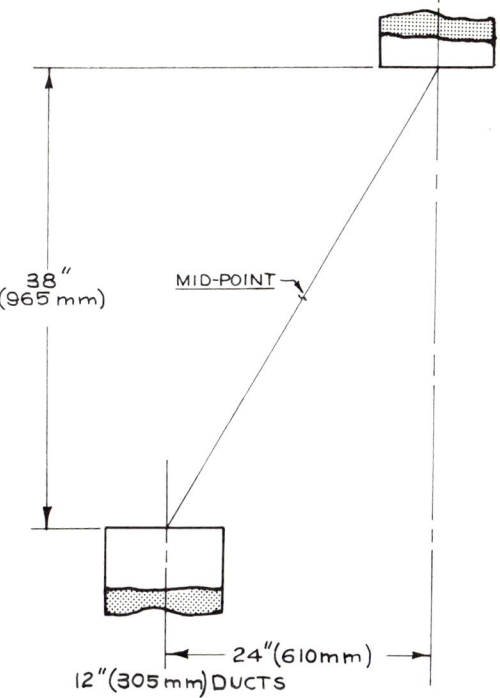

Fig. 5-32

chapter 6

ORTHOGRAPHIC PROJECTION

Probably the first drawings used to describe a job were crude clay scratchings, or paintings in the form of a picture. History relates that drawings of one type or another were used in building long before the time of Christ.

Objects to be constructed of sheet metal may be described to the workers by means of written or spoken language, pictures, or drawings. Each of these methods has certain advantages but each also has limitations. Perspective drawing or a picture, as the eye sees the object, is very descriptive, but it was soon discovered that it could be given many different interpretations, and that it did not convey exact shape, size, and function. With the coming of the Industrial Revolution in Europe and America, a system of projection drawing which did convey exact shape, size, and function, was studied and perfected. This system is called *orthographic projection*.

In practice, therefore, the advantages of the *written language* are exploited in written specifications, and the advantages of the *graphic language* are used to communicate by drawings.

6-1 Planes of projection

Orthographic projection is a form of multiview representation of an object. The lines of sight from the observer to the object are parallel lines. As shown in Figure 6-1, a *plane of projection* is an imaginary plane placed between the observer and the object. The plane of projection is perpendicular to the *orthographic projectors* (lines of sight). An *orthographic view* is thereby produced upon the plane.

6-2 Regular views

It is possible to completely enclose an object within six projection planes, the planes being either parallel or mutually perpendicu-

Fig. 6-1 An orthographic view.

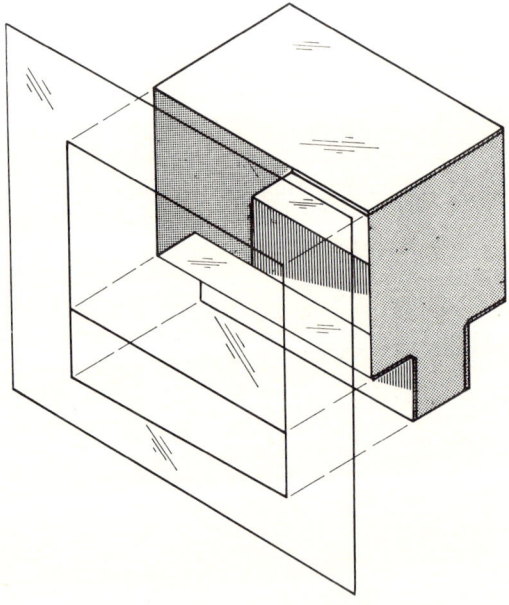

ORTHOGRAPHIC PROJECTION

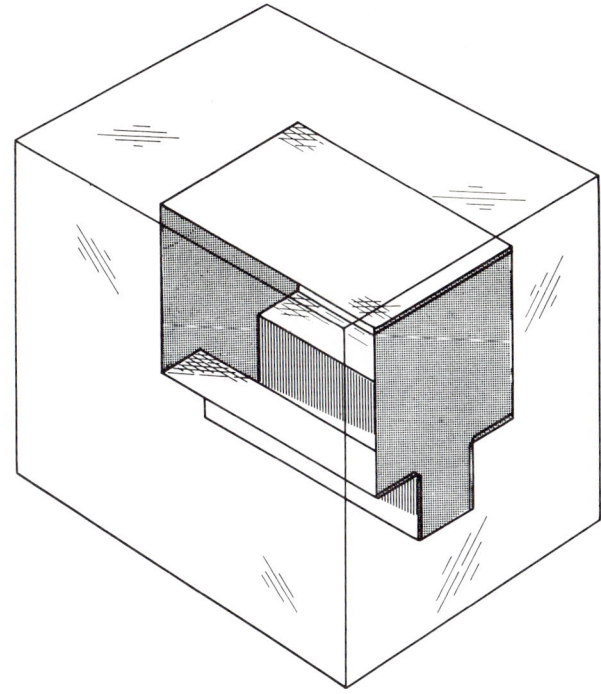

Fig. 6-2 Principal planes.

Fig. 6-3

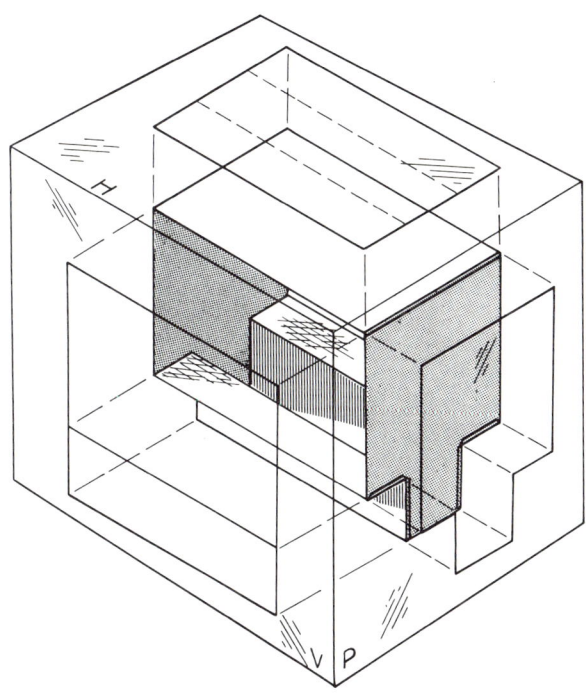

lar (boxed in by planes). This is illustrated in Figure 6-2. If orthographic projectors are extended to each of the six planes of projection, six different views of the object are obtained. These views are related to each other just as the planes are related. These six views are commonly called the *regular* or *principal* views of an object. They are named, according to the plane of projection involved, as follows:

 a. Front View or *Front Elevation.* This view is sometimes called a vertical projection because it is projected upon the vertical plane (abbreviated F or V, see Figure 6-3).

 b. Top View or *Plan* or *Horizontal Projection.* This view is projected upon the horizontal plane and is abbreviated H. (See Figure 6-3.)

 c and *d. Side View* or *Side Elevation, End View* or *End Elevation.* This view is sometimes called a profile projection because it is projected upon the right or left profile plane and correspondingly is named right or left (abbreviated P).

 e. Bottom View. This view is theoretically projected upon a horizontal plane, but, to avoid confusion, the plane is appropriately called the bottom plane of projection.

 f. Back View. This view is relatively unimportant and is seldom used.

6-3 Third-angle projection

Observe in Figure 6-3 that only two of the three possible principal dimensions of an object can be represented in a single view. Therefore, it becomes necessary to relate, according to a standardized arrangement, two or more views in order that a complete representation of shape and size, hidden as well as visible, be conveyed.

The relationship of the *six principal* views and the standard arrangement of them is shown in Figure 6-4*A*,*B*. This arrangement of views on a single paper plane is produced by "unfolding" or "flattening out" the projection

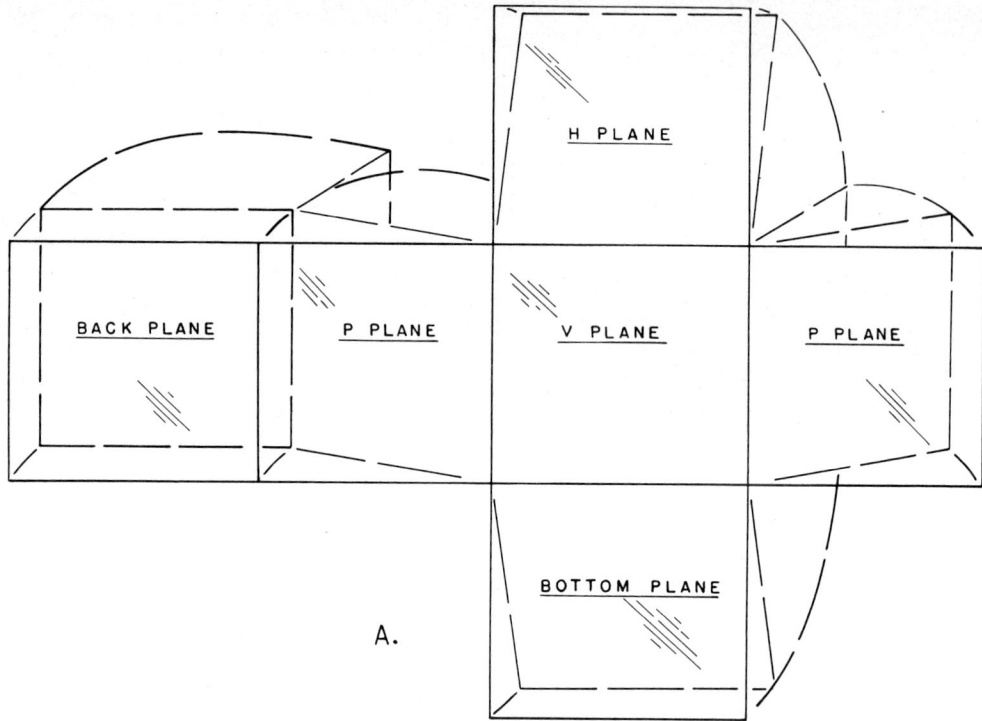

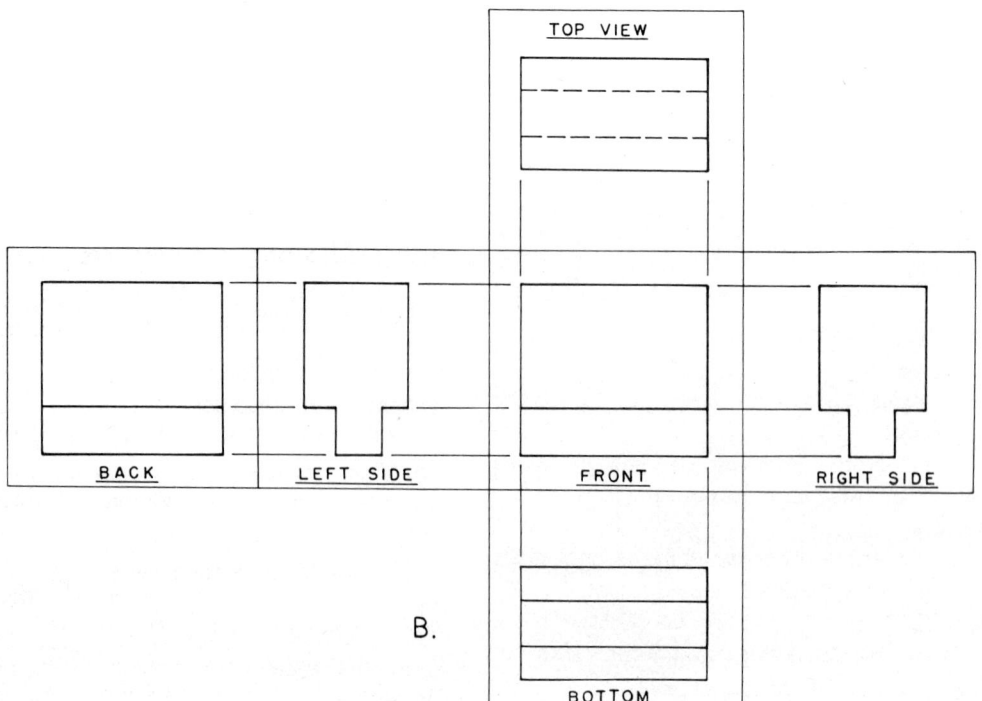

Fig. 6-4

ORTHOGRAPHIC PROJECTION

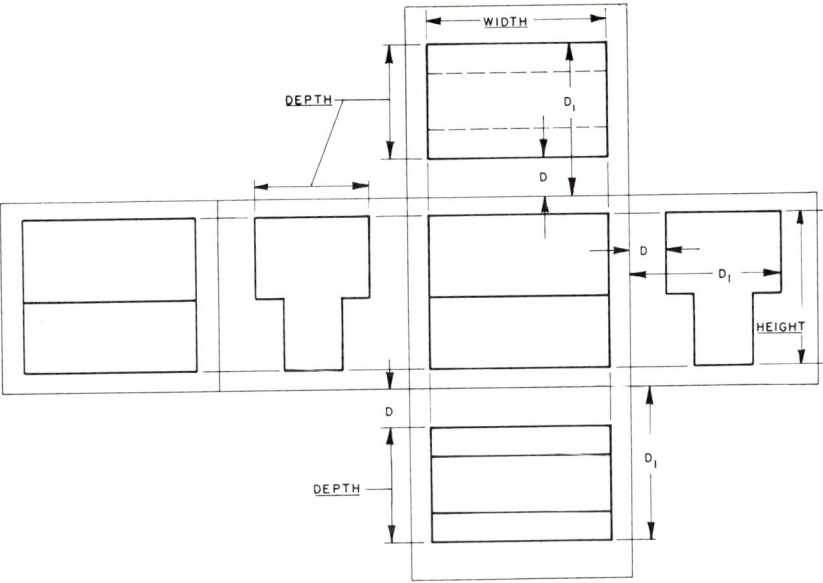

Fig. 6-5

box by using certain edges as imaginary "hinge lines." A common error of beginner and experienced draftsman alike, is to unfold the views using the back edges of the projection box as hinges. This results in reversed views. Pay particular attention to the hinge lines in Figure 6-4A, and avoid the above error by noting that the front faces of the object, as seen in the top, bottom, and side views of Figure 6-4B are *toward* the front view; *they are the closest faces to the front view*. The standard arrangement of views, as shown in Figure 6-4A,B, is approved by the American National Standards Institute and is the basis for all industrial drawing in the United States. The principal views *must* always occupy the following positions: the top view in line with and projected above the front view; the bottom view in line with and projected below the front view; the right-side view projected to the right of the front view; the left-side view projected to the left of the front view; the back view on the same projectors as side and front views but placed to the left of the left-side view.

Because of the relative positions of observer, projection plane, and object, the arrangement is called *third-angle projection*. In this system, all views are obtained by projecting the object *through the plane* to the observer.*

For practical purposes, depending upon the object, only two or three views are needed to convey complete shape and size, or to solve a development problem. If a side view is needed, only one is necessary since both contain the same information. However, one is usually better than the other. The same may be said for top and bottom views, and for front and back views.

Figure 6-5 shows six views of an object.

* *First-angle projection,* which is standard in many European countries, places the object in front of the plane, between observer and plane, so that the view is obtained by projecting the object "on the plane."

In the past, some sheet metal drawings have been "first angle" or a combination of both methods. While this may result in certain short cuts and space-saving techniques, it has also led to confusion on the part of the student or beginner. In this textbook all drawings are "third angle" in order to conform to American standards.

Notice that *height* is represented in the front, two side, and back views, and that it may be projected from one view to another. *Width* is shown in the top, front, bottom, and back views and again, except for the back view, projection is convenient. *Depth* appears in the top, bottom, and two side views and may be transferred from one to another, as shown in Figure 6-5, by using the "folding lines" as reference when transferring distances D and D_1.

6-4 Position of lines in space

Any two adjacent views of a line will completely describe the position of the line in space. With practice, every student of sheet metal drafting can learn to "read" two related views to the extent that the line can be completely visualized in space. As an example, study the two views of line AB of Figure 6-6A. The student should be able to hold up a pencil and place it in the same relative position in space as AB is described orthographically in the figure.

All lines parallel to the H plane are known as horizontal lines. AB in Figure 6-6A is but one example. The horizontal information is obtained from the front view where the line appears level and parallel to the edge view of H; therefore, the true length of the line (abbreviated TL) projects to the top view and may be measured there. For all horizontal lines, the true angle θ_v that the line makes with V can also be measured in the top view. This is true only for horizontal lines since for all lines other than *level* lines, the angle will project, in the top view, larger than its true value (horizontal lines are inclined to V and P).

Fig. 6-6 Regular views of lines.

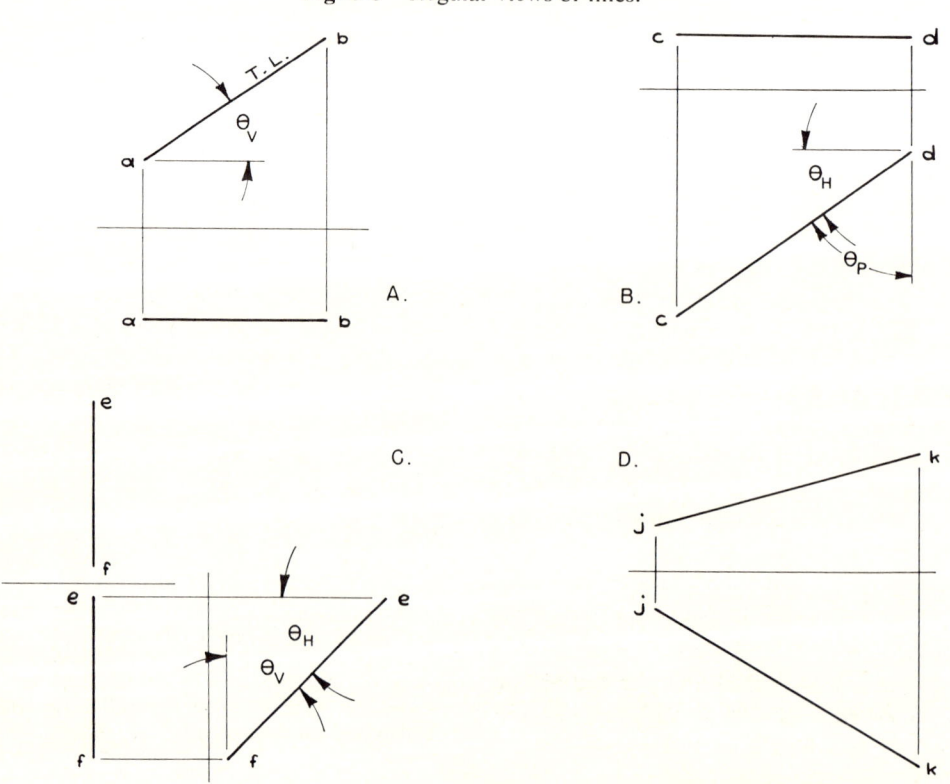

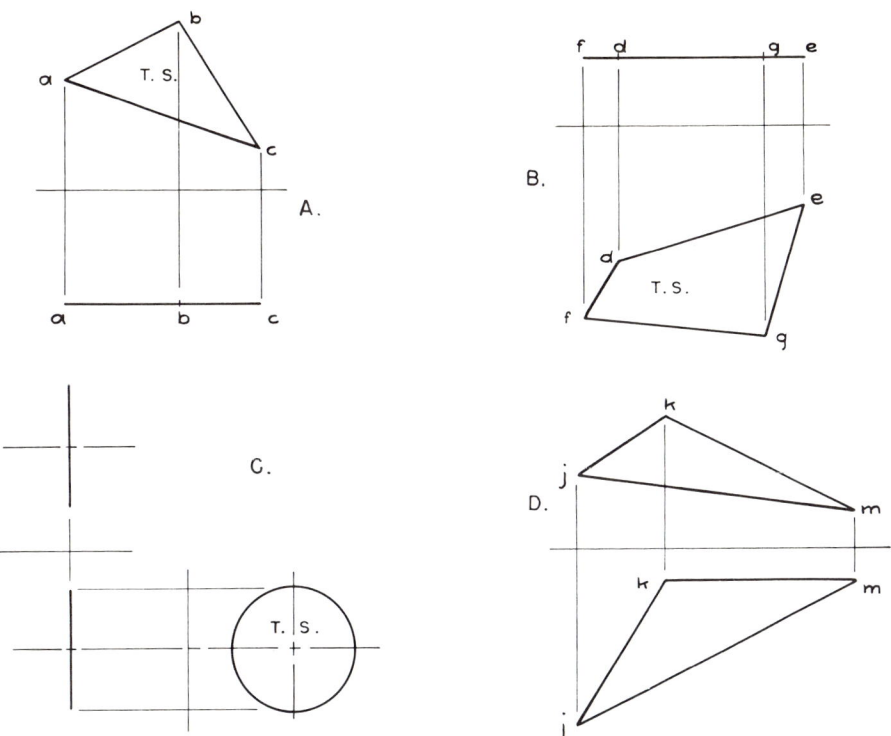

Fig. 6-7 Regular views of plane figures.

Notice that the "hinge" line between top and front views may be visualized as either the edge view of H, when viewed from the front, or as the edge view of V when viewed from the top.

A line parallel to V and inclined to H and P is known as a *frontal line*. In Figure 6-6B, the line CD is a frontal line. Note that: (1) The line appears parallel to V in the top view. (2) It makes the true angle θ_H with the edge view of H as seen in the front view. (3) The true length of the frontal line is *always* measurable in true value in the front view.

A line parallel to P and inclined to H and V is called a *profile* line. The top and front views of the profile line EF of Figure 6-6C both show the line to be parallel to the edge view of the profile plane. A side view is necessary if the true length of the line and the true angles θ_V, θ_H are desired.

A line which makes acute angles with all three principal planes is known as an *oblique* line. The line JK in Figure 6-6D is a typical oblique line. Notice that no true angles are measurable in either view nor does the true length of the line project in any regular view. The development of sheet metal patterns is primarily based upon determining the true lengths of oblique lines (which is discussed in section 7-4).

6-5 Position of planes in space

The plane surfaces of objects may be classified as *horizontal, frontal,* or *vertical, profile, inclined,* and *oblique,* depending upon whether they are parallel or inclined to the principal planes of projection.

For example, a horizontal surface is one

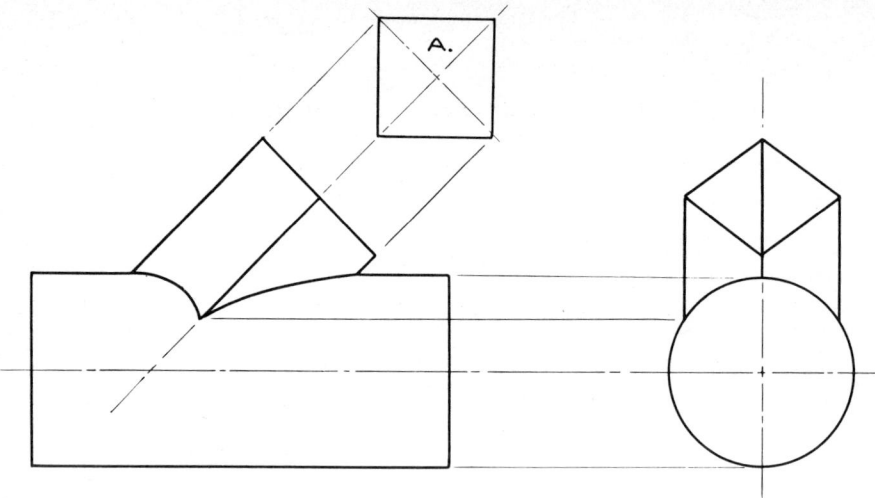

Fig. 6-8 Auxiliary view showing pipe shape.

which is parallel to the H plane and perpendicular to V and P. A horizontal plane figure, as *ABC* in Figure 6-7*A*, will project true shape and size (abbreviated: TS) upon the H plane.

A frontal plane figure, represented as *DEFG* in Figure 6-7*B* is parallel to V and perpendicular to H and P. The front view is the true size and shape of the figure.

A profile surface is shown parallel to P and perpendicular to H and V in Figure 6-7 *C*. A profile view (side view) shows the true size and shape of the surface.

An inclined surface is a plane which is perpendicular to one of the principal planes and is inclined to the other two. An inclined surface appears as an edge in one regular view and foreshortened in the other two.

An oblique plane is inclined to all three principal planes of projection. It will therefore appear foreshortened in all regular views (see Figure 6-7*D*). The development of sheet metal patterns is essentially the determining of the true size and shape of the inclined and oblique surfaces of the object (section 7-7).

6-6 Choice of views

The three basic dimensions of an object—height, width, and depth—may be determined from two related views. To describe the shape more adequately, three views are sometimes preferred. By taking both shape and size into careful consideration, most objects are best described by combining the desirable features of two regular views with the important features of partial and special views.

Such is the case in Figure 6-8. The front view of the intersection is desired because the elements of both pipes appear true length there. The side view is valuable because it indicates the shape of both pipes and helps in the plotting of the line of intersection. A special view *A* is needed in order to plot the line of intersection and to indicate the true shape and size of the square pipe. The purpose of this discussion of views is to point out the importance of sketching, selecting, planning view arrangements, and of analyzing the results thus obtained. To help in the selection of views and the positioning of the object on the drawing, remember the following axioms:

ORTHOGRAPHIC PROJECTION

a. **Elements** of a sheet metal object will project true length on a plane of projection only *when they are parallel to that plane.*

b. **Surfaces** of a sheet metal object will project true size on a plane of projection only *when they are parallel to that plane.*

c. The "girth," or true distance around a pipe, will show in a view in which the axis of the pipe appears as a point. This view is also the true shape of a *right section* of the pipe.

d. Two related views are necessary in order to obtain the true length of a line or the true size of a surface.

e. A circular cylinder will have, in all possible views, one dimension equal to its diameter.

6-7 The need for special views

The special view *A* of Figure 6-8 is necessary because that is the only view showing the girth and true shape of the square pipe. In most sheet metal problems, the true length of lines or the true size of surfaces do not show unless numerous special views are made. These special views are called *auxiliary views*. They are projected upon special planes called *auxiliary planes of projection*. Draftsmen must be thoroughly familiar with the basic principles involved in the construction and application of auxiliary views. The following chapter is concerned with the basic theory and the procedures for drawing and reading all required auxiliaries.

EXERCISES

Problems 1 through 8 (Figure 6-9). Using dimensions suitable for the size of paper available, complete the given views and draw three additional views of each object. (Omit the back or rear view.) Consider the objects as hollow or open. Sketch noncircular curves. Show all lines, both visible and hidden, in all views.

Problems 9 through 16 (Figure 6-10). Directions same as for problems 1 through 8.

Problems 17 through 22 (Figure 6-11). Using dimensions suitable for the size of paper available, draw three views which best describe the shape of each object. (Use the indicated view as front view.) Show all visible and hidden edges, including brake lines.

Problem 23

a. Construct a top, front, and right-side view of a *horizontal* line, *AB.* The line is to be 3¼ in [83 mm] long, making an angle of 30 degrees with V (*A* is to be to the left, and in front of *B*).

b. Construct the top, front, and left-side views of a *frontal* line, CD. The line is to be 2½ in [64 mm] long, making an angle of 30 degrees with H (it is to slope downward to the right from *C*).

c. Construct the top, front and right-side views of a *profile* line, EF. The line is to be 3 in [76 mm] long, making an angle of 60 degrees with H (*F* is above and behind *E*).

Problem 24. Draw top, front, and right-side views of a horizontal, regular, hexagonal plane figure with one long diameter of 3 in [76 mm] perpendicular to V.

Problem 25. Draw top, front, and right-side views of a frontal elliptical plane figure. The minor diameter equals 2 in [51 mm], the major diameter equals 3¼ in [83 mm]. The major diameter to slope downward to the right at 15 degrees with H.

Problem 26. Draw top, front, right-side, left-side, and bottom views of a circular, profile, plane figure. The diameter of circle is to be 2 in [51 mm].

Problem 27. Draw the top, front, and right-side views of an open right circular cylinder 2½ in [64 mm] in diameter, 4 in [102 mm] long, in such a position that the elements of the cylinder are true length in front and top views.

Problem 28. Draw the top, front, and right-side views of a right square pyramid whose base is 2 in [51 mm] square and whose altitude is 3½ in [89 mm]. Position the base at 30 degrees to H and at 90 degrees to P, with the apex above and in front of the base.

Fig. 6-9 Problems 1 through 8.

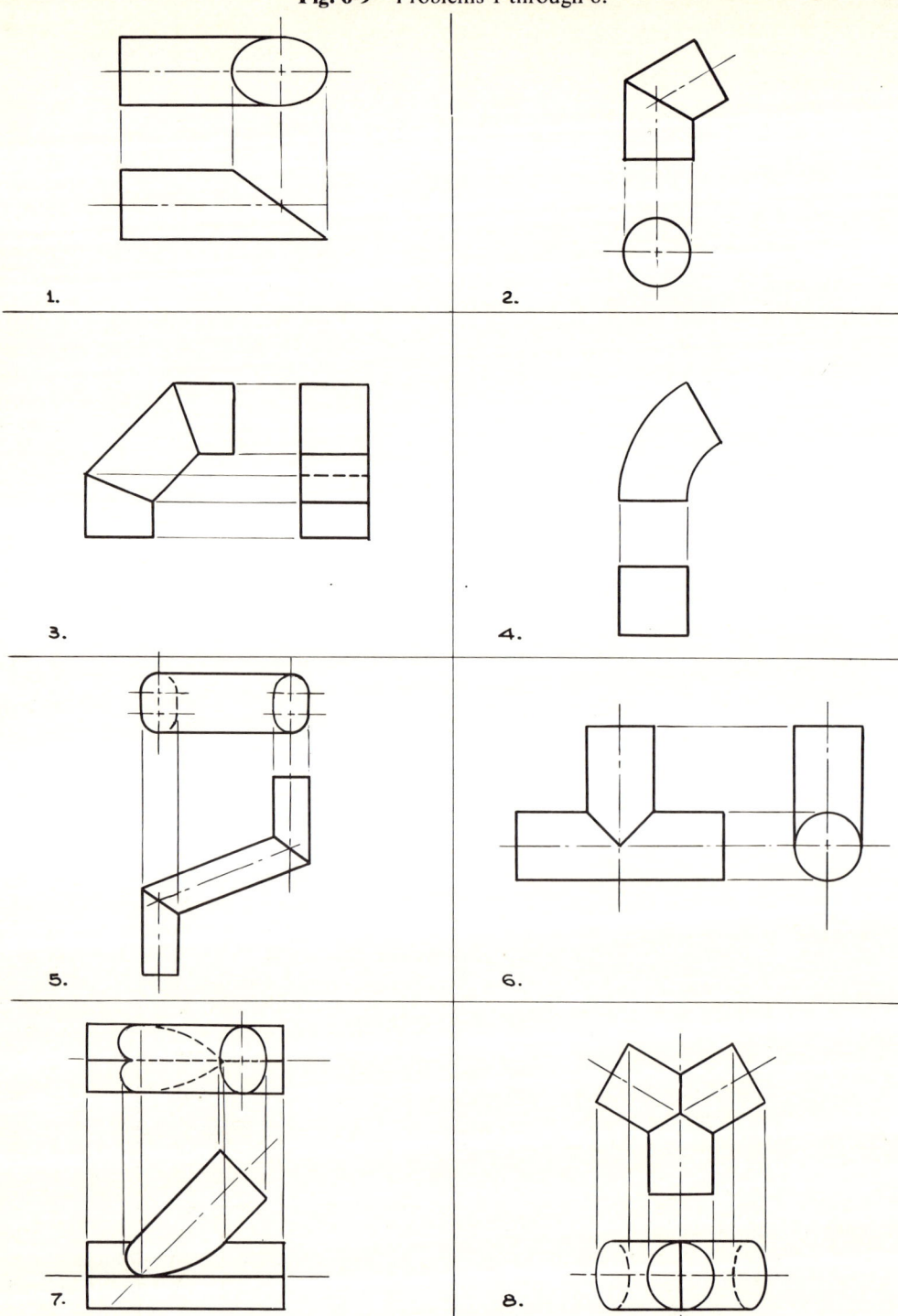

Fig. 6-10 Problems 9 through 16.

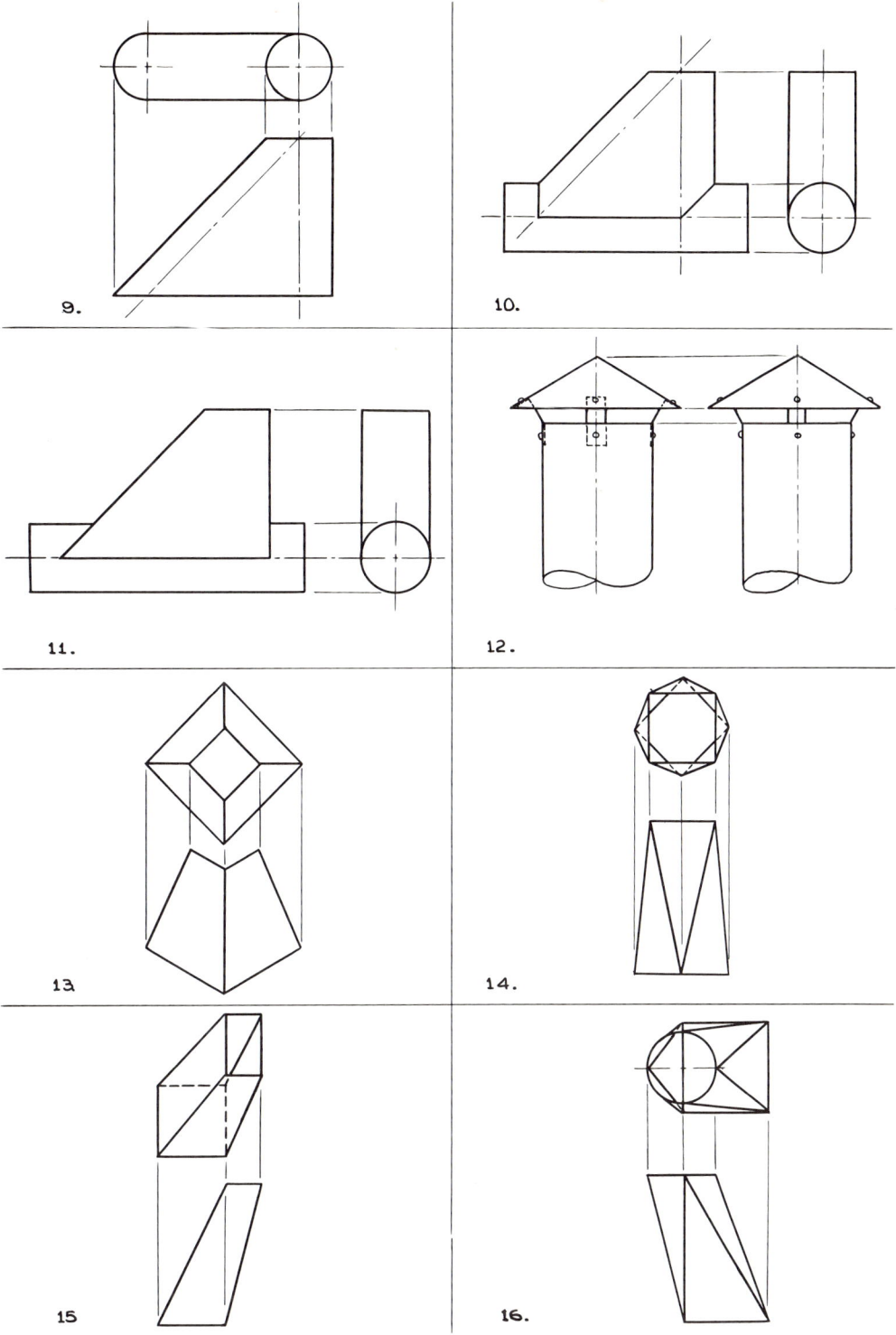

Fig. 6-11 Problems 17 through 22.

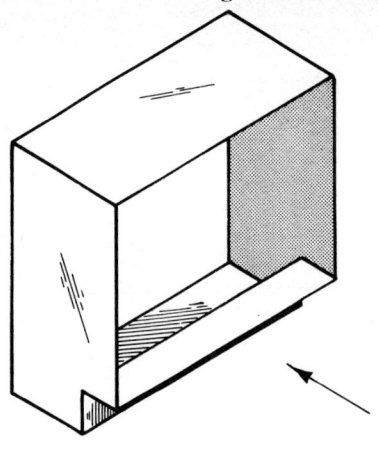

17.

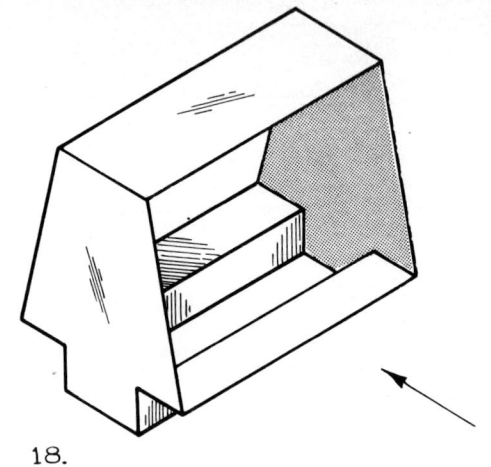

18.

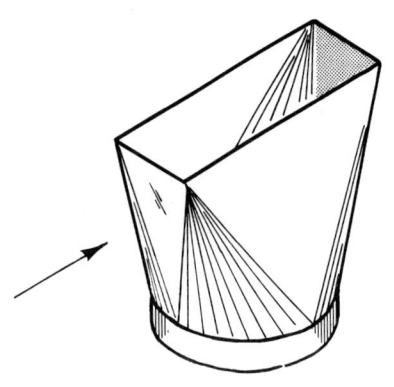

19.

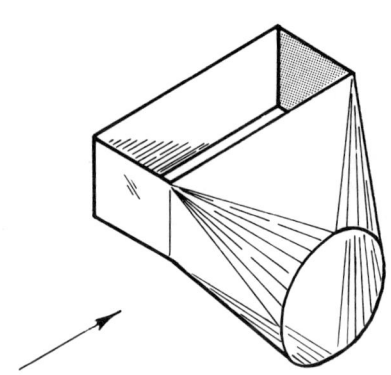

20.

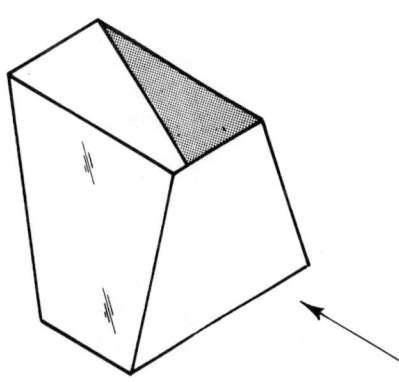

21.

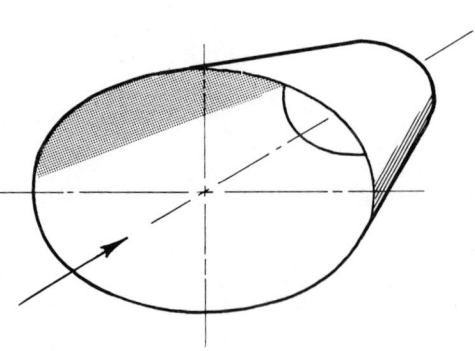

22.

chapter 7

AUXILIARY VIEWS

Primary or single auxiliaries are drawn with the auxiliary plane of projection perpendicular to one of the principal planes and inclined with the other two. Because there are three principal planes H, V, and P the primary auxiliaries may be projected from the top, front, and/or side views.

Secondary, successive, or oblique auxiliaries are drawn on auxiliary planes of projection inclined to all three principal planes. Secondary auxiliaries can only be projected from a primary auxiliary.

7-1 Primary auxiliary planes and views

A typical auxiliary plane of projection perpendicular to H and inclined to V and P is shown in the "exploded" pictorial drawing of Figure 7-1. The auxiliary plane is used, as shown in Figure 7-2, to produce the true size and shape of one of the flared sides of the exhaust hood. In orthographic projection two adjacent views are always projected upon *mutually perpendicular planes;* there-

Fig. 7-1 Auxiliary plane perpendicular to H.

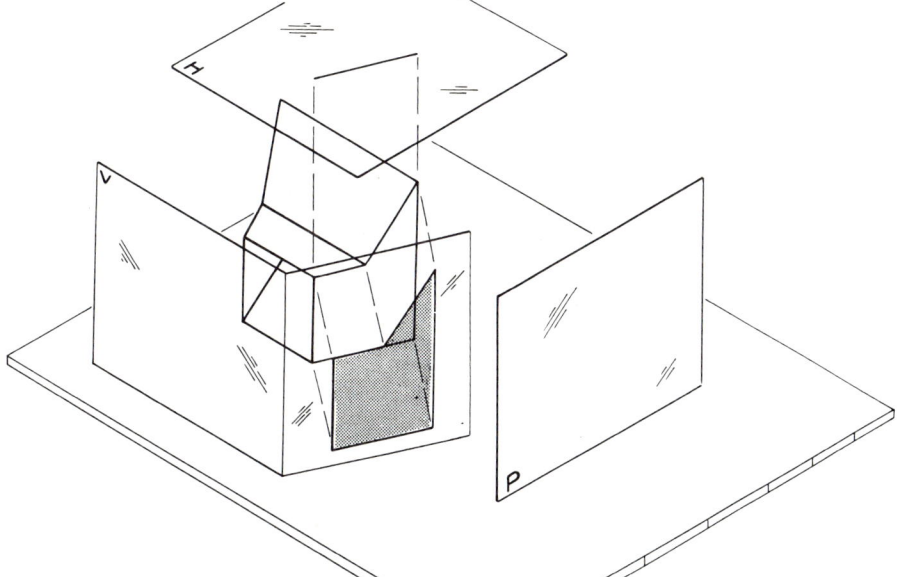

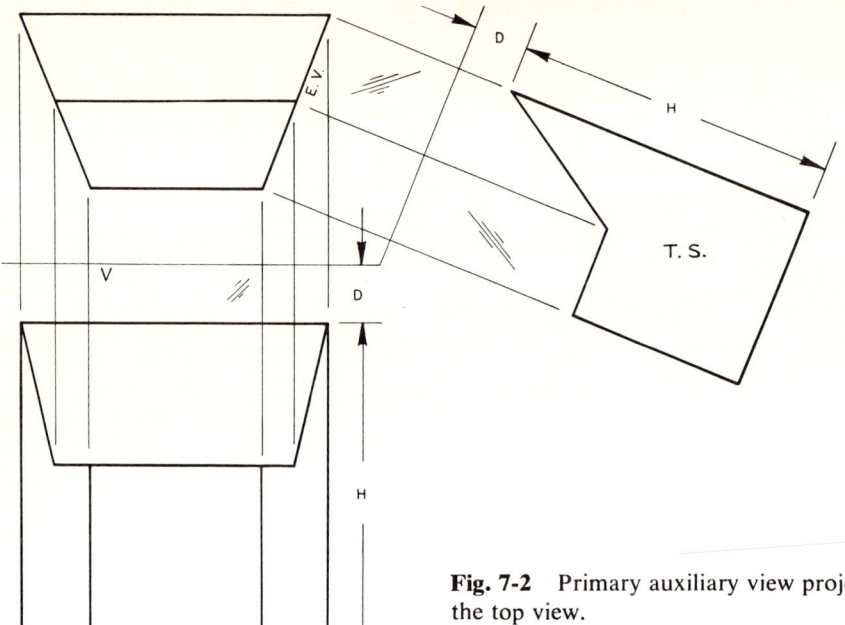

Fig. 7-2 Primary auxiliary view projected from the top view.

fore, all views projected upon planes perpendicular to H on the drawing are projected *from the top view*. Of the two dimensions in this type of auxiliary view, one is always *height,* the other is obtained by projection from the top view. How is height provided when making the auxiliary view? It is transferred from a view which contains it, such as the front view in Figure 7-2.

As an example of a primary auxiliary view projected from a front view, examine Figure 7-3 which shows, pictorially, the auxiliary

Fig. 7-3 Auxiliary plane perpendicular to V.

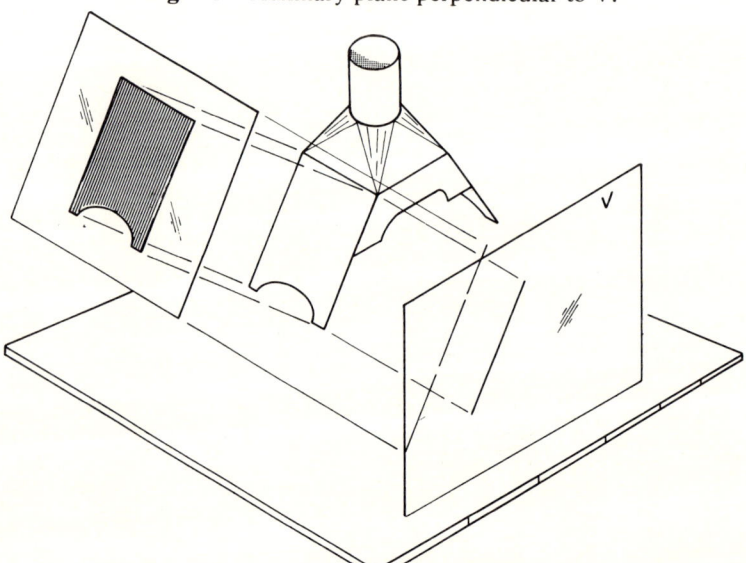

AUXILIARY VIEWS

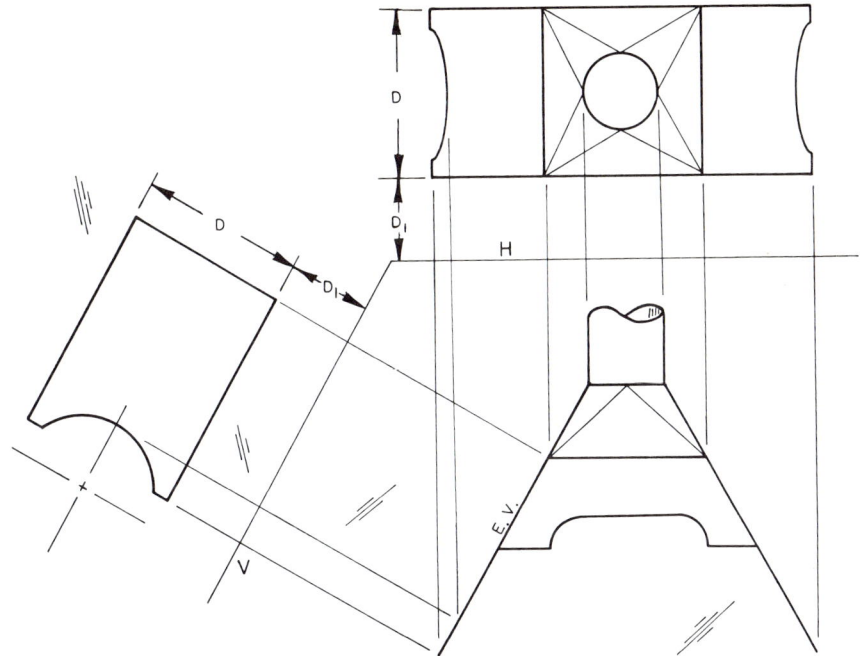

Fig. 7-4 Auxiliary view projected from front view.

Fig. 7-5 Auxiliary plane perpendicular to **P**.

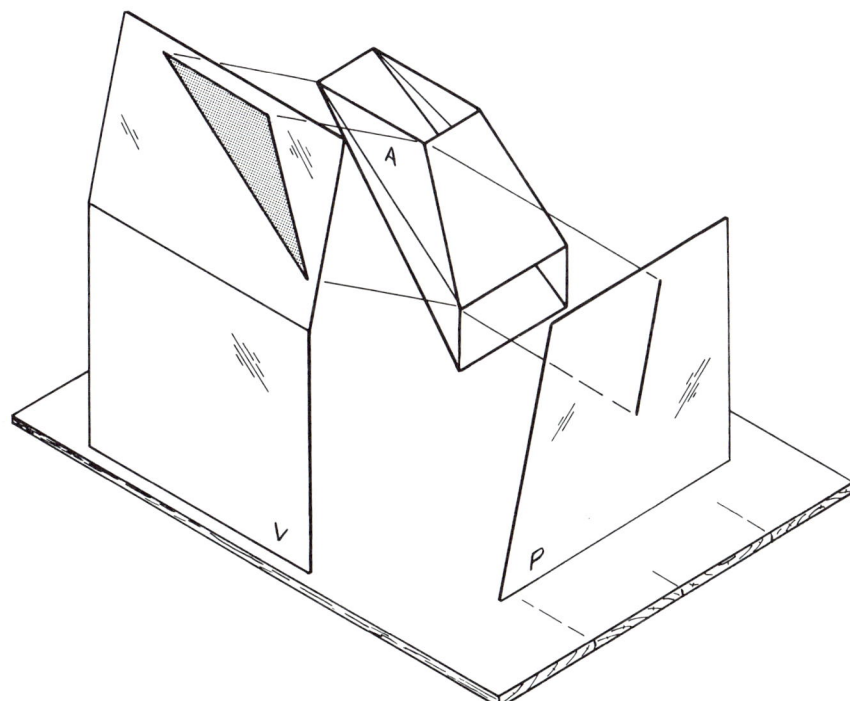

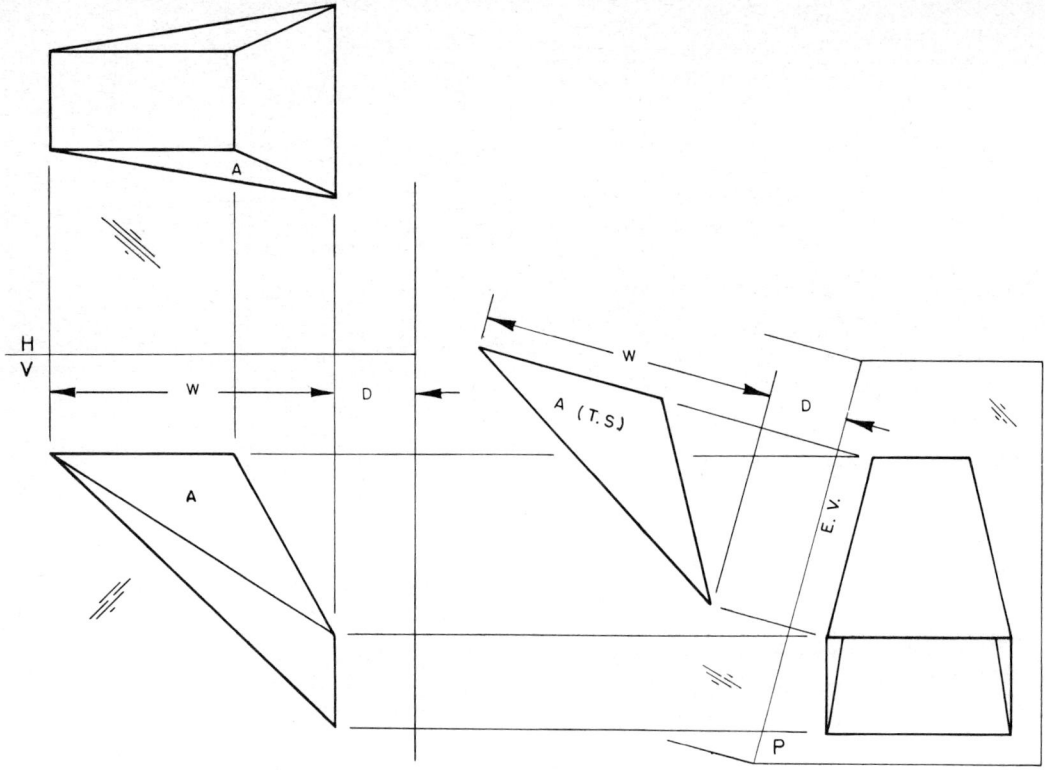

Fig. 7-6 Auxiliary view projected off side view.

plane of projection parallel to the side shield of the melting furnace hood and perpendicular to V. The auxiliary view in Figure 7-4 shows the side shield in its true shape and size. Notice that all primary auxiliary views projected from a front view (on planes perpendicular to V) have two dimensions, one resulting from projection off the front view, the other always being depth. Depth is transferable from top or side views.

The next example of a primary auxiliary view involves a plane of projection perpendicular to P and inclined to H and V. Such a plane is illustrated pictorially in Figure 7-5. Panel *A* of the reverse elbow is desired true shape and size. Because *A* is perpendicular to P and appears edgewise in the side view, the required auxiliary view is projected from the side view as shown in Figure 7-6. Notice that the transfer distance, for any auxiliary projected from a side view, is always *width,* as obtained in top or front view.

7-2 Oblique auxiliary planes and views

Because the relationship between the planes of adjacent views is always 90 degrees, it is impossible to project directly from a principal view to an oblique plane of projection. Therefore, an oblique auxiliary view must always follow a primary auxiliary view. The primary view is selected and arranged so that the oblique surface, whose true size and shape is desired, will ap-

AUXILIARY VIEWS

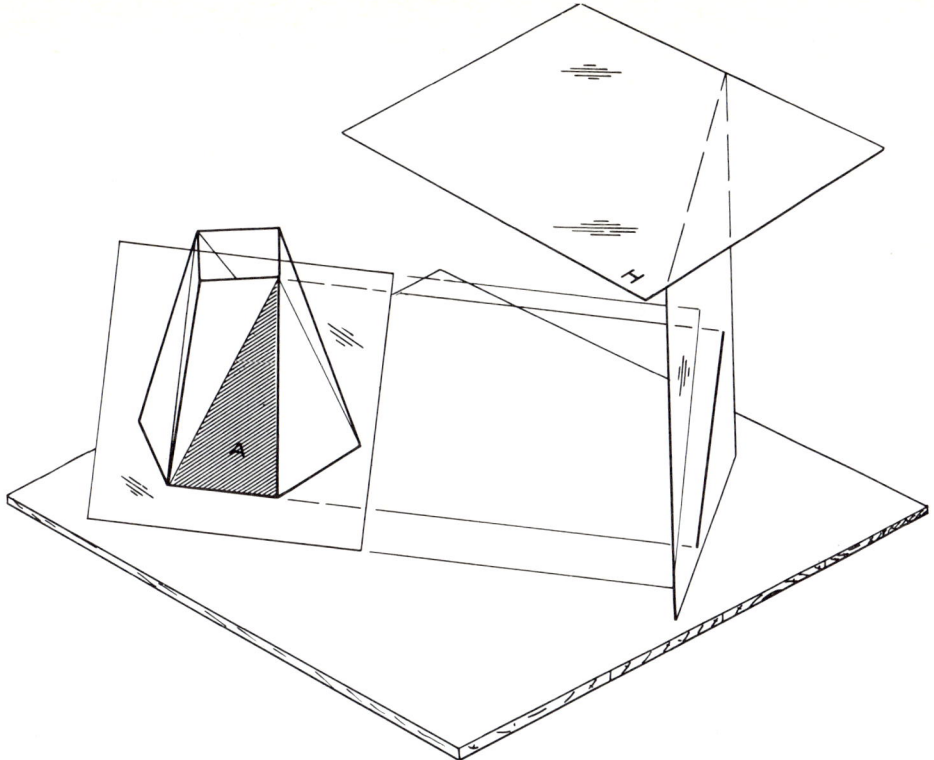

Fig. 7-7 Secondary auxiliary plane perpendicular to a primary auxiliary plane.

Fig. 7-8 Secondary auxiliary view projected from a primary auxiliary view.

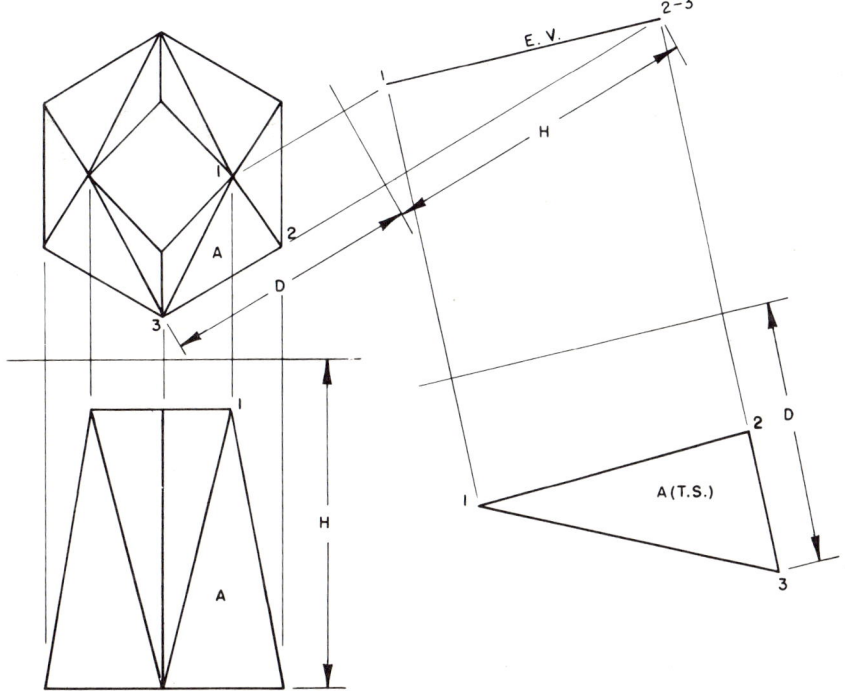

pear as an edge (see Figure 7-7). Then the second auxiliary can be made on a plane perpendicular to the primary plane and parallel to the desired surface. In effect, the second auxiliary is projected from an auxiliary view. A typical theoretical arrangement of planes is shown pictorially in Figure 7-7. The same problem, as it would be constructed on the drawing, is illustrated in Figure 7-8. The oblique view shows the true size and shape of surface *A* of the square to hexagonal transition.

7-3 Functions of auxiliary views

By means of auxiliary views, an object may be viewed from any desired direction. Thus, the drawing of auxiliary views should be recognized as a very important phase of the draftsman's and engineer's skill. Ability to select and construct the required auxiliaries depends on the draftsman's ability to recognize the direction of sight needed for each view. He or she must know just what can be accomplished with auxiliary views and how these views depend on each other.

The four main functions of auxiliary views are

    a. To find the true length of a line
    b. To produce the point view of a line
    c. To produce the edge view of a surface
    d. To find the true shape and size of a plane figure

Each operation above acts as a foundation upon which the next may be constructed (see Figure 7-9). For example, if a true length view of a line is given or obtained, then a point view of that line may be projected *from the true length view*. If an edge view of a plane is required, the point view of a line on that plane establishes the line of sight in which the entire plane appears as an edge (try this by sighting a triangle edge). In order to produce the true shape and size of a surface on a drawing, *an edge view of the surface must first be obtained*.

All of the basic concepts stated above are applied in Figure 7-8. For example, the line 3-2 is true length in the top view; and, an auxiliary view of 3-2 as a point produces surface 1-2-3 as an edge. A second auxiliary projected from the edge view, on a plane parallel to the edge view of the surface, produces surface 1-2-3 true shape and size.

7-4 To find the true length of a line

A line will project true length upon a plane of projection which is parallel to it. For this reason, as discussed in section 6-5, a horizontal line is always true length in the top view, a frontal line true length in the front view, and a profile line true length in the side view. An oblique line, because it is not parallel to any principal plane, projects foreshortened in all regular views. The true length of any oblique line may be found by pro-

Fig. 7-9

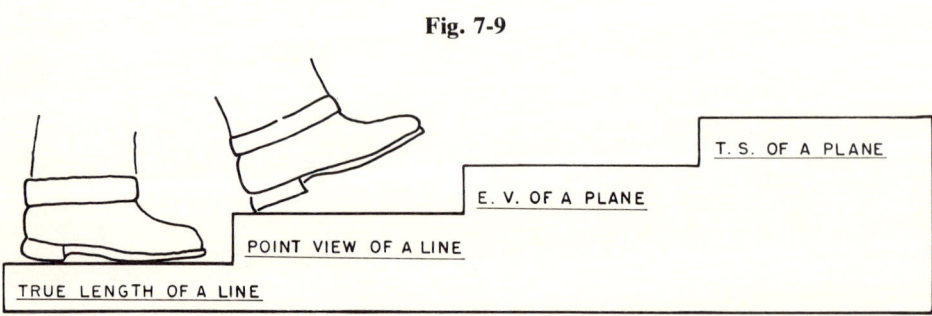

jecting the line upon an auxiliary plane which *is* parallel to the line, or by rotation, a method described in Chapter 8.

The auxiliary-view method is illustrated in Figure 7-10, H, V, and P. Consider the top and front views of line AB as given. The true length of AB is required. Three examples are illustrated. In example H the auxiliary plane is parallel to AB and perpendicular to H. The height measurements H_1 and H_2 are measured in the front view and transferred, by compass, to the proper projectors in the auxiliary view. The resulting view is the true length of line AB. Notice that height is measured in the front view, *down* from the H, V intersection and is measured in the auxiliary view *down* from the H, I intersection. In example V, the auxiliary plane is parallel to the line and perpendicular to V.

Notice that the transfer dimensions D and D_1 are *depth* dimensions brought from the hinge line between the view under construction and the "parent" view. In the third example P, the auxiliary view is projected on a plane parallel to the line and perpendicular to P. In order to project on a plane perpendicular to P, a side view is necessary. In all auxiliary views projected from a side view, the transfer dimension is *width*. The measurements W and W_1 are obtained in the front view.

It has now been demonstrated that the true length of a line may be obtained by projection from any view. The draftsman has a choice when selecting views in solving problems involving the true length of a line. A careful analysis of the entire problem before any drawing construction is made will pay in simplicity of solution and in time saved.

Fig. 7-10 True length of a line by auxiliary views.

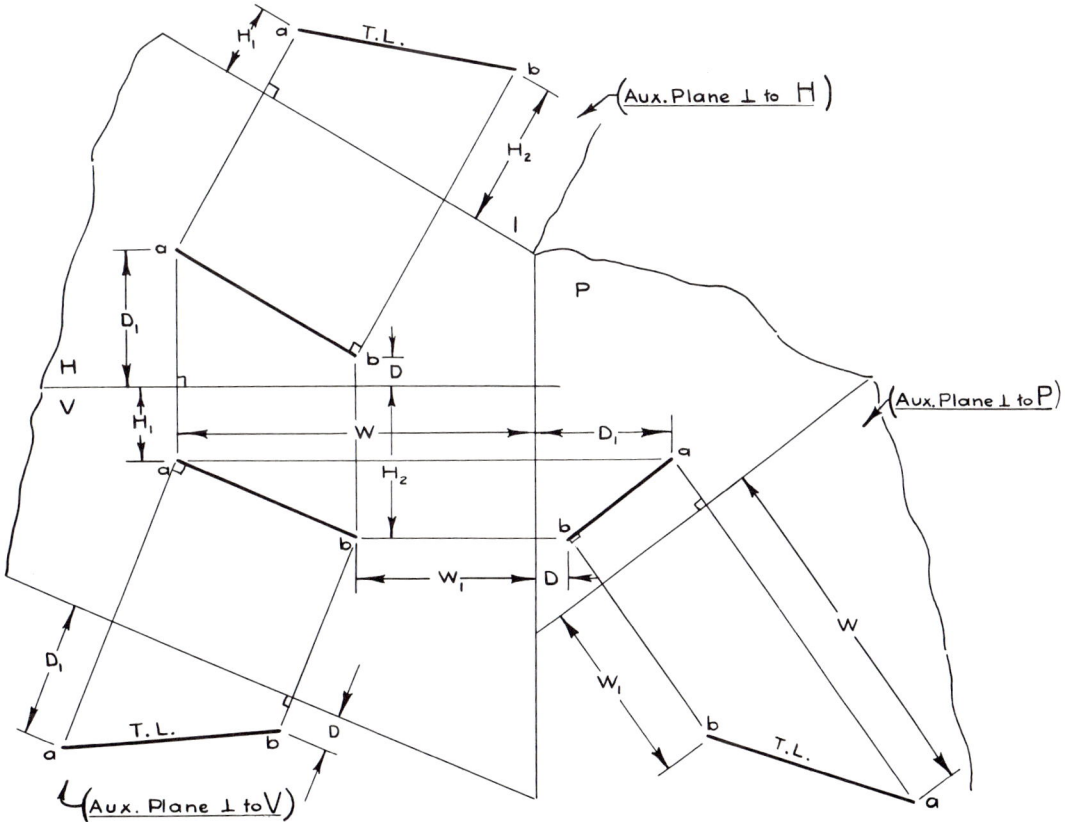

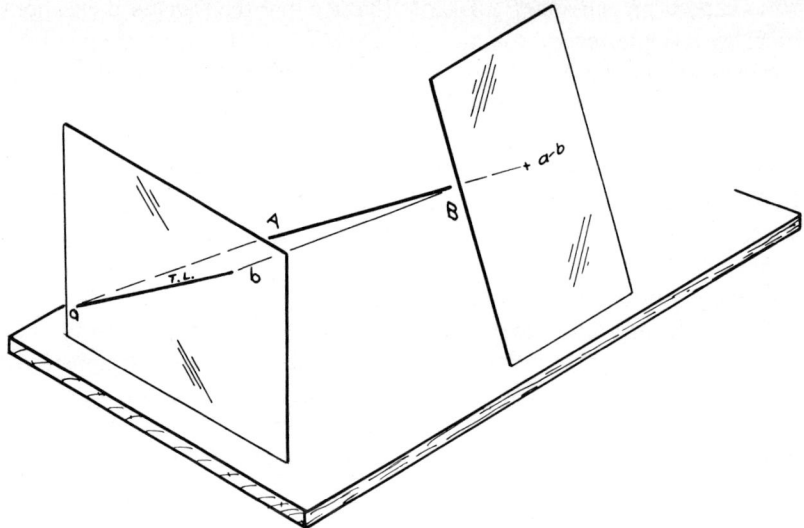

Fig. 7-11 Plane perpendicular to a line.

7-5 To produce the point view of a line

A line will project as a point on a plane of projection which is perpendicular to it. If the plane of a true-length view is parallel to the line, and the plane of a point view is perpendicular to the line, the two planes are perpendicular to each other, as shown pictorially in Figure 7-11. The views are directly related and it is easy to project from one to the other. The important basic fact to remember is that *a point view of a line is impossible to construct until a true-length view, to project from, is available*. As an illustration, in Figure 7-12, the top and front views of line AB are given. A point view is required. Before the point view can be made however, a true-length view of the line must be provided. The point view is projected as a second auxiliary from the first auxiliary. Note the two sets of reference dimensions used: D, D_1; and D_2.

As examples of problems in which the point view of a line is desired refer to the "direction line" on a plane, as described in section 7-6; also, to problem 9 at the end of this chapter, a technique is suggested for determining bend angle when "braking up" large cones. Because the only line common

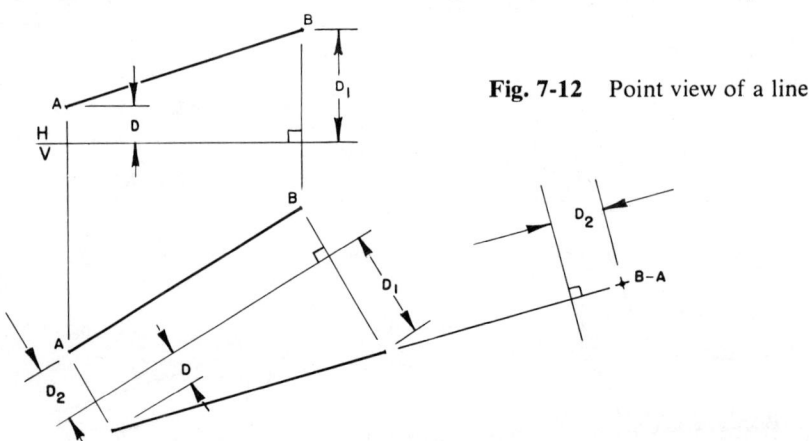

Fig. 7-12 Point view of a line.

AUXILIARY VIEWS

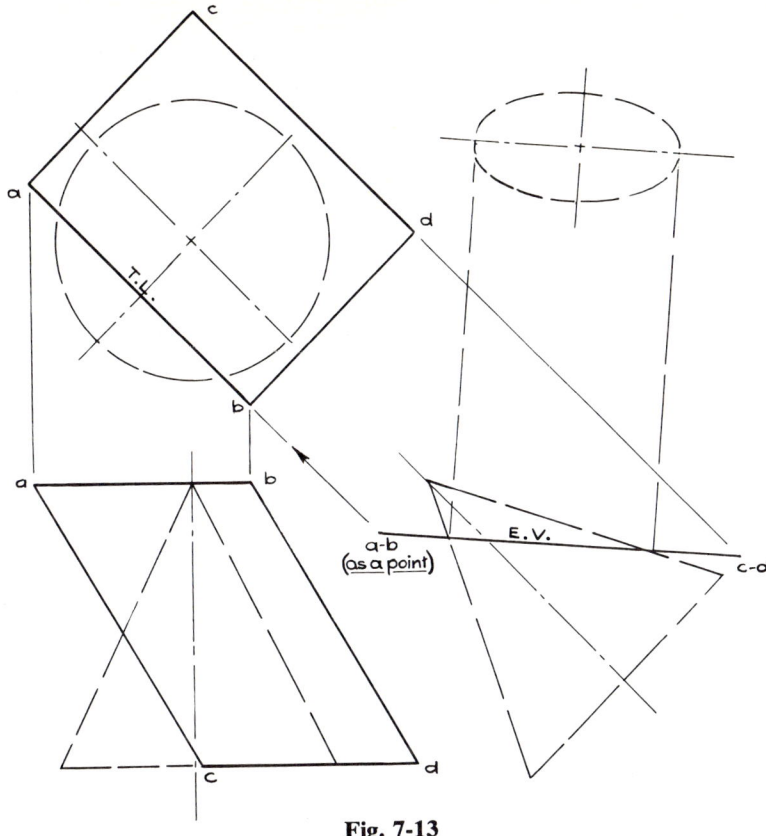

Fig. 7-13

to two planes is the line of intersection of the two planes, *a point view of the line of intersection results in both planes appearing as edges.* The true angle between planes may be measured in that view.

7-6 To produce the edge view of a plane

If any line on a plane is viewed as a point, the entire plane is seen as an edge. In Figure 7-13, two views of plane *ABCD* are given, and an edge view is required. The line *AB*, being true length in the top view, is used to establish the projection direction. One of the handiest tools of the draftsman is a direction line: a horizontal, frontal, or profile line placed upon a plane and used to establish projection direction or line of sight when an edge view is required. As an example, an edge view of the plane 1-2-3-4 is projected from the top view in Figure 7-14. Direction was established by means of the horizontal line 3-*X*. The same plane is again used in Figures 7-15 and 7-16 to illustrate use of frontal and profile direction lines. In Figure 7-15 the frontal direction line is 4-*Y*, arbitrarily placed on 1-2-3-4 parallel to V. In Figure 7-16 the profile direction line is 2-*Z,* arbitrarily placed on the plane parallel to P.

7-7 To find the true size and shape of inclined and oblique surfaces

A surface will project true size and shape upon a plane of projection only when the plane is parallel to it. Therefore, in order to

SHEET METAL DRAFTING

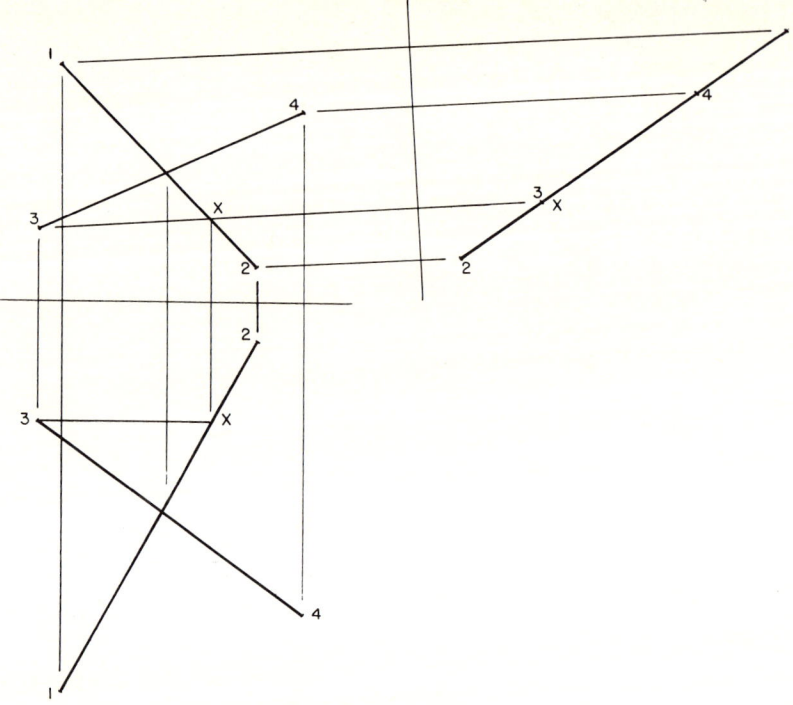

Fig. 7-14 Using a horizontal direction line.

Fig. 7-15 Frontal direction line.

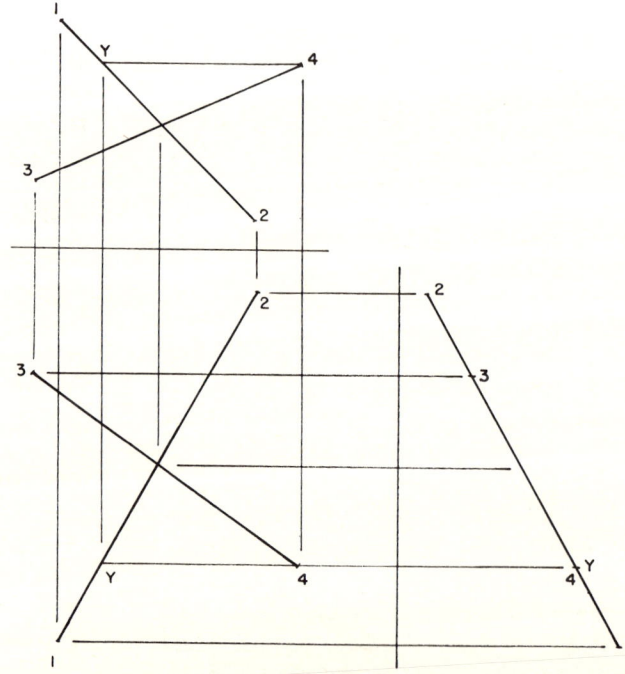

Fig. 7-16

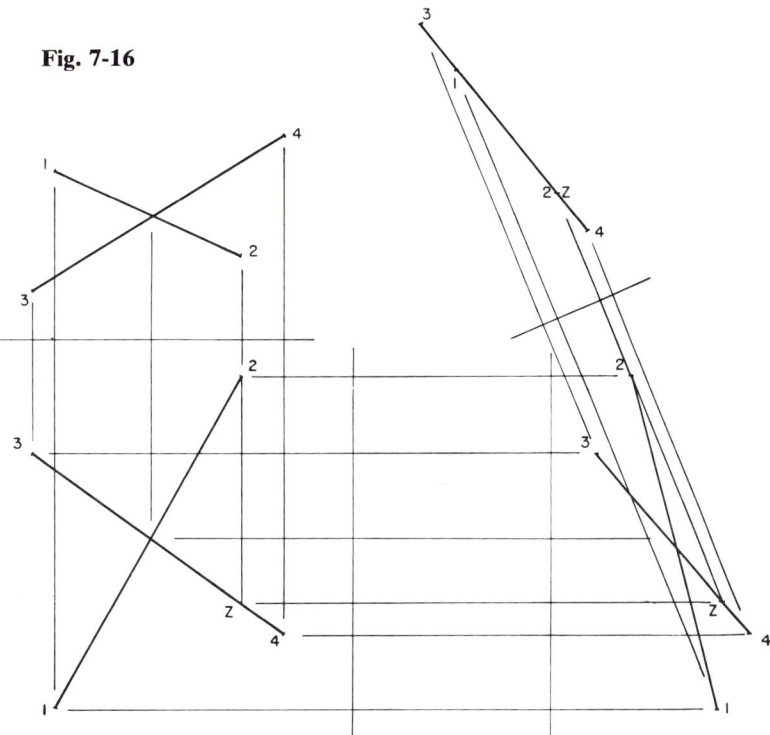

produce a true size and shape view, it is first necessary to have an edge view of the surface to project from. The auxiliary plane of projection is placed perpendicular to the plane of the edge view and parallel to the edge view of the surface. This *basic* concept was first illustrated in Figure 7-8. As a practical example, consider the design of a box with rectangular base and sloping sides. The required corner angles and sloping height may be found as shown in Figure 7-17. A partial front and side view of a corner, drawn with the specified box height and the vertical height of the sloping sides, establishes the *edge view* of the two sloping surfaces. From the edge views, auxiliaries are

Fig. 7-17 Given: box height H; side slope $\theta = 15$ degrees; end slope $\phi = 10$ degrees. Required: side corner angle α; end corner angle β; side true height e; end true height f.

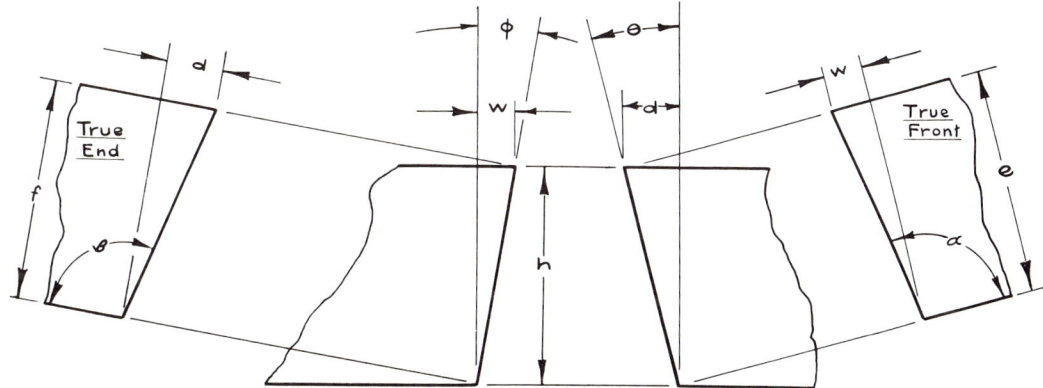

projected which show the true sloping height and the required corner angles (in fact, the true sloping heights are equal to the length of line representing the edge views). The true shape and size of an oblique surface may be found as shown in Figure 7-18. In Figure 7-18A the top and front views of a triangular pyramidal aluminum cap 1-2-3-4 are given. It is required to find the true shape and size of surface 1-2-3.

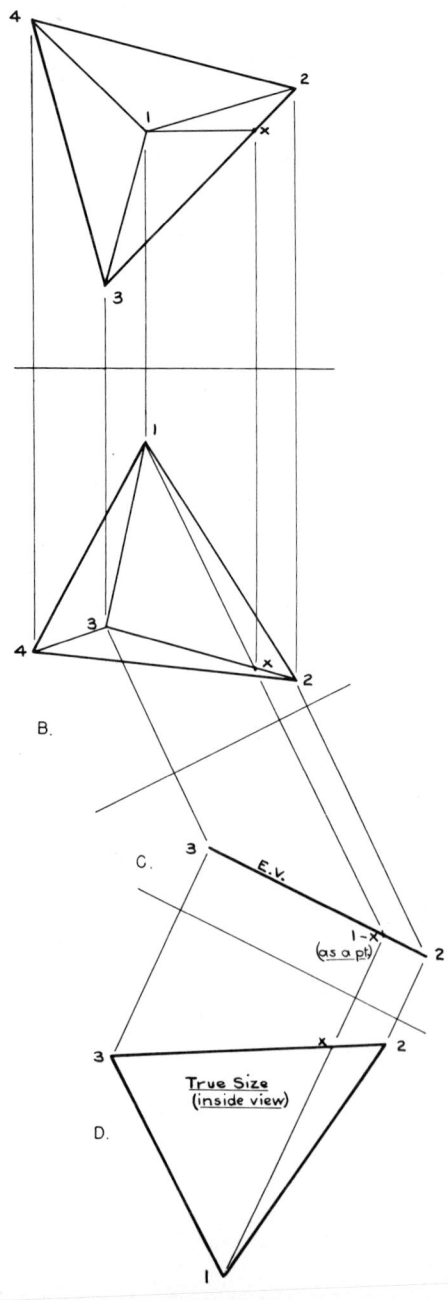

Fig. 7-18

AUXILIARY VIEWS

None of the edges of 1-2-3 are true length in the given views; hence they are not suitable as "direction lines." A *frontal direction line,* 1-*X,* is true length in the front view, therefore, a point view of 1-*X* may be constructed as shown in Figure 7-18C. This results in an edge view (EV) of 1-2-3. A second auxiliary view *projected perpendicular from the edge view* produces the required true shape and size view (TS), as shown in Figure 7-18D.

7-8 Procedure for making any auxiliary view

A careful study of Figures 7-8 through 7-19 might reveal that the following procedure can be established for the construction of any auxiliary view. This procedure is applicable, in logical sequence, to all problems involving points, lines, or surfaces.

a. Determine the Line of Sight. Practically all problems are easily solved if viewed in the proper direction. The draftsman's first concern then is to determine which line of sight is needed (see Figure 7-19).

b. Place the Necessary Reference Lines (sometimes called folding or hinge lines). In making any auxiliary view three views are involved, the view under construction, the previous or parent view, and the "second previous" view. A pair of reference lines are needed. One is placed between the view under construction and the previous view; the other is placed between the first and second previous views (see Figures 7-8 through 7-20).

c. Transfer Points to the View under Construction. All points in the new view may be located by transferring measurements from *the second previous view* to the proper projector in the new view area. The plane of the new view and the plane of the second previous view are both perpendicular to the plane of the parent view; therefore, the distances away from the folding line in the new view are equal to distances away from the folding line in the second previous view (see compass points in Figure 7-19).

d. Determine Visibility and Complete the View. After the points for the view have been transferred, there remains only to "read" the given views to determine which points are connected, which are visible, and which are hidden so as to produce the visible or hidden edges for the new view. Figure 7-20 is a good example of the application of the preceding steps. Starting with a top and front view of a sheet metal box, the drawing consists of successive auxiliary views made with lines of sight selected at random. The above procedure is used to produce each of the views of the box and to determine visibility. Numbering the corners may help in keeping tab of all points in all the views.

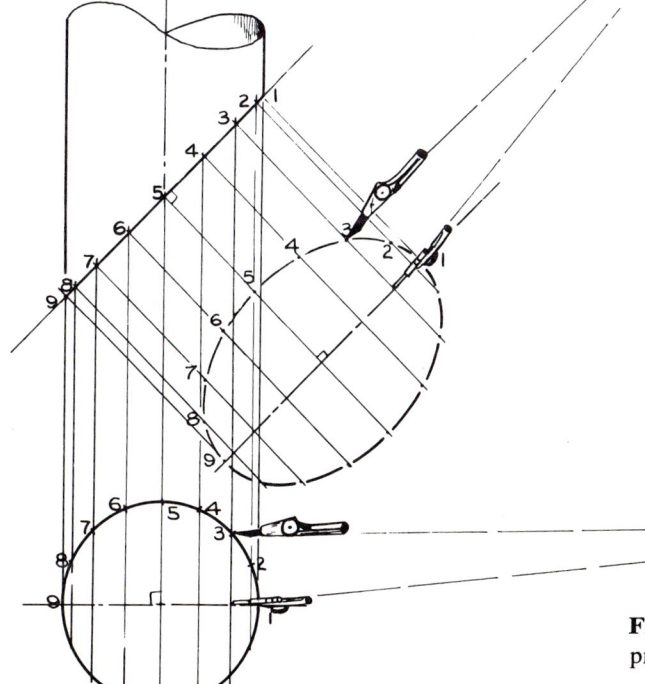

Fig. 7-19 To establish curves, project and transfer points.

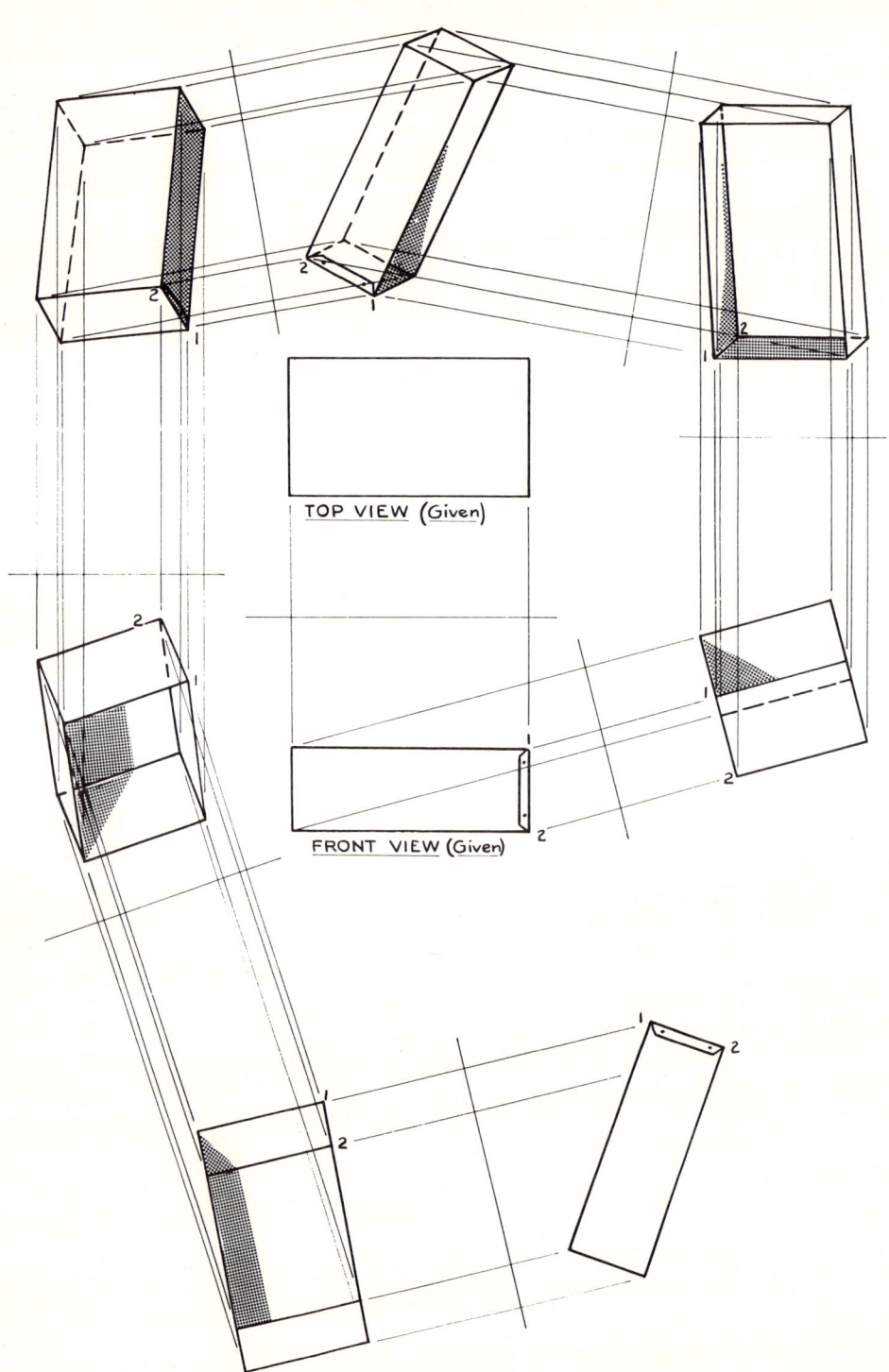

Fig. 7-20 Successive auxiliary views.

EXERCISES

Single or Primary Auxiliaries

1. Given the top and front views of the collector hopper for the sandblasting cabinet in Figure 7-21; using a scale suitable for the size of paper available, draw the top, front, and right-side views. Add auxiliary views to show the true shape and size of the back, front, and left panels of the hopper.
2. Refer to Figure 7-22. Draw the top, front, and right side views of the observation port for the sandblasting cabinet in Figure 7-21. Add an auxiliary view which will show the true shape and size of the safety glass needed. Add an auxiliary view to show the true shape and size of panel A.
3. Refer to Figure 7-23. Given the top and front views of the arm shelf for the sandblasting cabinet in Figure 7-21. Draw the top, front, and right-side views. Add an auxiliary view to show the true shape and size of shelf bottom. Add an auxiliary view to show the true shape and size of one side wing.
4. Using a scale suitable for the size of paper available, draw the pipe and roof intersection shown in Figure 7-24. Make an auxiliary view which will show the true shape and size of the roof opening.
5. Refer to Figure 7-25, and follow the directions of problem 4.
6. Construct the top and front views of the elliptical cylinder shown in Figure 7-26. Draw an auxiliary view to show the true shape and size of a right section of the cylinder. Project the right section, as taken at mid-point A, to the front and top views.
7. Construct the top and front views of the hollow cone shown in Figure 7-27. Pass a plane, 1-2-3-4, perpendicular to V and parallel to a cone element, AB, so that 3-4 is a diameter of the base. Make an auxiliary view which shows the true parabola thus cut. Project the line of intersection of plane and cone to the top view. *Hint:* Use a series of elements on the cone which pierce the plane, and project these elements and their piercing points to the auxiliary and top views.
8. Using as large a scale as possible for the drawing paper available, draw the plan and elevation views of the building shown in Figure 7-28. By means of auxiliary views, where necessary, determine the true length of each line and the total linial footage of
    a. Gutters (all eaves)
    b. Hip finish B
    c. Hip finish C
    d. Ridge roll
    e. Valley finish
9. Conical hoods, hoppers, separators, etc., are sometimes "bent" rather than rolled. If the brake is calibrated, it is helpful to know the approximate amount of bend at each brake line. The true angle between bent surfaces may be calculated by means of spherical trigonometry, the mathematics becoming more involved than experimental bending in the shop. By means of an auxiliary view, drawn to any convenient scale, the true angle between bent surfaces may be quickly and easily determined regardless of the size of the project. Draw the top and front views of the conical reducer shown in Figure 7-29. On the basis of twenty-five brake lines for the entire surface, lines V-2 and V-3 are at 15 degrees to V-1 in the top view. Draw an auxiliary view which will show the true angle between surfaces V-1-2 and V-1-3 (see section 7-5). Measure the angle.
10. A rectangular box with sloping sides and ends is required drawn to the following specifications: bottom = 8 × 14 in [203 × 356 mm]; vertical height of sides and ends = 3 in [76 mm]; ends slope outward 28 degrees from vertical; sides slope outward 17 degrees from vertical. With a partial front and side view and suitable partial auxiliaries, determine the amount of metal to be added to the base for each side, the amount to be added for each end, the angle at which to cut the side corners, and the angle at which to cut the end corners.

Fig. 7-21 Sandblasting cabinet. (*Leiman Bros., Newark, N.J.*)

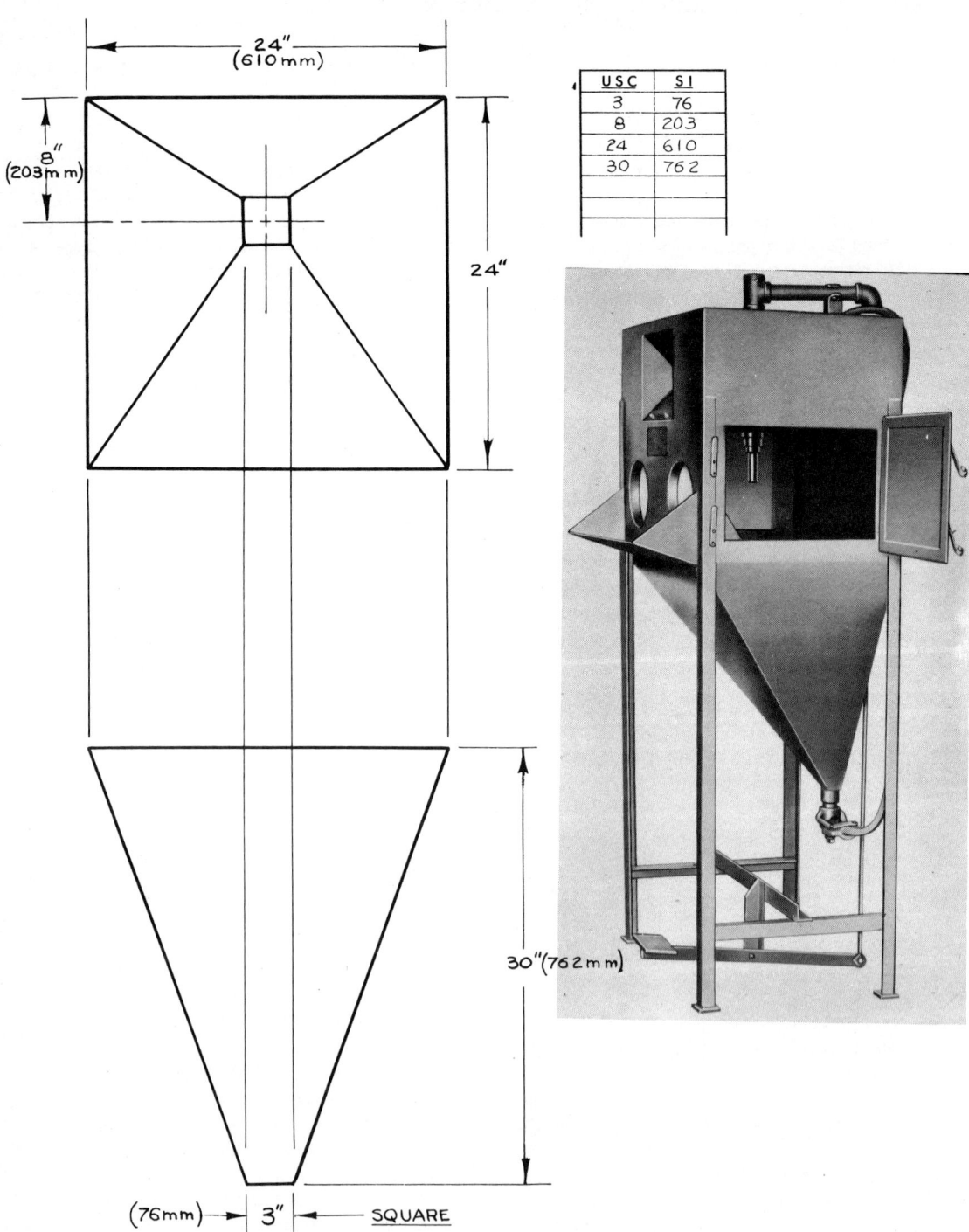

USC	SI
3	76
8	203
24	610
30	762

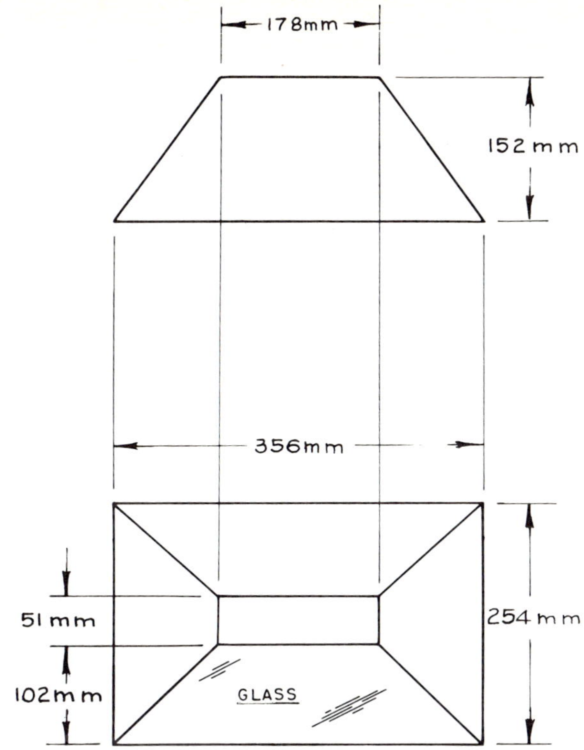

AUXILIARY VIEWS

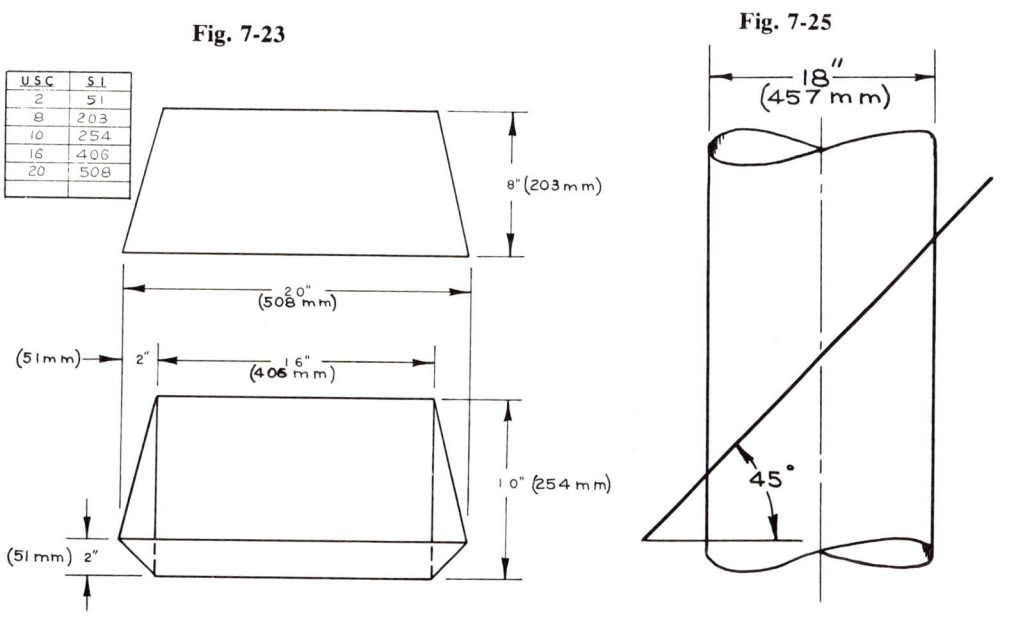

Fig. 7-22

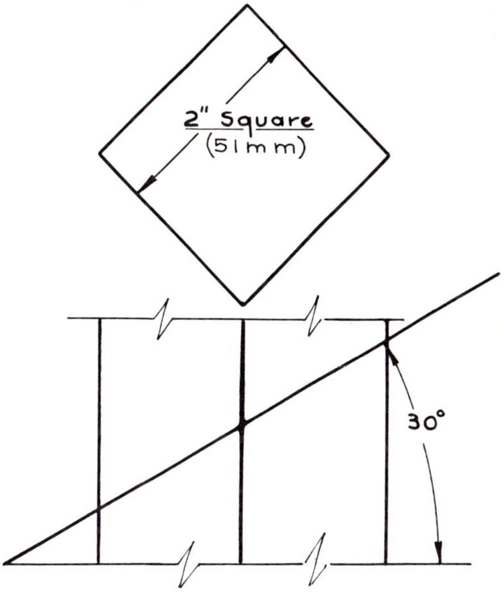

Fig. 7-23

Fig. 7-24

Fig. 7-25

SHEET METAL DRAFTING

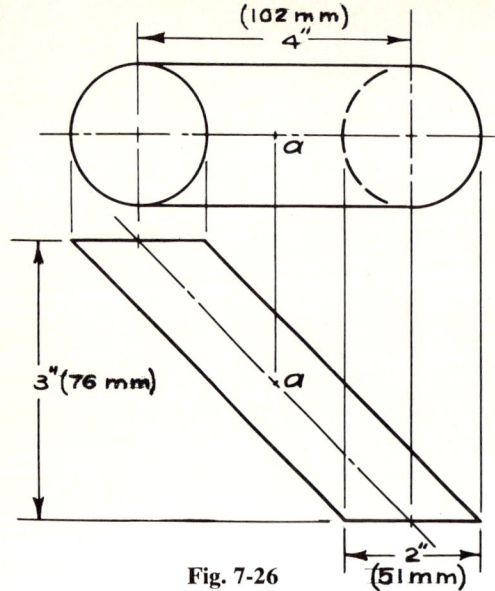

Fig. 7-26

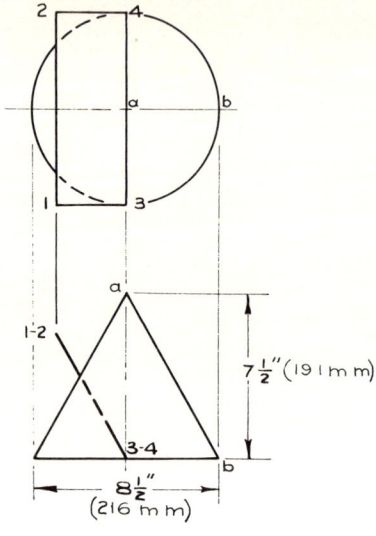

Fig. 7-27

Fig. 7-28

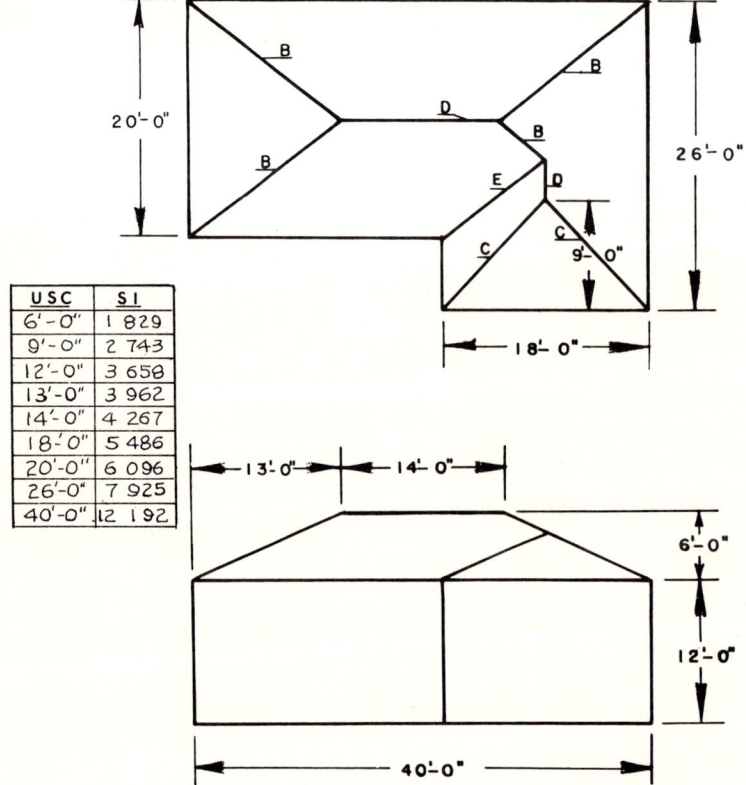

USC	SI
6'-0"	1 829
9'-0"	2 743
12'-0"	3 658
13'-0"	3 962
14'-0"	4 267
18'-0"	5 486
20'-0"	6 096
26'-0"	7 925
40'-0"	12 192

AUXILIARY VIEWS

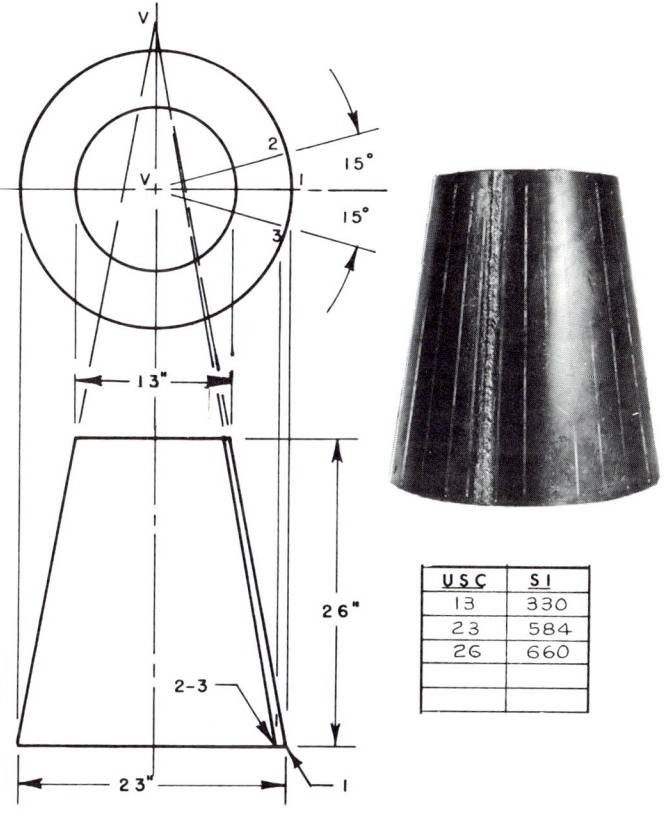

Fig. 7-29 (*Naylor Pipe Co., Chicago, Ill.*)

USC	SI
13	330
23	584
26	660

Secondary Auxiliaries

11. Redesign the arm shelf for the sandblasting cabinet of Figure 7-21 as per the top and front views of Figure 7-30. Draw the auxiliary views necessary to secure the true size and shape of one side wing.

12. Draw the top and front views of the roof shown in Figure 7-31. Draw a view which will show the true shape and size of the roof opening required for an 18-in [457-mm] square stack as indicated. Use visible and hidden lines to show the roof and stack intersection in the front view.

13. Draw the top and front views of the roof in Figure 7-31. Change the roof direction from N60° W to N45° W. Establish a view which will show the true shape and size of the roof opening when using an 18-in [457-mm] circular stack, as indicated in Figure 7-31. Use visible and hidden lines to show the roof and stack intersection in the front view. What are the true dimensions of the roof surface?

14. Draw the plan and elevation views of the building shown in Figure 7-28. Draw partial auxiliary views to determine the true angles so that sheet copper may be bent to fit the following: hip finish *B*; valley finish *E*; hip finish *C*.

    (*Hint:* see section 7-5. Only the line of intersection as a point and one additional point from each roof plane is needed to establish the edge view of both roof planes.)

15. Refer to Figure 7-32. Given the top and front views of the center line of a circular tube 12 in [305 mm] in diameter which intersects a floor (horizontal plane) at *A* and a wall (vertical plane) at *B*. Draw the auxiliary views which are necessary in order to complete the top and front views

of the tube. Show the true shape and size of the elliptical openings in the floor and wall. By scaling the drawing, determine the over-all length of tubing necessary.

16. Refer to Figure 7-32. Given (1) the top and front views of the center line of a 12-in [305-mm] square duct; (2) the duct intersects vertical and horizontal planes as indicated; (3) the duct is so positioned that *one diagonal of a right section* is parallel to the V plane. Draw the auxiliary views necessary to complete the top and front views of the duct. Show the true shape and size of the openings in the vertical and horizontal surfaces. By scaling the drawing, determine the over-all length of duct required.

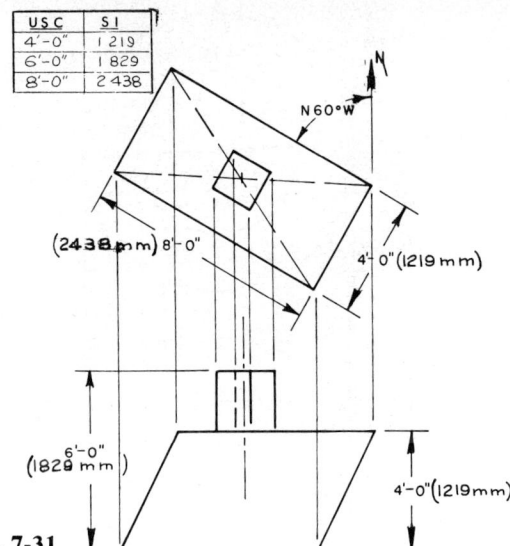

Fig. 7-31

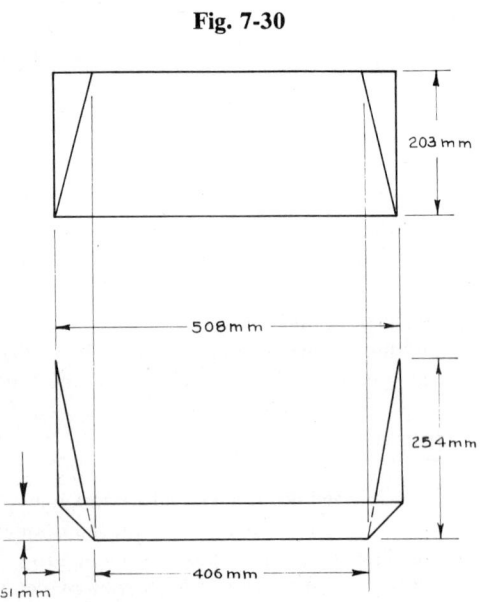

Fig. 7-30

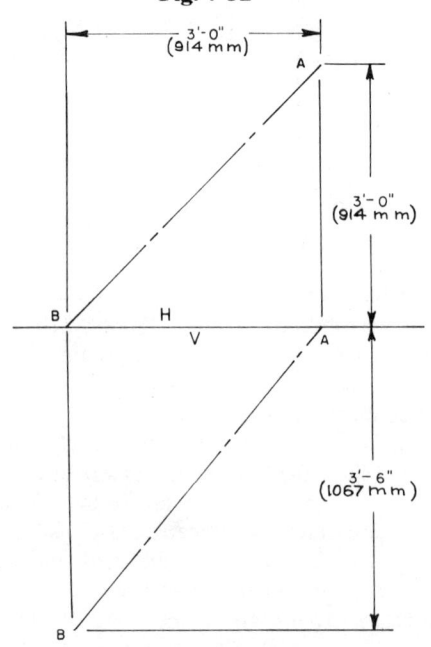

Fig. 7-32

chapter 8

ROTATION

The auxiliary-view method of solving problems theoretically assumes that the object remains in one position while the observer moves in order to secure the desired view. It is logical to assume that the same results can be achieved by rotating the object; the observer need not change position and thus the problem is solved by using only the given views. This method is known as *rotation*. Some problems are most easily solved by rotation, others by auxiliary views. For a great many problems it is desirable to combine both methods. The principles and the application of *rotation* will be discussed in this chapter.

Fig. 8-1 Principles of rotation.

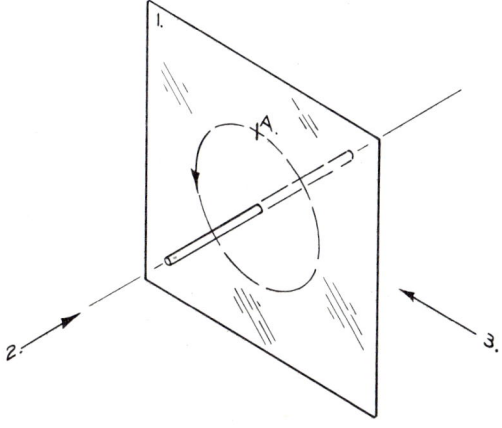

8-1 Fundamentals of rotation

Rotation, which may be observed in the everyday operation of mechanical equipment, is based upon natural mechanical phenomena. The facts of rotation are so obvious and so easy to comprehend that many students fail to realize their importance. For example,

  a. *A point rotated about an axis will generate a plane perpendicular to the axis.*

  b. *A view which shows the axis as a point will show the plane true shape and size, and the path of rotation will appear as a circle.*

  c. *A view which shows the axis true length will show the path of rotation as a line perpendicular to the axis* (the edge view of the plane of rotation; see Figure 8-1).

8-2 To find the true length of a line by rotation

If a line is rotated about an axis which intersects the line, the line generates a *cone of revolution*. The axis of rotation may be real or imaginary. As an example, the line *AB* in Figure 8-2 is rotated around the imaginary axis *AC*. The resulting cone of revolution (a right circular cone) is indicated by dash lines. Because all the elements of a right circular cone are of equal length, the true length of *AB* is the same as the true length of element *AE*. Because *AE* is parallel to the V plane, *AE* shows true length in the front view.

The examples in Figure 8-3 show that the true length of any line may be conveniently found by rotation. Any axis which is parallel to any plane of projection may be used. The line is rotated until it too lies parallel to the same plane of projection. In Figure 8-3, the true length of *EF* is found by three rotations, 1, 2, and 3, each assuming a different axis parallel to a different plane, V, H, and V, respectively.

8-3 The sheet metal draftsman's true-length diagram

In developing, by triangulation (Chapter 11), a pattern for a sheet metal fitting, the draftsman often requires the true length of many lines. If the lines are rotated and projected within the given views, the views very soon become so line-heavy that it is difficult

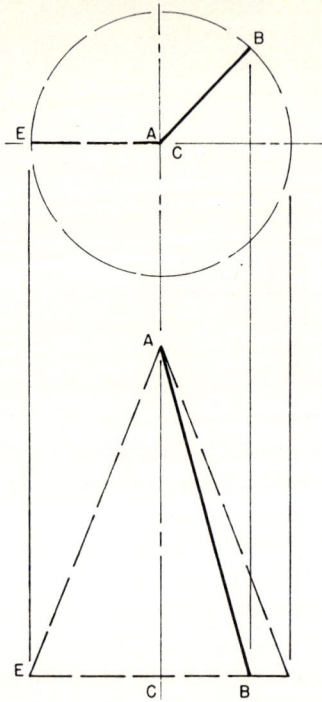

Fig. 8-2 Cone of revolution.

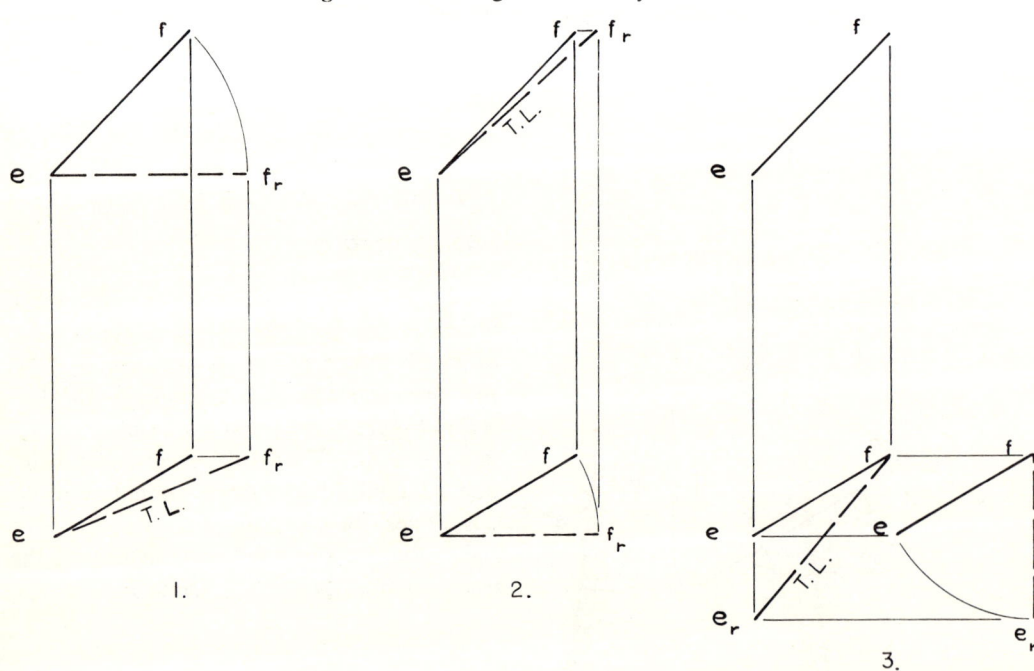

Fig. 8-3 True length of a line by rotation.

to keep the correct measurements separated. To avoid this many draftsmen construct a right triangle which is commonly called a *true-length diagram*. Basically this diagram represents rotation but it is removed from the original views. For example, to develop the pattern for the transition shown in Figure 8-4, the true lengths of *A*-1, *A*-2, *A*-3, and *A*-4 are needed. If each numbered point were to be rotated on a vertical axis through *A*, the difference in height between *A* and the numbered points would remain as originally shown in the front view. Therefore, a true-length diagram for each of the required lines may be made. Project the height to one side and transfer, with a compass, the top view of each line to a base line. The hypotenuse of the right triangle thus formed represents the true length of the line.

8-4 To find the true shape and size of a surface by rotation

As stated in section 7-6, a direction line may be used to establish the edge view of a surface. Using rotation and the direction line as an axis, the entire surface may be shown true shape and size in one of the given views.

As an example, Figure 8-5 shows the top and front views of a square tapered stack base, intersecting a ¼ pitch roof. Only the true shape and size of one panel (1-2-3-4) need be found because the four panels of the stack base are alike. Edge 2-3 is true length in the front view. A point view of 2-3 results in an edge view of the entire panel. Use the point view of 2-3 as an axis and rotate points 1 and 4 until the edge view lies parallel to V. Project points 1 and 4, revolved, to establish the rotated position of each point in the front view. Using long dashes, connect the rotated points 1 and 4 with 2-3 on the axis to show the true shape and size of the panel.

(Since the tapered base is part of a pyramid, other methods, described in Chapter 10, for developing a panel pattern could be used. For simplicity and accuracy, however, the method shown in Figure 8-5 is an excellent one.)

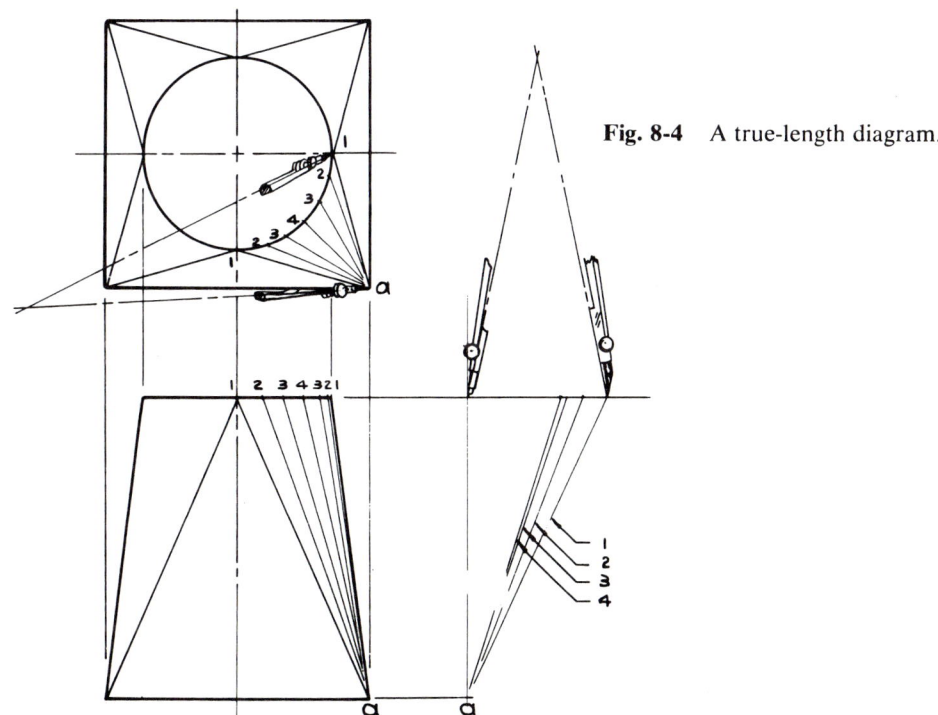

Fig. 8-4 A true-length diagram.

Fig. 8-5 Obtaining the true size and shape of a surface by rotation.

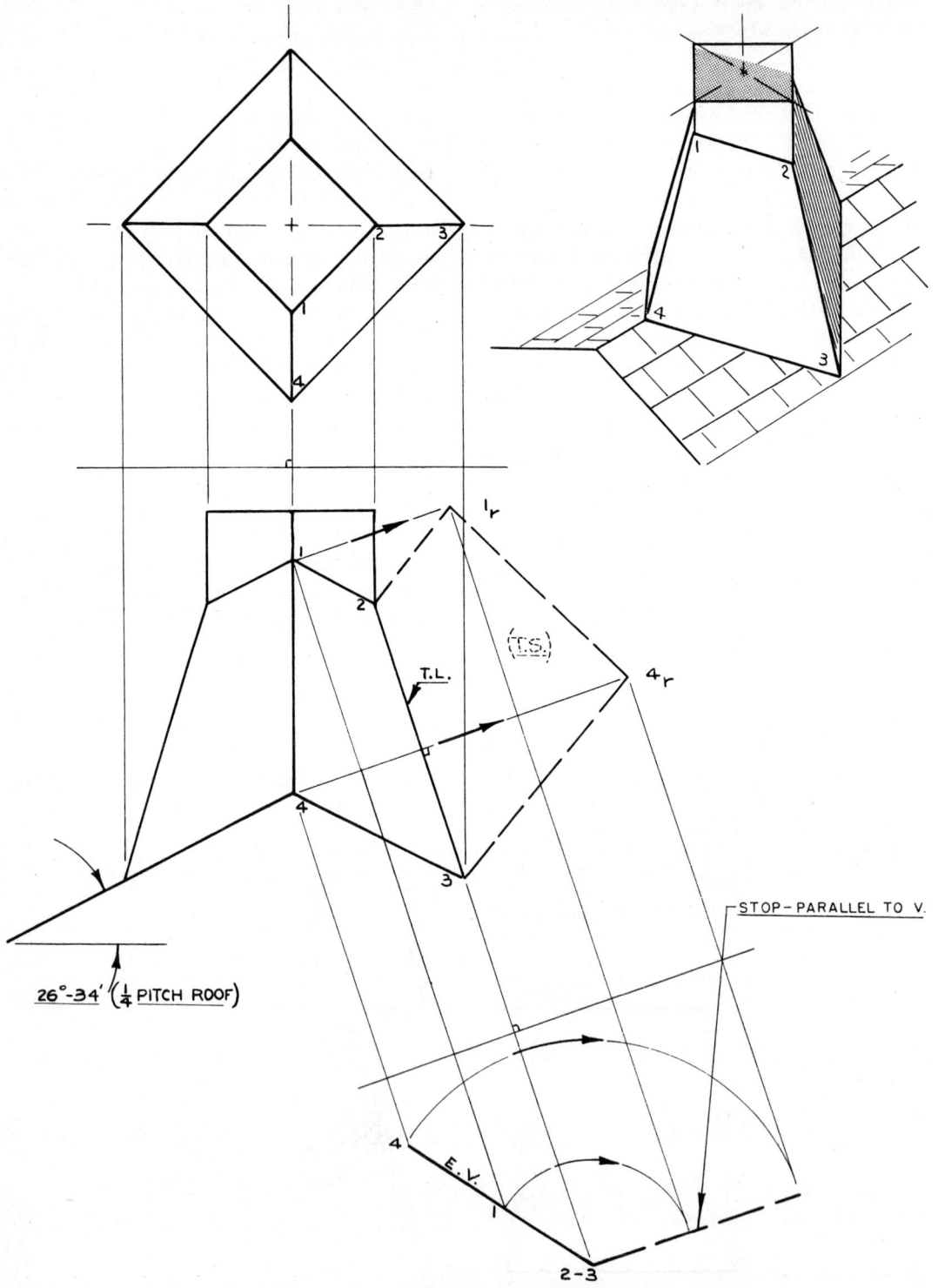

EXERCISES

1. Using as large a scale as possible, draw the views of the building in Figure 7-28. By means of rotation, determine the true length of the hip rafter *B*, the valley rafter *E*, and the hip rafter *C*.
2. Construct a true-length diagram for the above problem.
3. Draw the top and front views of the arm shelf in Figure 7-30. By means of an edge-view auxiliary and rotation, determine, in the front view, the true shape and size of one side wing. By means of an edge-view auxiliary and rotation, determine, in the front view, the true shape and size of the bottom of the shelf.
4. Draw the top and front views of the elliptical cylinder shown in Figure 7-26. Take a right section through point *A* and revolve the right section about an axis perpendicular to V until the right section is parallel to H. Using the four-center method described in section 5-14, construct an ellipse in the top view to represent the rotated right section.
5. Draw the top and front views of the roof in Figure 7-31. Intersect the roof with a stack 18 inches [457 millimeters] square, as shown. By means of an edge-view auxiliary and rotation, determine in the top view, the true shape and size of the opening required.

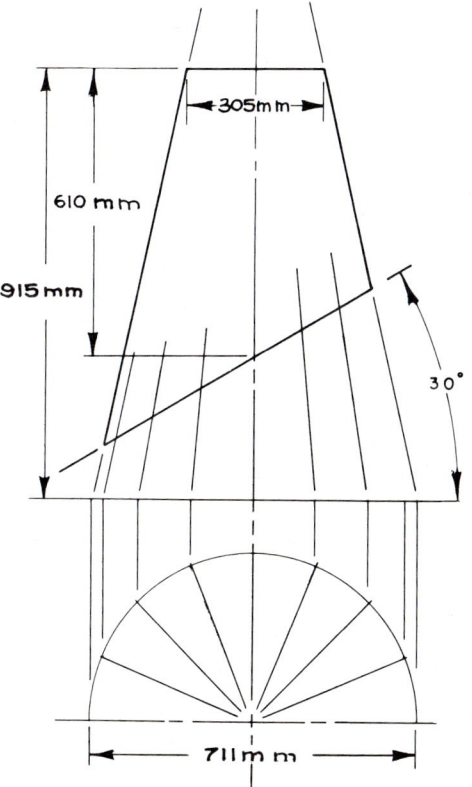

Fig. 8-6 Roof collar.

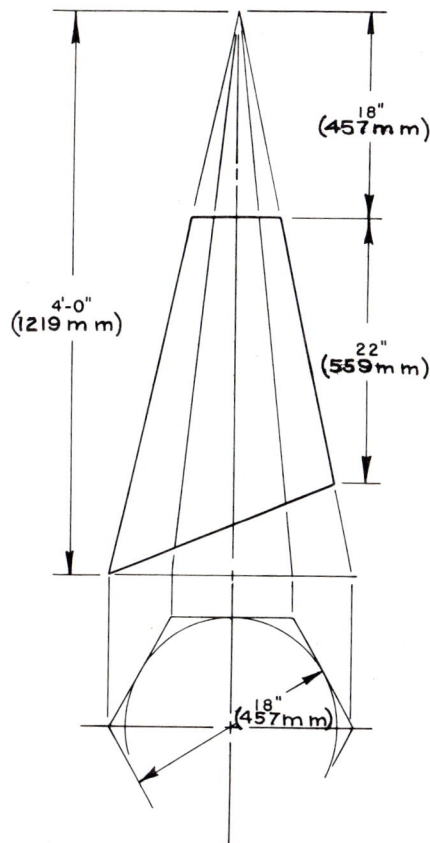

Fig. 8-7 Roof collar.

6. Draw the top and front views of the roof collar shown in Figure 8-6. By means of rotation determine the true length from apex *A* to each of the collar points
7. Draw the top and front views of the roof collar shown in Figure 8-7. By means of a true-length diagram, determine the true length of each of the collar corners from roof surface to the top of the collar.
8. Draw the top, front, and right-side views of a right square pyramid whose base is 2 in [51 mm] square and whose altitude is 3½ in [89 mm]. Draw the pyramid in such a position that one triangular side is true size and shape in the front view and the base appears as an edge in the side view.
9. Refer to Figure 8-8. A chute is required to load grain from the elevator bin to the hopper car. If the chute is to telescope, what are the minimum and maximum lengths required to reach the eight hopper ports with the car positioned as shown?

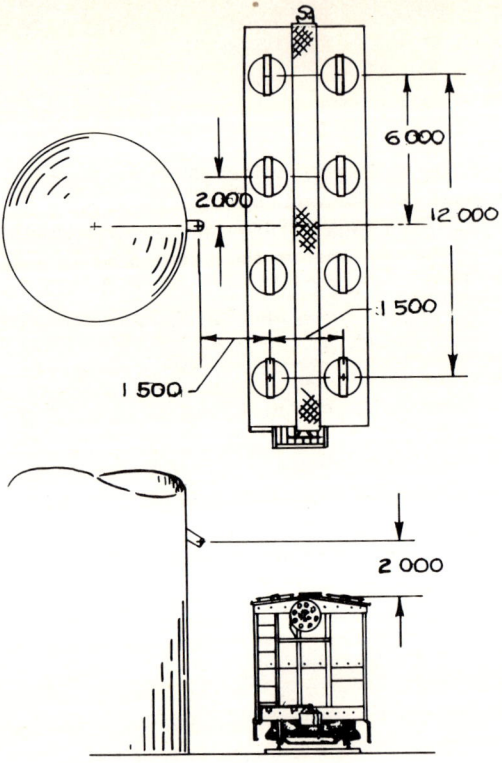

Fig. 8-8 Grain chute for elevator bin.

DIMENSIONED IN MILLIMETERS

chapter 9

PARALLEL-LINE DEVELOPMENT

A geometric object, of such a shape that straight lines may be drawn upon its surface parallel to its axis, may be classified as a *parallel form*. The lines are called *elements* of the object (section 5-1). Good examples of parallel forms in sheet metal are pipes, cylinders, and ducts. To *develop* the pattern of a sheet metal object means to unroll or unfold the surface of the object and thereby determine the true shape and size of the metal needed (Figure 9-1). The patterns for parallel forms are developed by the parallel-line method.

9-1 Essential views

To develop the pattern of a parallel form certain views are required (section 6-6): (1) A view is required in which the elements of the form show true length. For example, in Figure 9-2, the corners of the prism appear to be true length in the front view. The true lengths are transferred from the front view to the pattern. (2) The *girth* or true distance around the object must be known, or must show in a view. In Figure 9-3 the object is an elliptical cylinder. The *girth* appears in

Fig. 9-1 Developable parallel-line surfaces.

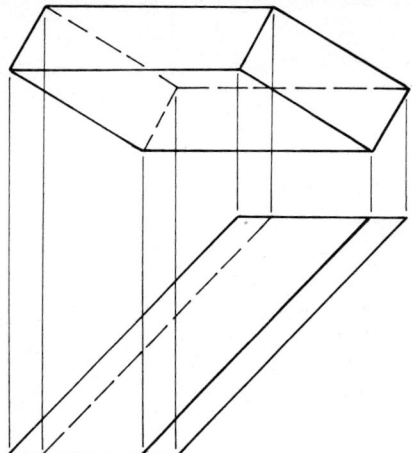

Fig. 9-2 Oblique prism.

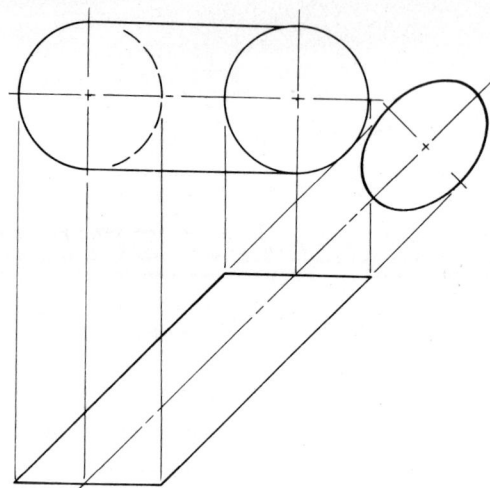

Fig. 9-3 Oblique cylinder.

the auxiliary view where the axis of the cylinder appears as a point. (This view is also the true shape of the right section of the pipe.) (3) A development area is needed where the pattern can be made. The object will unroll or unfold along a straight line perpendicular to each element of the object. The resulting view is called the "development" or the "stretchout" (Figure 9-4).

Fig. 9-4 Stretchout of cylinder.

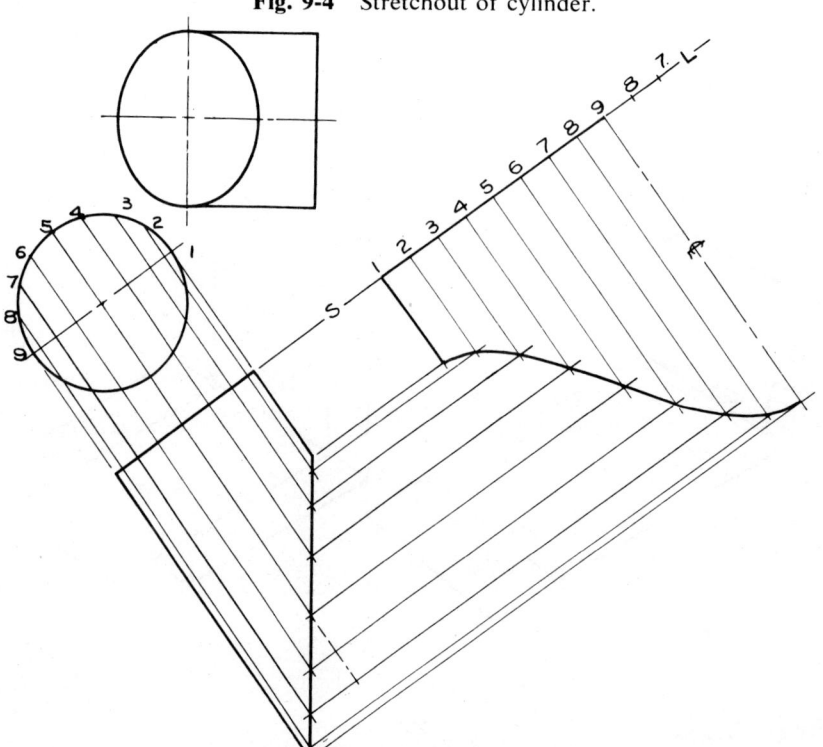

9-2 The stretchout line

In parallel-line development, the object unfolds or unrolls along a straight path called the *stretchout line*. The line is equal, in length, to the girth of the object. In the case of circular pipes, the stretchout distance is equal to πD (section 5-9), and is most accurately obtained by computation or from the sheet metal shop circumference rule. Figure 9-4 illustrates the step-off method of obtaining the total stretchout distance by transferring the spaces between elements obtained from the "girth view." The object in Figure 9-4 has one base which is right circular. Notice that the stretchout line is a projection of the edge view of that base. It is logical to place the stretchout line thus so that the true length of each element may be projected, rather than transferred, onto the development. The stretchout line need not be placed at one end of the object. On the contrary, it is preferable, in case both ends of the object are oblique to the elements, to place the line between ends, much in the manner of a string drawn taut about the object to obtain its girth (see Figure 9-5).

PIPE AND PLANE INTERSECTIONS

9-3 To develop cylinders or prisms with oblique ends

Many problems of parallel-line development involve the intersection of a cylinder or a prism with planes which are not perpendicular to the axis of the cylinder or prism. As discussed in section 7-6, a view which shows the intersecting plane as an edge will determine the length of each element of the cylinder or prism (two horizontal planes did this on the prism in Figure 9-5). As another example, Figure 9-6 shows a cylinder intersecting a floor at 30 and a wall at 60 degrees. Although Figure 9-6 represents the normal position of the pipe and planes, Figure 9-7

Fig. 9-5 Stretchout of prism.

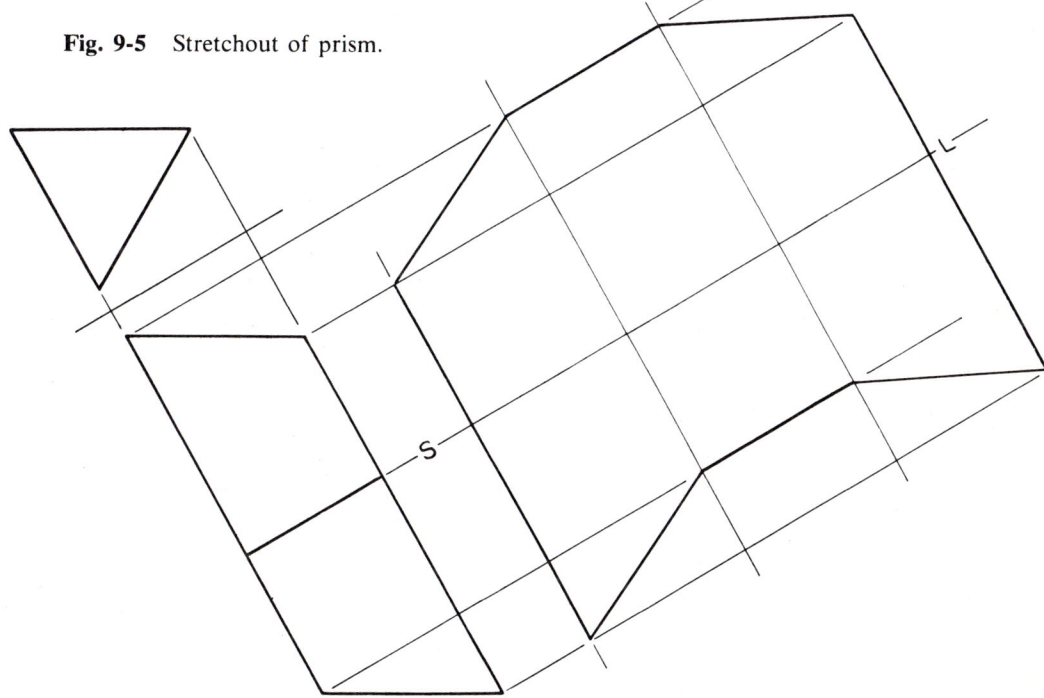

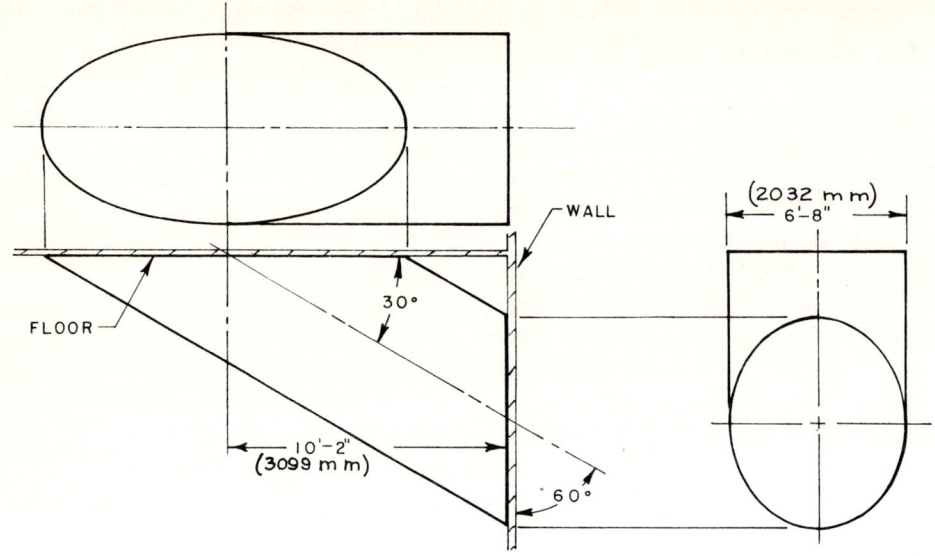

Fig. 9-6

Fig. 9-7

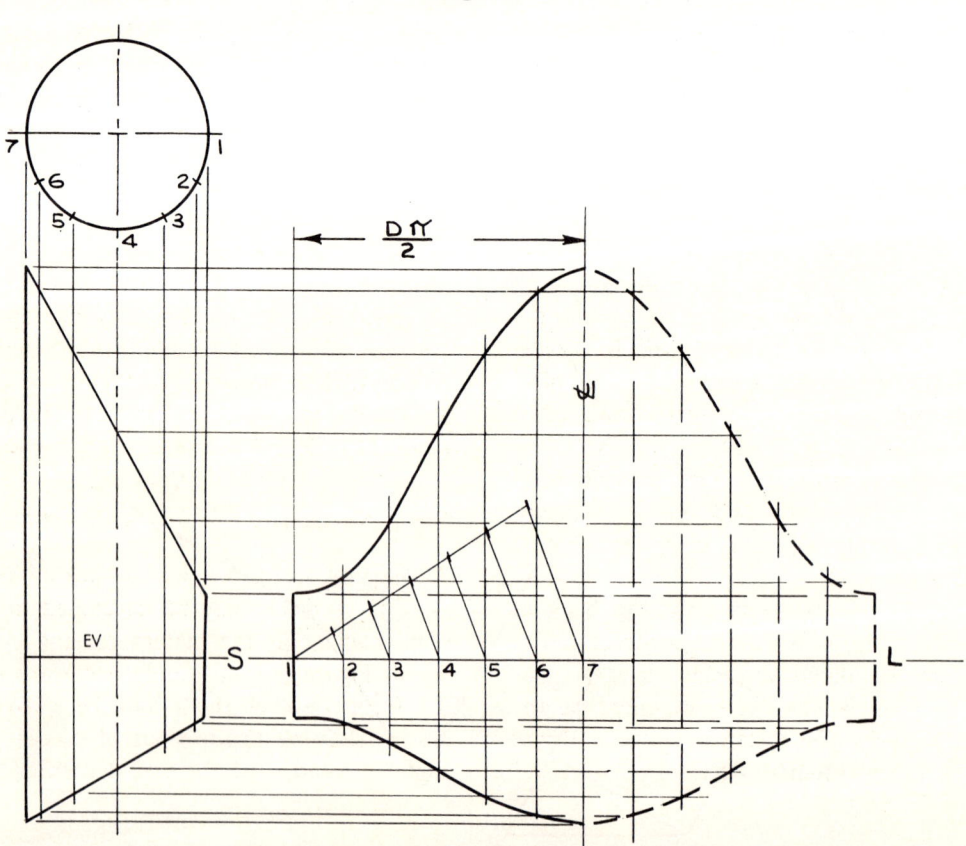

shows the problem arranged for easier development. The top view then furnishes the girth, the front view the true length of the elements, and the stretchout can be made to the right of the front view.

To make the development, the front half of the top view is divided equally into six spaces. Each numbered point represents an element on the cylinder surface. The edge view EV of the right section is placed at random in the front view and extended to serve as a stretchout line SL. Because the object is symmetrical, only a half-pattern need be developed. Points 1 and 7 are located on the stretchout line at a distance of one-half the cylinder circumference, $\pi D/2$, apart. Divide the 1–7 distance into six equal spaces and erect all elements perpendicular to the stretchout line. Project the true length of each element from the front view to the properly numbered element in the development, thus locating seven points on each curve. Sketch and draw the curves (section 3-14). The pattern thus developed represents the cylinder as though it had actually been unrolled along a path perpendicular to its axis, for a distance equal to one-half its circumference. Half-patterns are sometimes called "flop patterns." Patterns or templates are usually developed "inside up" with the seams along the shortest element. For a symmetrical object and a half-pattern, no attention need be paid to developing them inside up.

ELBOWS

9-4 Elbow nomenclature

Before drawing the front view of any elbow, an understanding of the terms applied to its various parts will be helpful.

The angle of bend of an elbow is measured in degrees and refers to the amount of "flow direction change" produced by the elbow (see Figure 9-8*A*).

The center-line radius is the distance from the center point to the center line of the elbow (Figure 9-8*B*). For elbows used in exhaust or blowpipe systems this is usually equal to twice the diameter of the pipe. These are sometimes called "long-radius elbows."

The throat radius regulates the amount of throat space available and is related to center-line radius and pipe diameter (Figure 9-8*C*).

The heel radius determines the back of the elbow and is equal to the throat radius plus the pipe diameter (Figure 9-8*D*).

The pipe diameter is represented on the front view by the distance between throat and heel arcs (Figure 9-8*E*).

Miter lines are shown at *F* in Figure 9-8. They represent seams between elbow sections.

The end sections are labeled *G* in Figure 9-8. Every elbow has two end sections. These are cut to the same curve as the center sections. They may be alike or may have seam relationship reversed, as desired.

The center sections are labeled *H* in Figure 9-8. Their pattern is equal to the patterns of the two end sections. The number of center sections for any elbow is always two less than the total number of sections (pieces). The angular bisector *J* in Figure 9-8 divides the center sections into two parts identical with end sections.

9-5 Necessary views

The patterns for the pieces of any elbow of uniform diameter may be developed in the same manner as the pattern for the cylinder in Figure 9-7. It is only necessary that the front view of the elbow be correctly drawn, showing the number of pieces, the angle of bend, and the true length of the

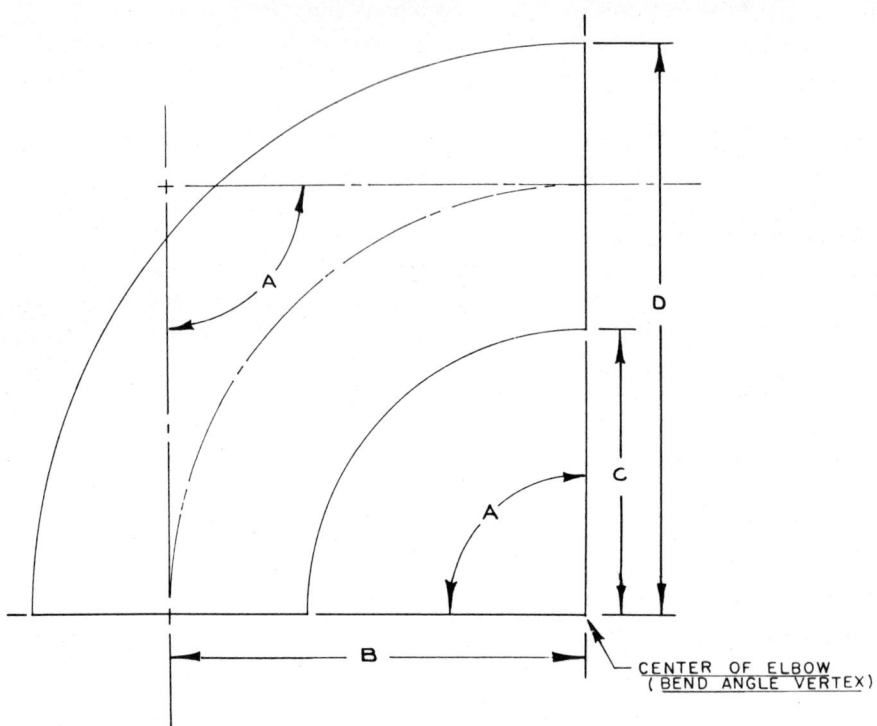
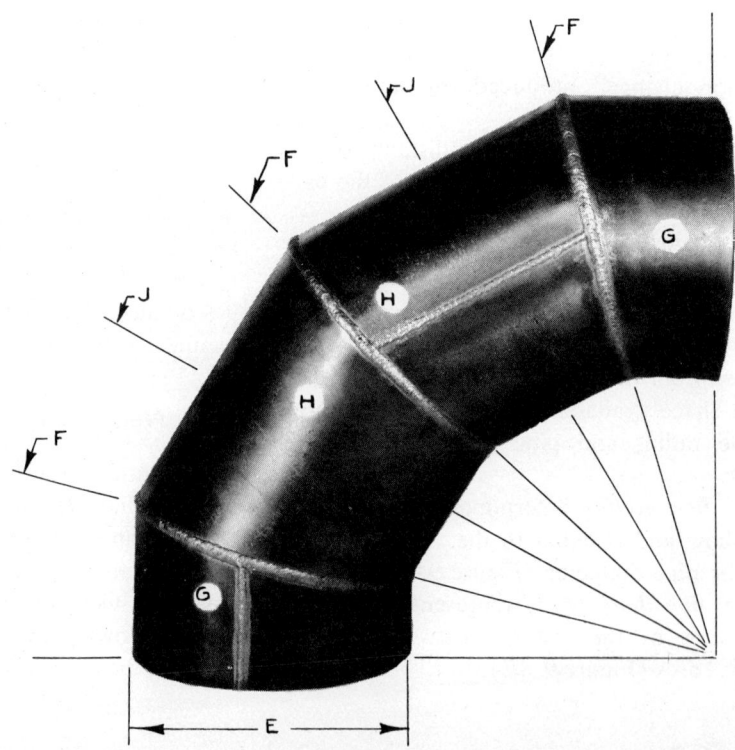

Fig. 9-8 Elbow nomenclature. (*Naylor Pipe Co., Chicago, Ill.*)

PARALLEL-LINE DEVELOPMENT 103

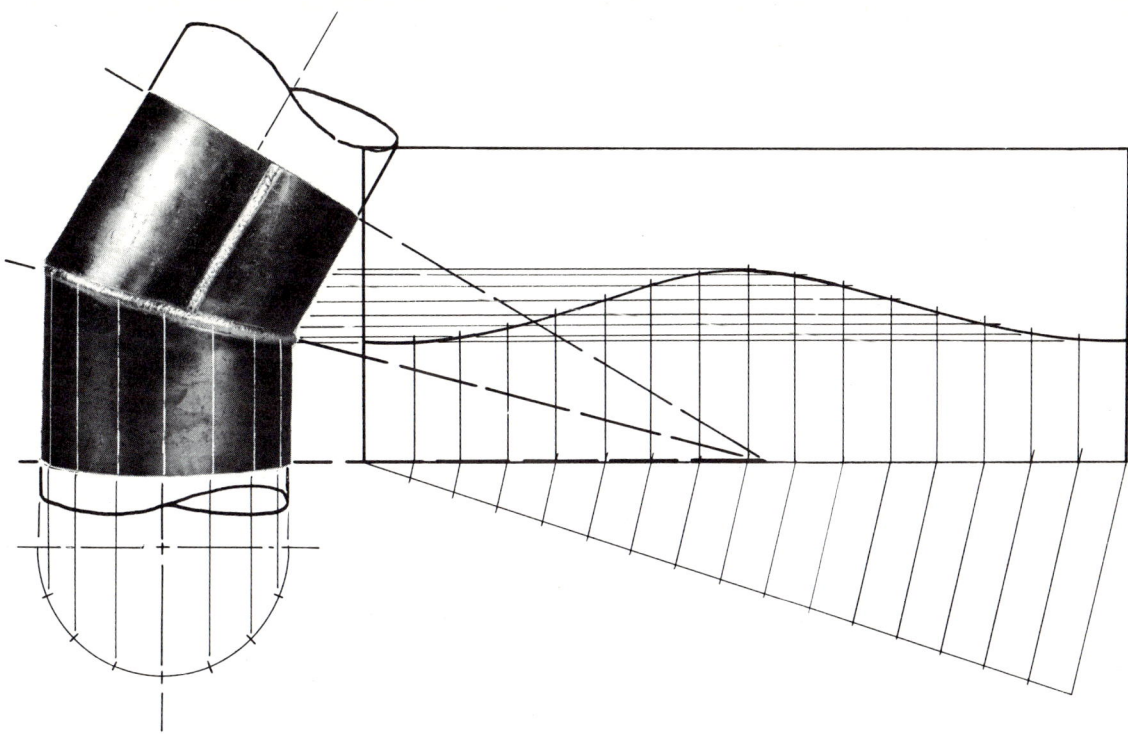

Fig. 9-9 Two-piece elbow. (*Naylor Pipe Co., Chicago, Ill.*)

Fig. 9-10 Common mistake in elbow design.

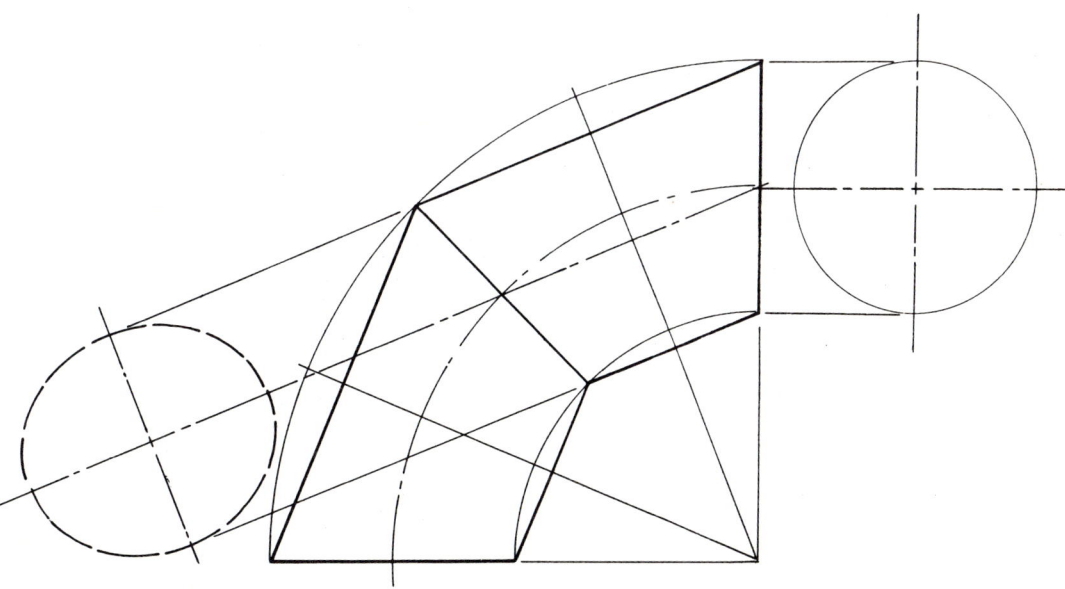

elements for each piece. A top or bottom view, showing the *girth,* is convenient but not always necessary, depending on the shape of the pipe.

9-6 Two-piece elbow

The front view of a two-piece, 30-degree elbow, as shown in Figure 9-9, is constructed as follows: (1) Draw the base and terminal lines for the angle of bend (30 degrees). (2) Bisect the angle of bend $30/2 = 15$ degrees; the bisector becomes the miter line of the front view of the elbow. (3) Draw the heel, center line, and throat arcs, as determined from job and pipe specifications. (4) Draw the two elbow sections *perpendicular* to the base and terminal lines of the elbow. Notice that the elbow pieces are parallel to the joining pipe sections. A common mistake of engineers and draftsmen is to design an elbow, as shown in Figure 9-10, which requires a different-shaped cross section in order to match the circular pipe, a shape which probably is neither desired by the designer nor required by the job. This error is more likely to occur in designing elbows of more than two pieces. *Always have the end sections parallel to the joining pipes* to avoid this.

9-7 Elbows of three or more pieces

The following procedure may be applied to any elbow, regardless of the number of pieces, the angle of bend, or the shape of the pipe. (1) Draw the base and terminal lines equal to the angle of bend (*ABC* of Figure 9-11). (2) Draw throat, center line, and heel arcs. (3) Divide the angle of bend into the required number of spaces. The center pieces of any elbow are twice as large as the end pieces, therefore, allowing one angular space for each end and two spaces for each center piece, as shown in Figure 9-11, the

Fig. 9-11 Development of a fishtail pattern.

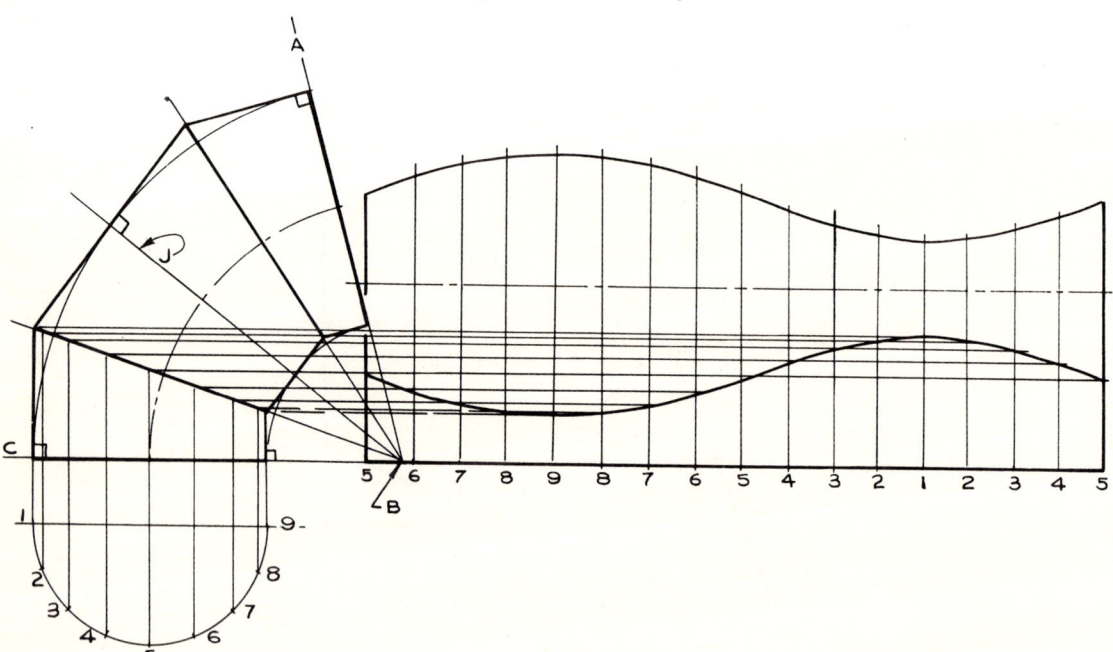

PARALLEL-LINE DEVELOPMENT

angle of bend may be spaced using the following formula:

No. of spaces = (no. of sections × 2) − 2

No. of degrees per space = $\dfrac{\text{angle of bend}}{\text{no. of spaces}}$

(4) Complete the front view by erecting the sides of the end pieces *perpendicular to the base and terminal lines* and the sides of the center pieces *tangent to throat and heel arcs* and *perpendicular to the center section midline J* (Figure 9-11).

The example in Figure 9-11 is a three-piece 75-degree elbow. The spacing would be computed as follows:

(3 × 2) − 2 = 4 spaces

$^{75}\!/_4$ = 18.7 degrees per space

The first and last spaces are always used as end sections and the remaining spaces are used in pairs to make up the center sections. It therefore should be obvious which lines become the miter lines and which the midlines of center sections.

For the four-piece elbow in Figure 9-8, the spacing would be

(4 × 2) − 2 = 6 spaces

$^{90}\!/_6$ = 15 degrees per space

9-8 Seam arrangement

When making the stretchout of an elbow curve, three different start and stop points are practical as shown in Figure 9-12. With the curve arranged as shown at *A*, the seams of the finished elbow are staggered (end sections seamed in throat and center sections seamed on heel). This arrangement saves metal. If the curve is started as in Figure

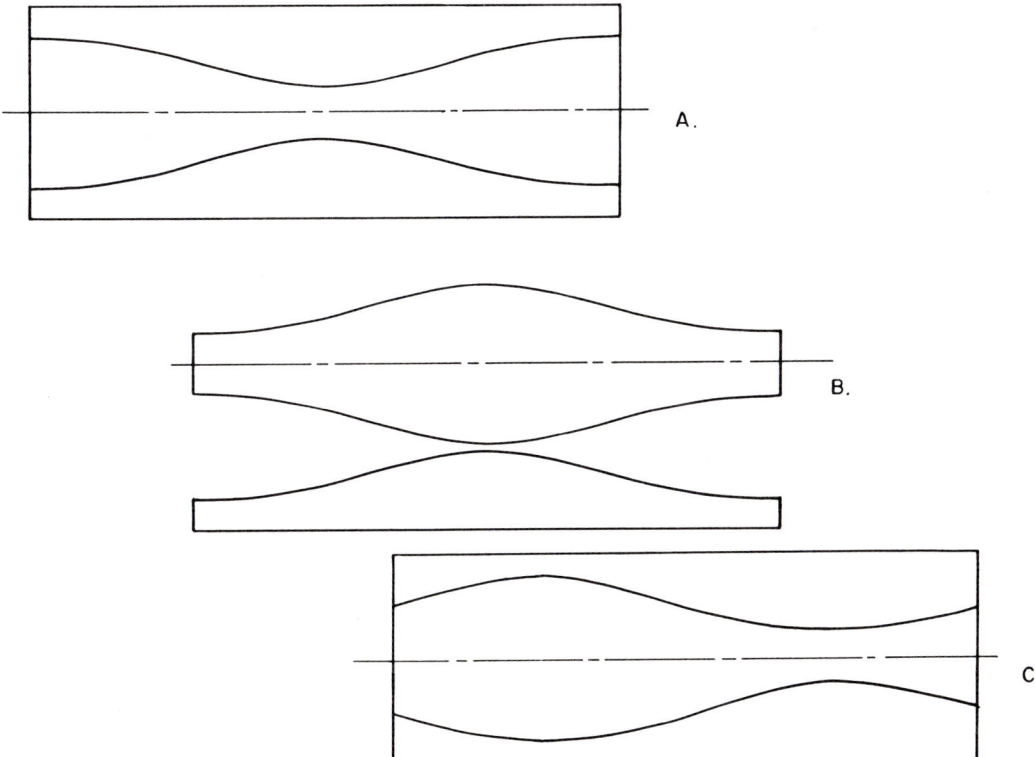

Fig. 9-12 Pattern arrangements.

9-12B the seams for all sections are as short as possible and are all throat seams. This arrangement, however, is more wasteful in cutting. Arrangement C presents several advantages over A and B. The metal is conserved, the seams are on the sides of the sections, they are all the same length, and they may be aligned or staggered left and right when forming the metal. Because of its appearance, this arrangement is called a "fishtail pattern." The fishtail pattern for the center section of the elbow in Figure 9-11 was made by repeating the pattern of the end section so that it was symmetrical on a stretchout center line.

9-9 Duct elbows

Duct elbows, used in the air-conditioning industry, are discussed as a part of this chapter because the basic principles of parallel-line development apply to some fittings of this type.

A curved duct elbow has four parts, namely, the throat (Figure 9-13A), the heel (Figure 9-13B), and the two cheeks (Figure 9-13C).

To draw the front view of a curved duct elbow similar to that shown in Figure 9-13: (1) Construct the angle of bend DEG with the center of the elbow at E. Make the throat radius, when possible, equal to the largest cross-section duct dimension (for an 8- by 10-inch [203- by 254-millimeter] duct, the throat radius is to be 10 in [254 mm]). (2) Draw the throat and heel arcs, which complete the front view. Only the front view is needed but, for clearness, a partial bottom view showing the girth is helpful. With the addition of proper allowances for seams, the pattern for the left and right cheeks is as shown in the front view. The patterns for the

Fig. 9-13 Rectangular 90-degree duct elbow.

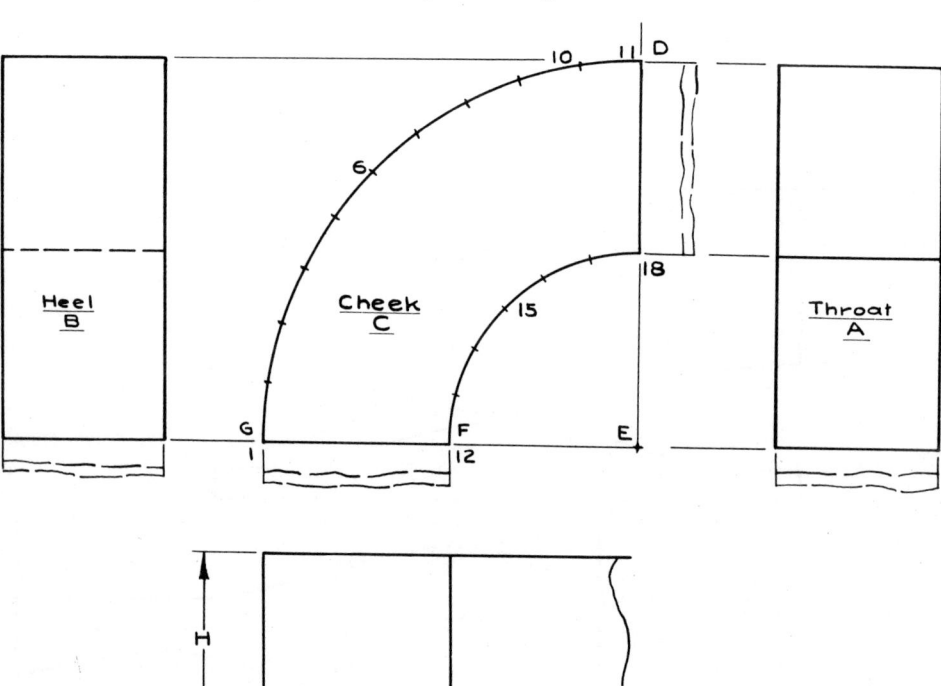

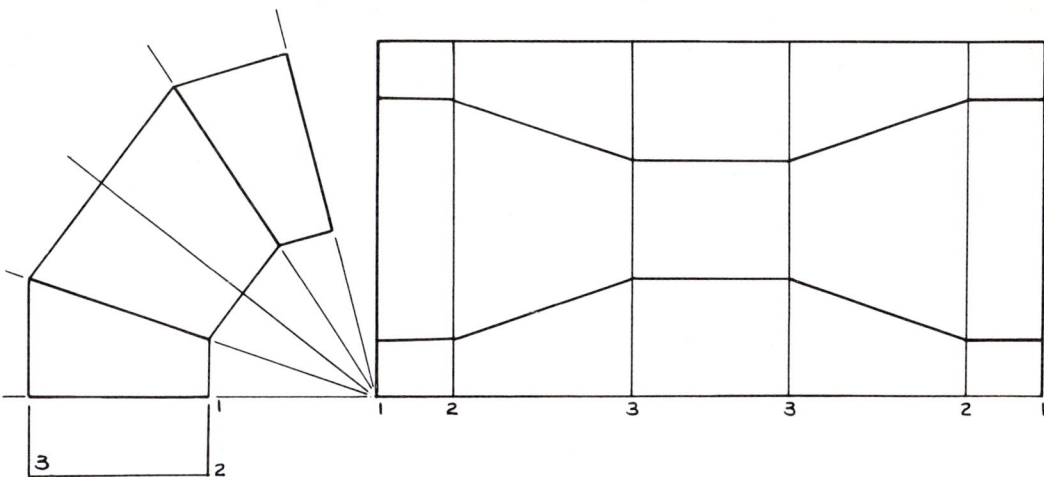

Fig. 9-14 Three-piece 75-degree duct elbow.

throat and heel pieces are rectangular strips. The width of the strips is as specified for the duct size and is indicated in the partial bottom view (Figure 9-13, H plus seam allowances). The length of the strips may be obtained from the front view by spacing the heel and throat arcs uniformly into a sufficient number of points, then by stepping the spaces off on the stretchout (1 to 11, 12 to 18, Figure 9-13). Since the chordal distances are slightly shorter than the circular distances, the circular distance may be accurately computed, as discussed in section 5-9.

Duct elbows of any number of sections may also be developed from front views similar to those in Figures 9-8, 9-9, 9-11. In order to relate the true length of the various elements or corners to the stretchout, provide a partial bottom view, which also shows the girth. For example, Figure 9-14 illustrates the development of a three-piece 75-degree duct elbow.

9-10 Transitional or change duct elbow

If the duct elbow produces a change in duct dimensions, the elbow is called a *transitional* or *change elbow*. The only change elbows developable by parallel-line methods are those in which the cheek measurement changes, the heel and throat strips remaining a uniform width. The others are developed by triangulation as discussed in Chapter 11. The elbow in Figure 9-15 is a 60-degree square-to-rectangular change. The front view of a change elbow is drawn as follows: (1) Draw the angle of bend (*ABC* in Figure 9-15). (2) Draw the throat arc *DG*; if possible, make the throat radius equal to the larger duct dimension (*DB = ED*, Figure 9-15). (3) Locate the heel radius and center, by bisecting the heel angle *EFH* (the bisector crosses the base line at *D*). (4) Locate the tangency point *K* by erecting a perpendicular to *FH* through *D*. (5) Draw heel arc and heel flat to complete the front view.

For patterns, the cheeks are as shown in the front view; the heel and throat strips are as wide as shown in the partial bottom view and their lengths can be computed, using the angle of bend, or the step-off method (see section 5-9). The patterns thus drawn are net and additional metal would have to be allowed for seaming or for bend allowance in heavy material.

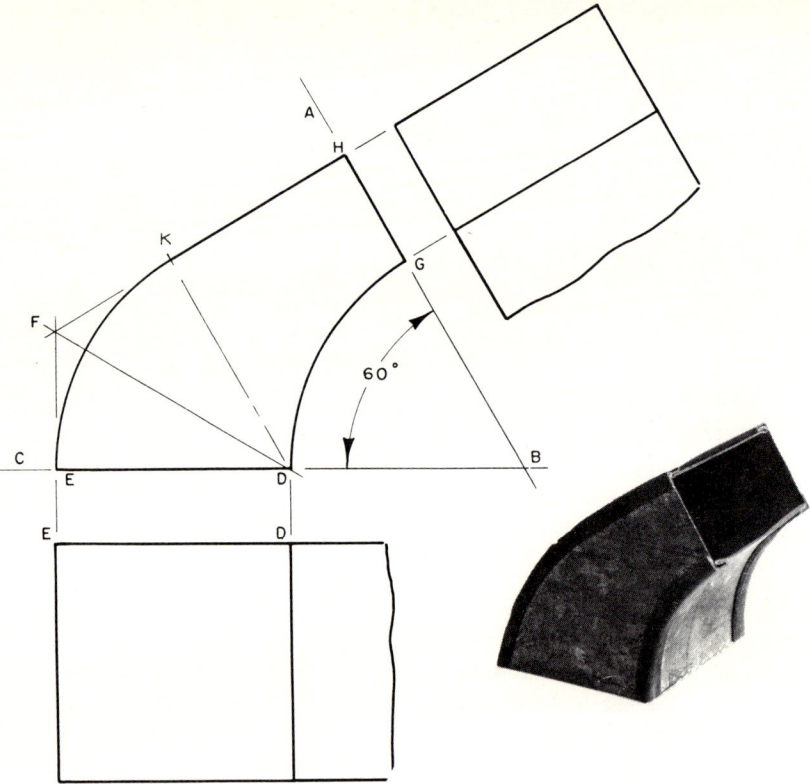

Fig. 9-15 Transition elbow.

9-11 Compound curved duct elbow

Fig. 9-16 Compound curved duct elbow.

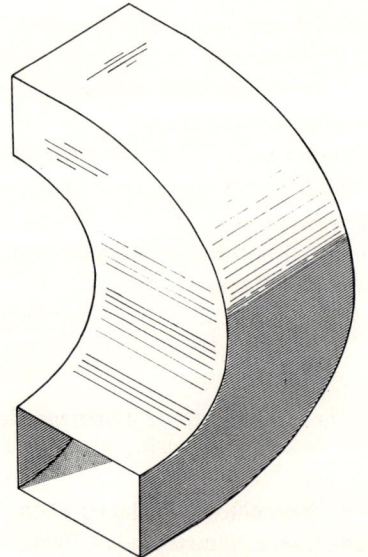

The fitting, shown in Figure 9-16, is a compound curved duct elbow; so named because it contains a 90-degree bend angle in two directions. The fitting is used to connect offset ducts at 90 degrees to each other. The elbow is made up of four pieces of metal. Note that each piece connects a vertical plane of one pipe to a horizontal plane of the other. This results in each piece being a combination of cheek and heel, or a combination of cheek and throat. If the two ducts joined by the fitting are alike in dimensions, and if the available space will permit, it is possible to design the fitting so that only two patterns are needed, one a cheek-and-heel pattern, and the other a cheek-and-throat pattern. Two pieces are cut from each pat-

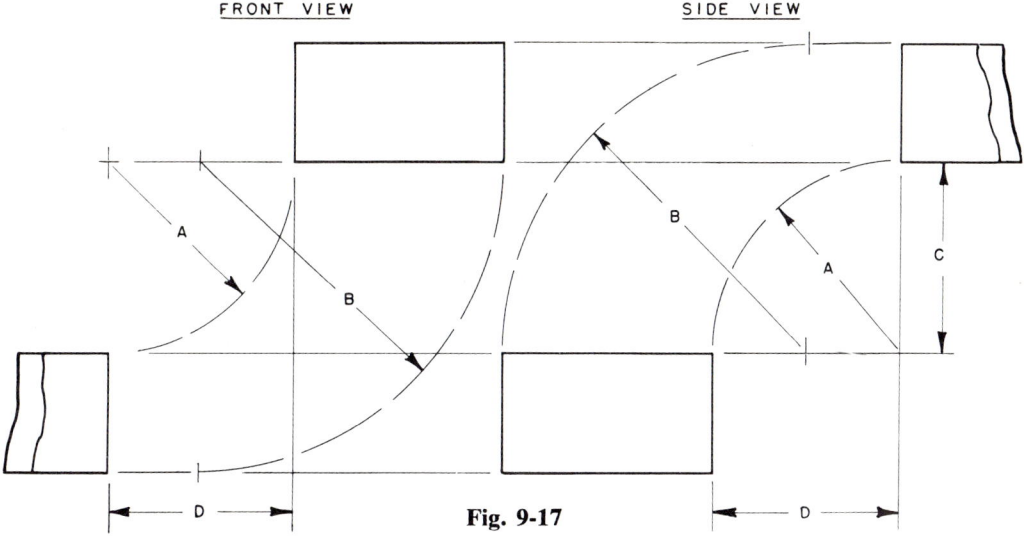

Fig. 9-17

tern to make up the elbow. To do this, several important factors must be considered when selecting and making the views of the fitting.

a. Two views are necessary.

b. The amount of offset should show in both views.

c. End-to-side measurement of the duct should equal the offset space between ducts.

If the above design factors are adhered to, the throat and heel radii can be the same in both views.

As an example, examine Figure 9-17. The two views selected are the front and side views—because the offset distance shows in both views. The specifications for the job at hand will dictate the offset distance required, but the end-to-side measurement of the duct distance D can be made equal to the space between ducts C so that the same heel and throat radii may be used in each view, as previously mentioned. The heel and throat radii of the front view represent the edge view of two metal strips; likewise, the heel and throat radii of the side view represent the edge view of two surfaces of the fitting. In each case, the surfaces represented are *quarter cylinders*. (The heel strips have additional flat metal at both ends; throat strips, at one end only.) Since the surfaces involved are quarter cylinders, parallel-line methods are used to develop the patterns for each strip.

The views of the fitting, in Figure 9-18, show elements added to the heel strip. The elements appear as points on the edge view of the strip (right-side view). The elements are equally spaced and should be numbered in this view. The true length of each of the elements is seen in the front view. Develop the heel strip just as any cylinder could be developed, by stepping off the distances between elements, as taken from the side view, then projecting the true lengths of the elements from the front view. Notice that the stretchout of the heel strip, as shown in Figure 9-18, is inside up. Two pieces are cut from the pattern, with allowances for seams, and *rolled in the same manner*.

In order to avoid confusion, separate illustrations have been used. Figures 9-17, 9-18, and 9-19 illustrate view construction, heel stretchout, and throat stretchout, respectively. In practice, one drawing would suffice to develop both the heel and throat patterns. The throat pattern is developed in the same manner as the heel strip and is

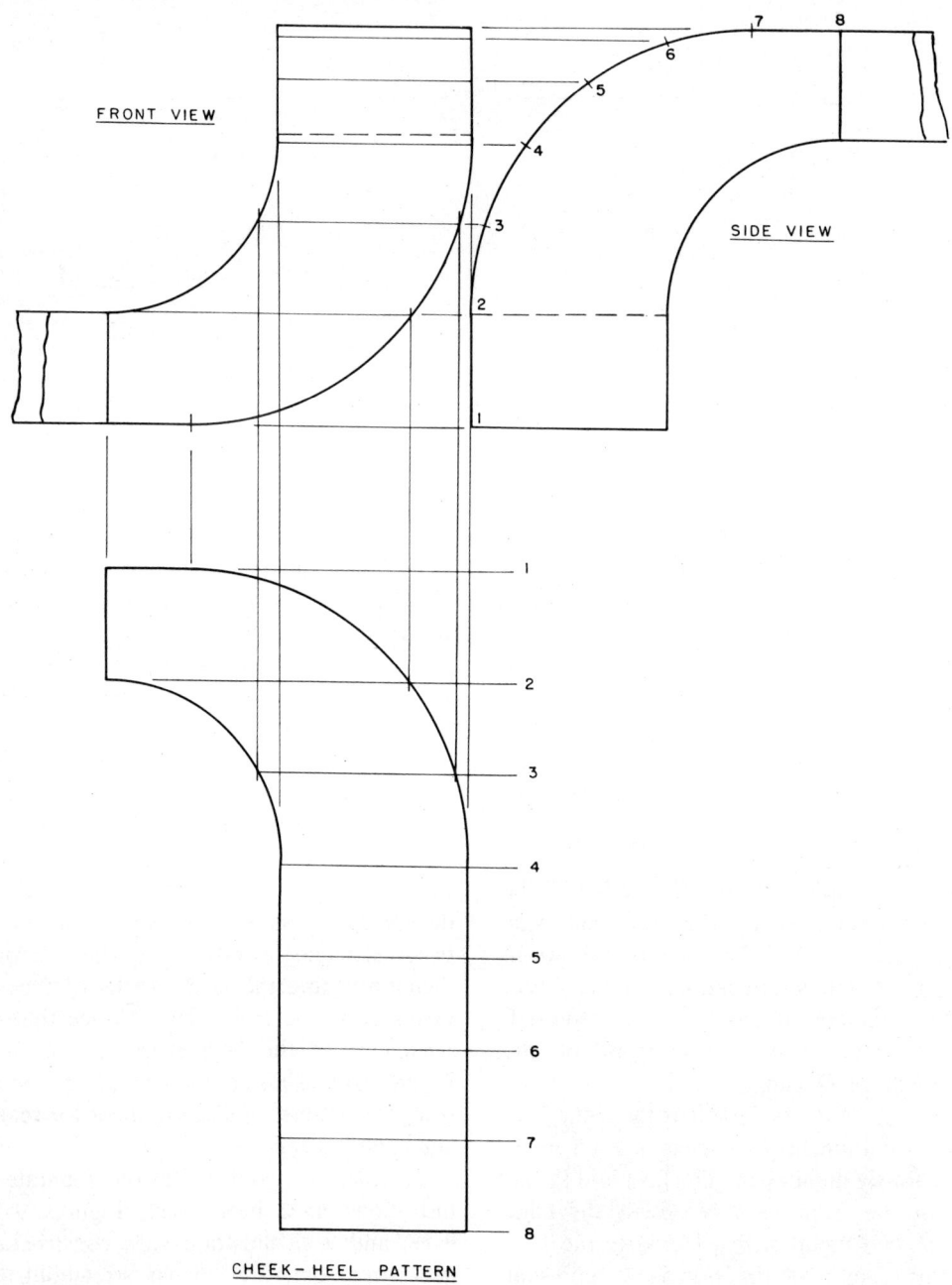

Fig. 9-18

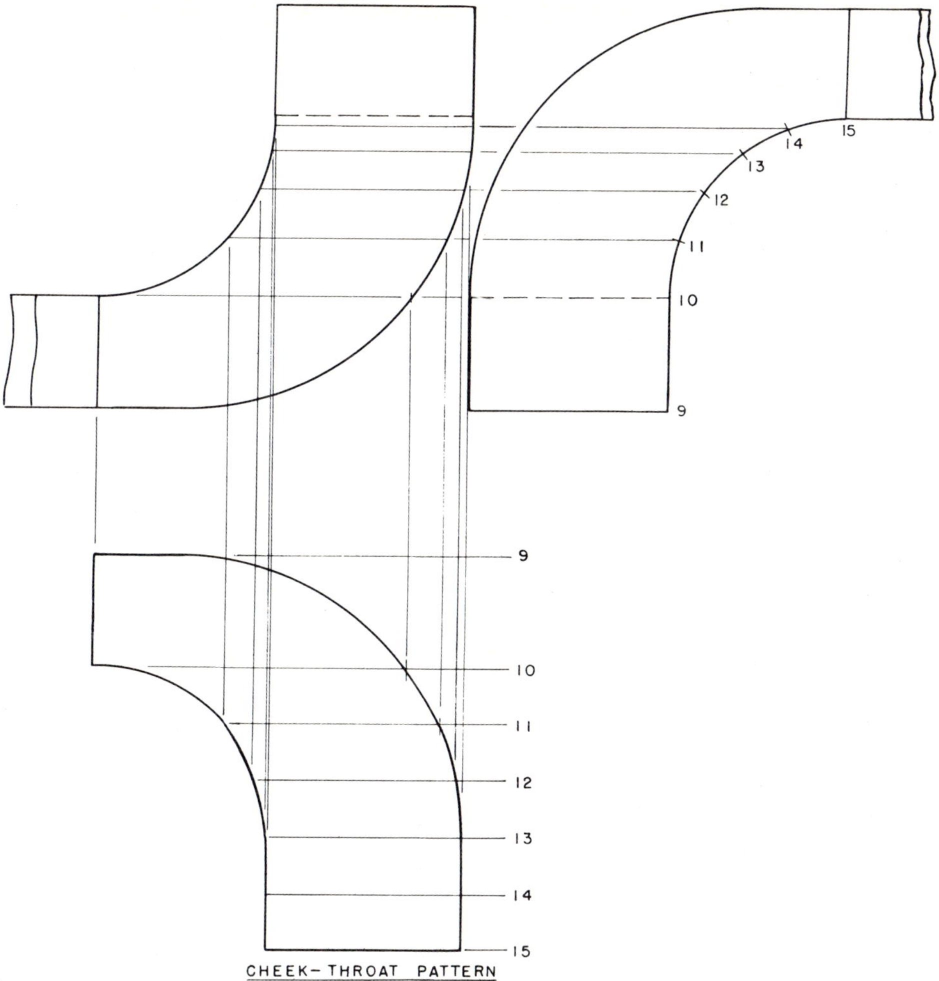

PARALLEL-LINE DEVELOPMENT

FRONT VIEW SIDE VIEW

CHEEK—THROAT PATTERN

Fig. 9-19

Fig. 9-20 Compound curved transitional duct elbow.

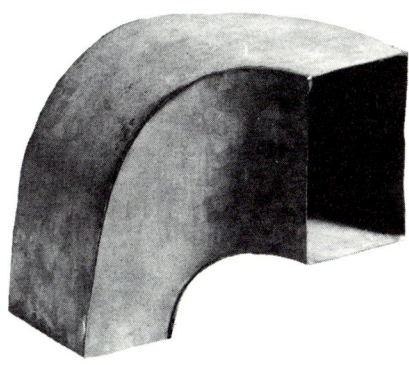

shown in Figure 9-19. Again notice that the throat pattern is inside up, and two pieces are cut and rolled alike.

9-12 Compound curved transitional duct elbow

A compound curved transitional duct elbow, shown in Figure 9-20, is similar to the fitting in Figure 9-16 except that the duct dimensions change.

To draw the two necessary views of Figure 9-21, again select those in which the offset distance will be shown. The specifications for the job at hand dictate the offset distance required for the fitting. The remaining dimensions necessary for the completion of each view are the distance *D* which locates, in each view, the end of the fitting; and the distance *E* between pipes. *D* is made equal to *E* in order to use a quarter cylinder throat radius. The heel radius *A*, in the top view, is obtained from the distance *X* (measured from the inside edge of the fitting mouth to the outside edge of the fitting, as shown in Figure 9-21). The heel radius *B* in the front view is obtained in the same manner (measured from the inside edge of the fitting mouth to the outside edge of the fitting) this being distance *Y* in the front view.

The development of the four patterns necessary is shown in Figure 9-22. Elements are placed on the four surfaces and num-

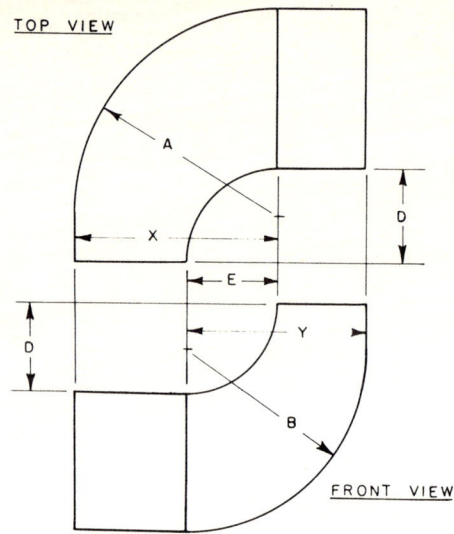

Fig. 9-21

Fig. 9-22

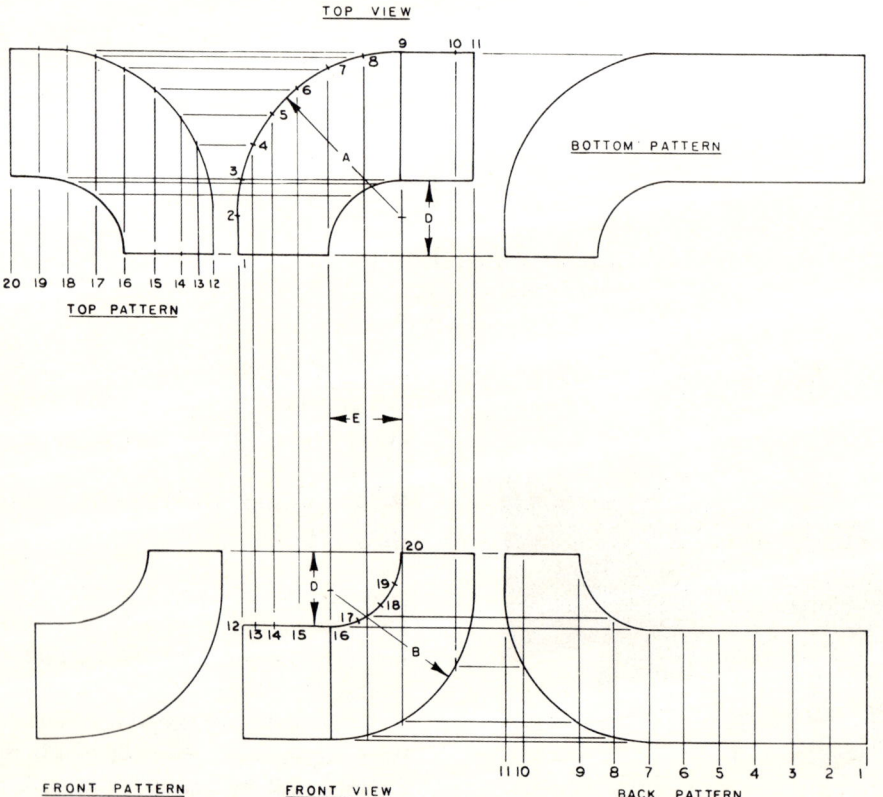

bered in the point views. Elements on the top and bottom surfaces appear as points in the front view and show true length in the top view; therefore, the stretchouts of these are projected from the top view. Elements on the front and back pieces appear as points in the top view and show true length in the front view; therefore, the stretchouts of the front and back strips are projected from the front view. Again notice that the patterns for all pieces have been developed inside up to facilitate bending.

OFFSETS

9-13 Offset design factors

An offset fitting or double elbow, as shown in Figure 9-23, is sometimes called a three-piece reverse elbow. The offset is used to divert the flow of gases or other fluids around an obstruction and to cause the flow to be moved out of line a specified distance (*A* in Figure 9-23). *Simple or primary offsets* contain offset measurement in one direction only (*A* in Figure 9-23). *Compound offsets* have two offset measurements (*A* and *B* in Figure 9-24).

The first major consideration in offset design is the pipe shape. Offsets may be designed for round, square, rectangular, elliptical, oblong, or any other shaped pipes. For primary offsets, the front view is the same, regardless of pipe shape. For example, the front view of the offset shown in Figure 9-23 remains the same for all pipe shapes, as shown in the top view.

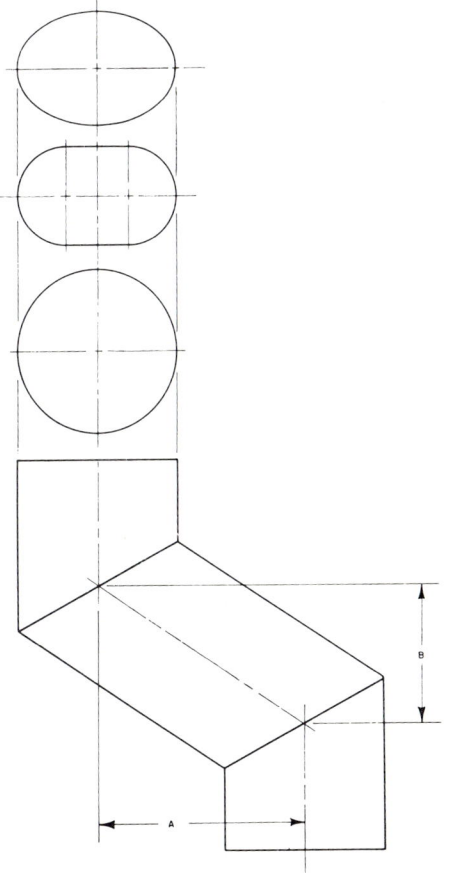

Fig. 9-23 Three-piece offset.

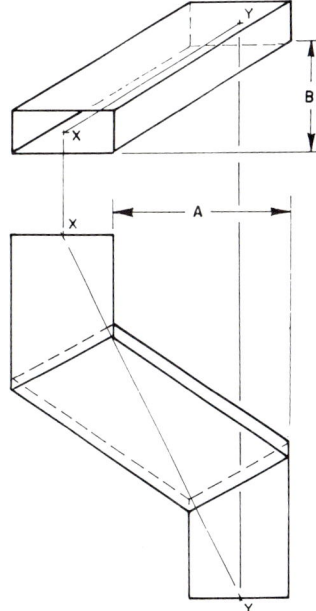

Fig. 9-24 Compound offset.

The most important measurement involved in an offset design is the amount of offset which is established by the specifications for the fitting. Another measurement affecting the design is the "run" (*B* in Figure 9-23). Possible variation in run allows the designer a measurement of such a nature that the fitting is practical to make and yet long enough to minimize friction of flow diversion, especially if high velocities are involved.

9-14 Three-piece offset layout

Several interesting geometric constructions may be used when drawing the front view of an offset, depending on which measurements are given and what features are desired in the completed fitting.

When the total run of the center section is not critical, the front view may be designed as shown in Figure 9-25. (1) Space the center lines, as in *A*, the specified offset distance apart. (2) Decide upon a suitable run, and draw the offset center line *XY* (Figure 9-25B). (3) Bisect angle *XYZ* to establish a miter line through *Y*. (4) The upper miter line through *X* is established in the same manner (parallel to first). (5) Construct a partial bottom view (of the pipe end only). (6) Complete the front view by projecting the contour elements from the bottom to the front view (Figure 9-25C). The contour elements of each piece are parallel to the center line of that piece. Notice that the total run of the center piece is now slightly more than the original run of center line *XY*. The length of the end sections can be made suitable for the job at hand, preferably with both top and bottom end pieces alike.

Fig. 9-25

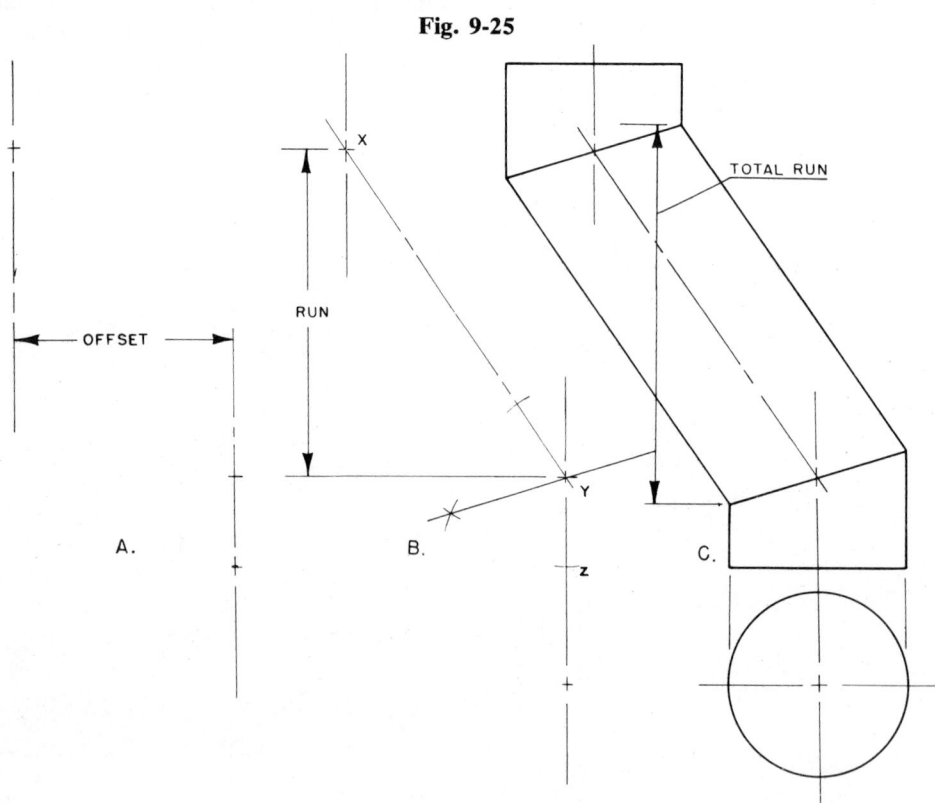

PARALLEL-LINE DEVELOPMENT

A method which results in the same front view as just described is illustrated in Figure 9-26. (1) Center lines are determined as before (Figure 9-26A). (2) A partial bottom view (of the pipe end only) is drawn. (3) Using the bottom view as a diameter, spheres are drawn with their centers at X and Y (Figure 9-26B). (4) The contour elements of the fitting are drawn tangent to the spheres. (5) The front view is completed (Figure 9-26C) by connecting the element intersection points, as shown, to form the miter lines of the view.

The most common front view used, when designing offsets, is one in which the entire fitting, including end sections, is drawn within a specified run distance. (1) Draw the center lines, spaced offset distance apart, as shown in Figure 9-27A. (2) Determine the total length of the fitting, thus locating D and F (Figure 9-27B). (3) Draw the construction line DF and locate the mid-point E. (Point E is the center of the fitting, since it represents one-half the offset and one-half the run.) (4) Construct the perpendicular bisector of DE, and the perpendicular bisector of EF. The perpendicular bisectors become the miter lines of the fitting. (5) Construct a partial bottom view (end of pipe only) and complete the front view by projection. Observe that a line through E, perpendicular to the center section elements, divides the fitting into two similar elbows. This is indeed a reverse elbow of three pieces, in which the center section is twice as long as either end section. Also notice that, since three pieces

Fig. 9-26

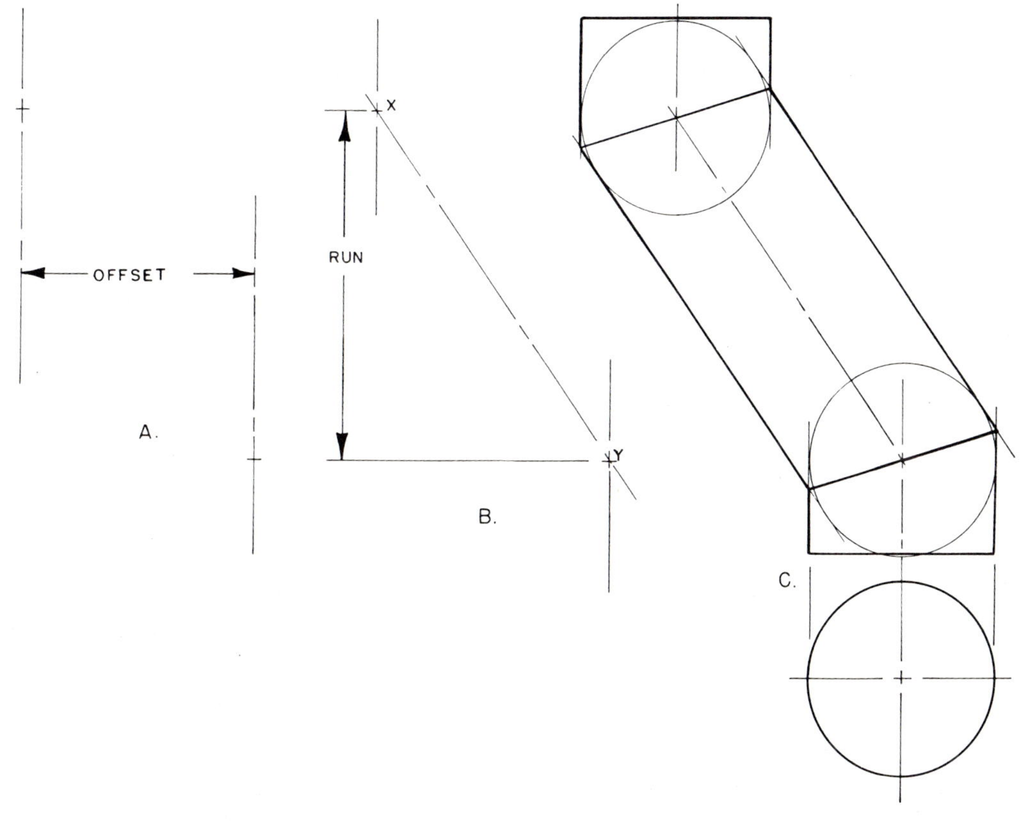

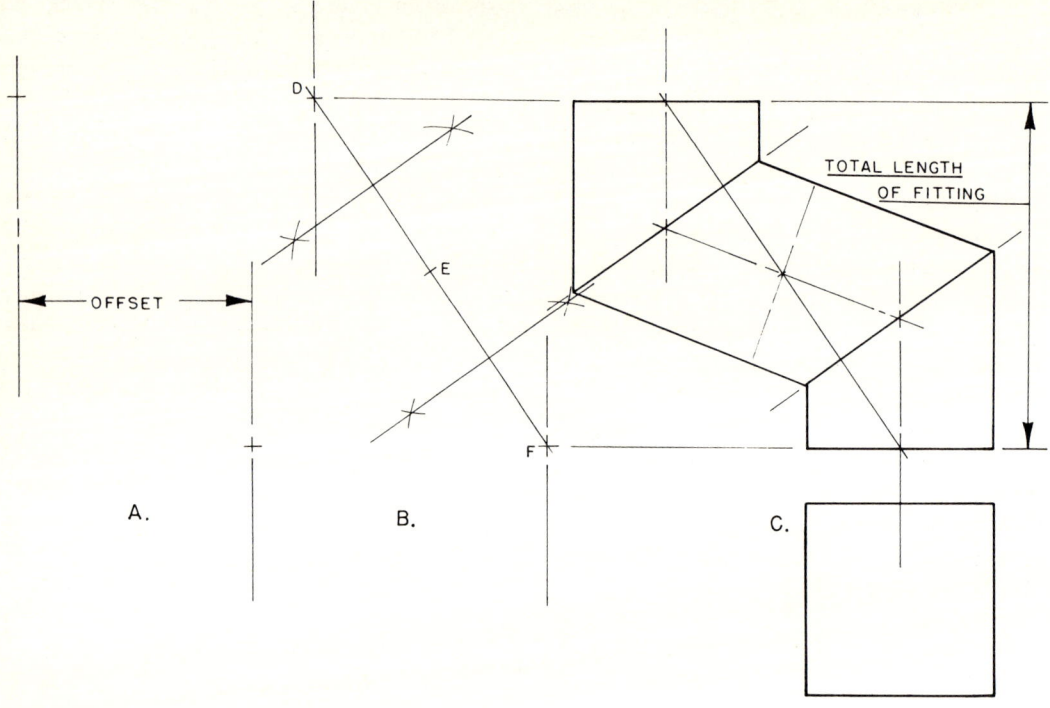

Fig. 9-27

go into the original construction from *D* to *F,* the offset angle of the center piece is somewhat larger than might be originally anticipated or desired. In cases where the offset or run measurements are long, it is common practice to use a couple of two-piece elbows and straight pipe, as shown in Figure 9-28.

When only the center section is desired within specified limits, many of the previous

Fig. 9-28

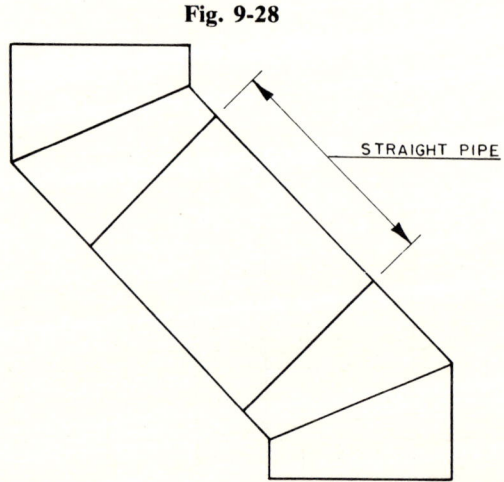

PARALLEL-LINE DEVELOPMENT

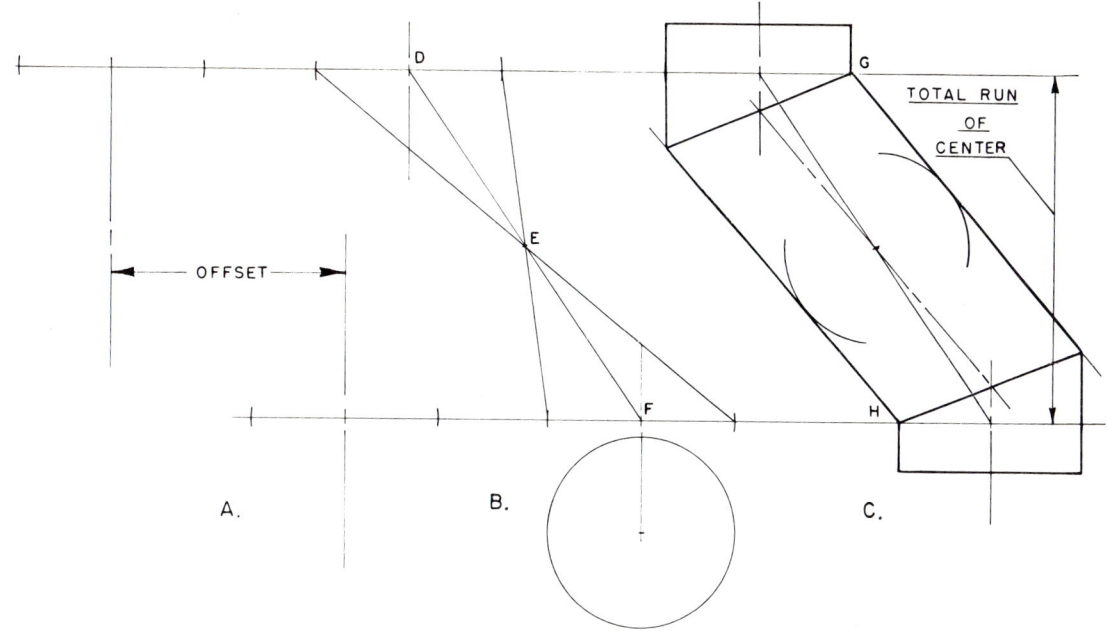

Fig. 9-29

procedures, with a few changes, may be used. Figure 9-29 shows the layout of center lines and maximum run desired for center section. (1) Again, draw the construction line *DF* and locate the mid-point *E*. (2) Using the bottom view as a diameter, a sphere is drawn with center at *E*. (3) The contour elements of the center section are drawn tangent to the sphere through points *G* and *H*. (4) The miter lines are de-

Fig. 9-30

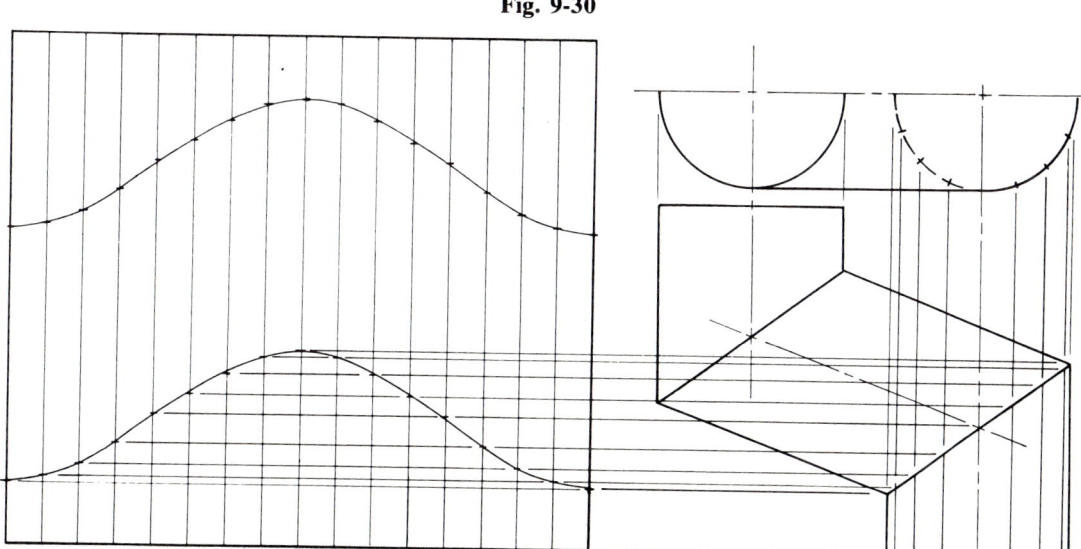

Fig. 9-31 Typical fittings used in a drier installation. (*Pittsburgh Electrodryer Corp., Pittsburgh, Pa.*)

termined by bisecting the throat angles, or by extending the contour elements of the end sections to meet those of the center section. Notice that the total run of the center section fits the limits established for that portion of the fitting.

9-15 Three-piece offset development

The development of a pattern for each piece of the offset is essentially the same as the pattern development described earlier. Figures 9-4, 9-5, 9-7, 9-9, 9-11, 9-12, and 9-14 all show various kinds and pieces of elbows developed. Apply the same general principles to the development of the offset of Figure 9-30.

9-16 Curved duct offset

Figure 9-31 is a photograph of a practical application of two duct elbows, a transitional elbow, and a curved duct offset

on a drier installation. For drawing a front view of a curved duct offset, the most important measurements are the offset distance A (Figure 9-32) and the throat radius B (Figure 9-32). The run distance C can be

Fig. 9-32

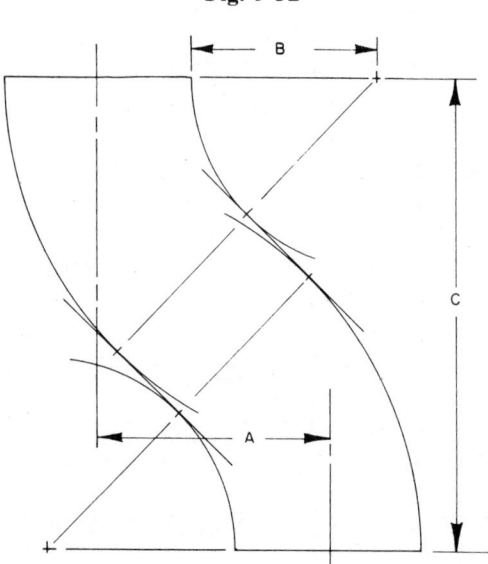

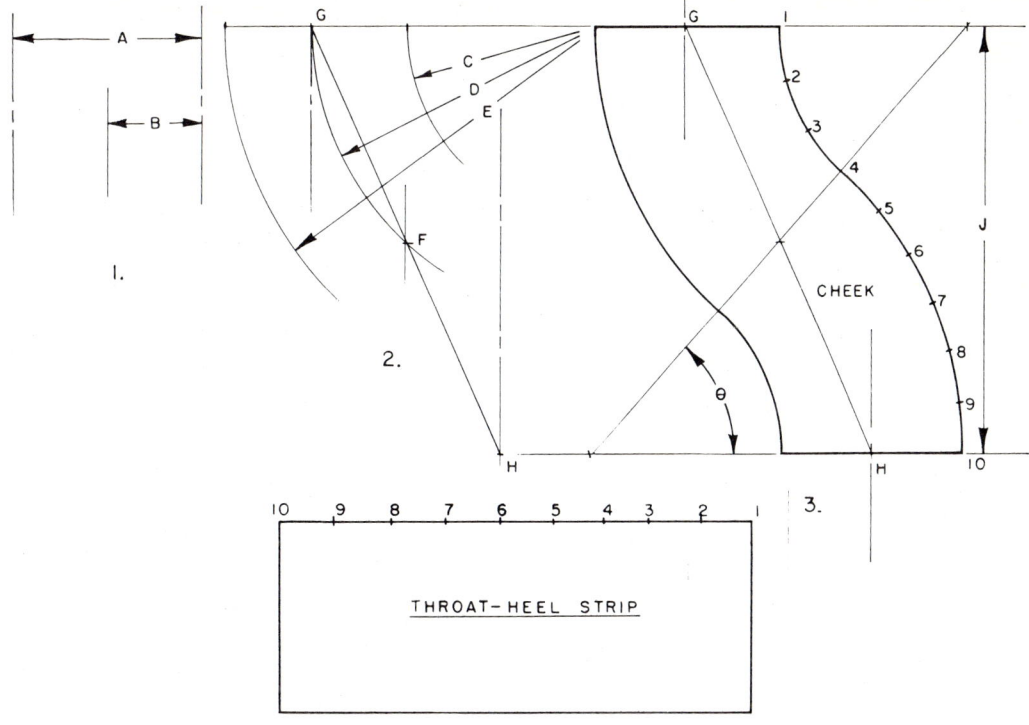

Fig. 9-33 Curved offset design.

varied to suit the job at hand. The throat radius and offset specified regulate the minimum possible run for the fitting. When using the minimum run a continuous curve results as shown on the offset of Figure 9-31. As the run measurement is increased, the amount of straight duct in the fitting increases. If specifications call for a large, long fitting, it may be more practical to make up the job as two conductor elbows and a center section of straight duct.

To draw the front view of any continuous curved offset, (1) Draw center lines representing the amount of offset (A, Figure 9-33). (2) Draw the bisector of the offset distance B. (3) From specifications, determine the duct size and throat radius. (4) Draw throat arc C, center arc D, and heel arc E (Figure 9-33). Center arc D intersects the bisector at F. (5) A straight line from G through F locates H, thereby establishing the minimum run J.* Complete the *reverse* or *ogee* elbow by using the same radii as were used for the top half of the fitting. (7) the two cheek pieces are cut as shown in the front view (plus an allowance for seams). (8) Develop a heel-throat strip, using duct width (plus seam allowances) for the width of strip, and a throat-heel step-off for the length. (Developed circular distances are discussed in section 5-9.) Two strips are required per fitting. (The ogee elbow represents a practical application of the ogee construction described in section 5-16.)

* The minimum-run distance can be computed, using the following formula:

$$\text{Run} = \sqrt{(2D)^2 - (2D-A)^2}$$

where
D = center-line radius
A = offset distance

9-17 Compound offsets

In a compound fitting, two offset measurements are specified, as shown in A and B of Figure 9-24. The two measurements produce an oblique center line XY. Even though XY, in the top view, represents the true amount of offset, the line XY projects foreshortened in all regular views. In previous problems, the axes and elements of pipes were true length in one of the regular views (usually the front view). In order to develop the surfaces, it is necessary to have a view in which the elements are true length. Therefore, fittings with an oblique axis must be developed either from an *auxiliary view* (Chapter 7) or else by *rotation* (Chapter 8).

Auxiliary-view Method

1. Draw the top view (Figure 9-34), using measurements A and B and arranging pipes C and D as required. Complete the top view by drawing the required corners of the oblique portion of the fitting.

Fig. 9-34 Compound offset development.

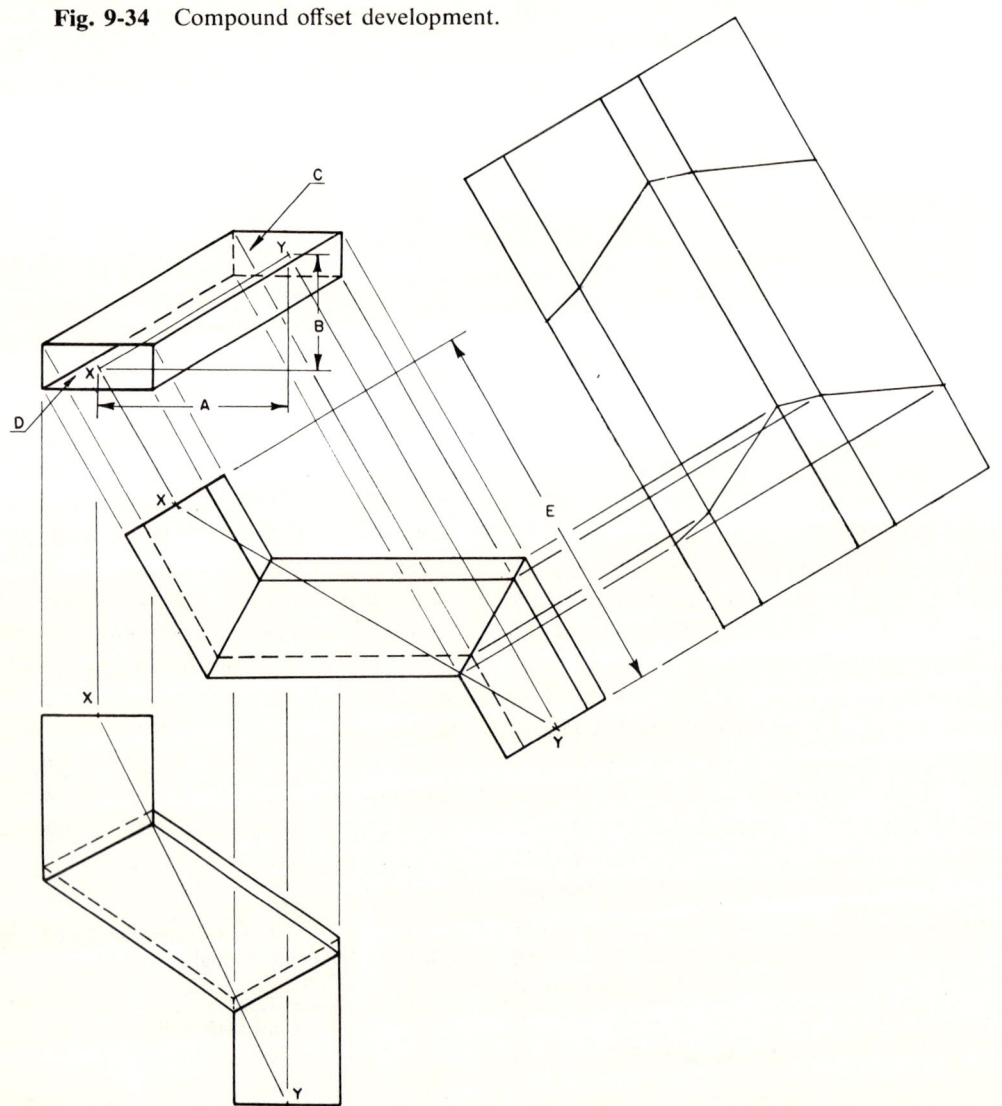

PARALLEL-LINE DEVELOPMENT

2. Select a line of sight perpendicular to *XY* in the top view (Figure 9-34).

3. Project the corners of the top view to the auxiliary view.

4. In the auxiliary view, lines perpendicular to the projectors establish the run distance *E* (Figure 9-34).

5. Complete the auxiliary view, using methods described in section 9-14. The miter planes will show as edges in this view.

6. A front view (included in Figure 9-34) is not needed for development (but may be required on an installation drawing).

7. The development for each part of the fitting is illustrated in Figure 9-34.

Rotation Method

1. The offset measurements *A* and *B* in Figure 9-35 are specified. By construction, find the true offset *F*.

2. After the right triangle has been constructed to determine *F*, transfer this triangle to a rotated position, such that *F* becomes a frontal line.*

3. Complete the top view, arranging pipes *C* and *D* positioned as required.

4. Complete the front view, using methods described in section 9-14. In this view, all pipe elements or corners are true length.

Rotation makes it easier for the beginner to visualize the fitting, since only one offset measurement *F* is shown on the view. The

---

* The value of *F* may be computed, using the formula

$$F = \sqrt{A^2 + B^2} \quad \text{(Figure 9-35)}$$

Fig. 9-35

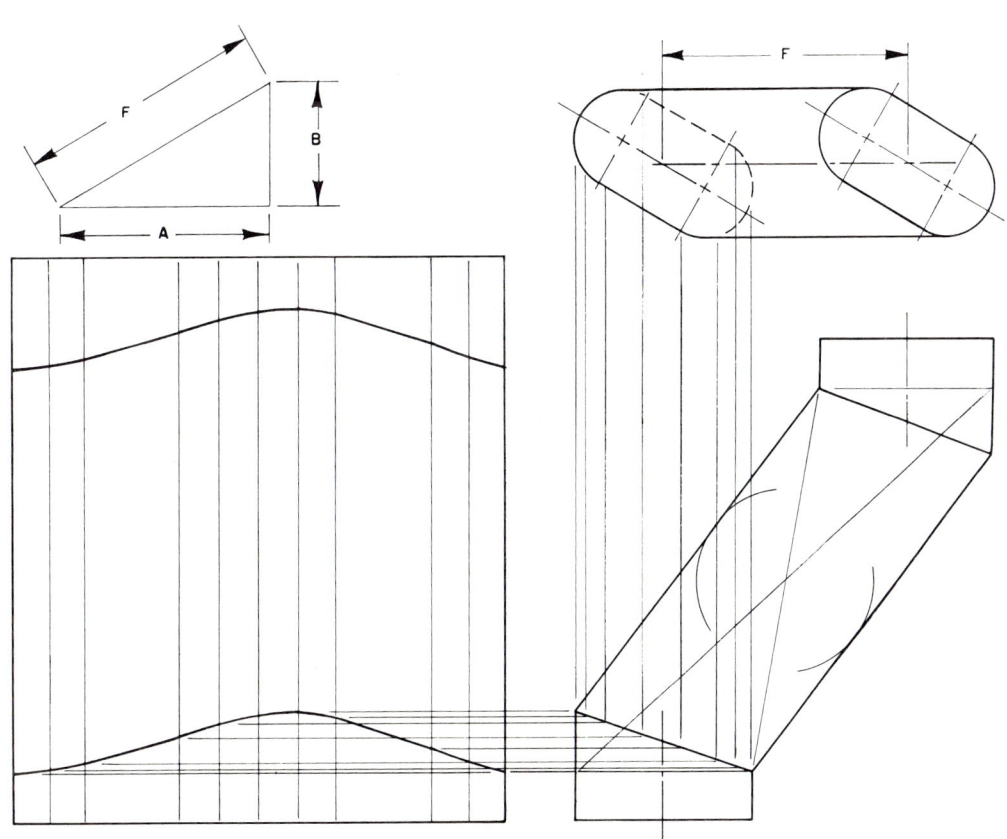

view occupies a more normal position on the drawing board. It should be noted by now that compound offsets actually exist only for square, rectangular, elliptical, or oblong pipes; that round offsets may be constructed as primary offsets, using the true distance F as determined by specifications, construction diagram, or computation. (Even offsets involving all shapes of pipe may be thought of as primary offsets if the connecting pipes are considered rotated to a new position.)

INTERSECTIONS

9-18 Tees, Ys, breeching

Because of their appearance, sheet metal fittings involving the intersection of main and branch pipes are generally called tees, Ys, breeches, etc. A typical tee joint is shown in Figure 9-36.

The *line of intersection* between the two pieces must be determined before the stretchout of each piece can be made. Joints may be analyzed according to those factors which affect the appearance of the line of intersection in various views. For example, in a view made on a plane parallel to the axes of both pipes, the line of intersection will appear as a straight line when all intersecting surfaces are flat (planes) as shown in Figure 9-37. For round, elliptical, or oblong pipe, the line of intersection will appear as a straight line when the pipes are of *equal size* and *identical shape,* and have

Fig. 9-36 A 90-degree tee joint. (*Naylor Pipe Co., Chicago, Ill.*)

Fig. 9-37 Intersection of square and hexagonal pipe.

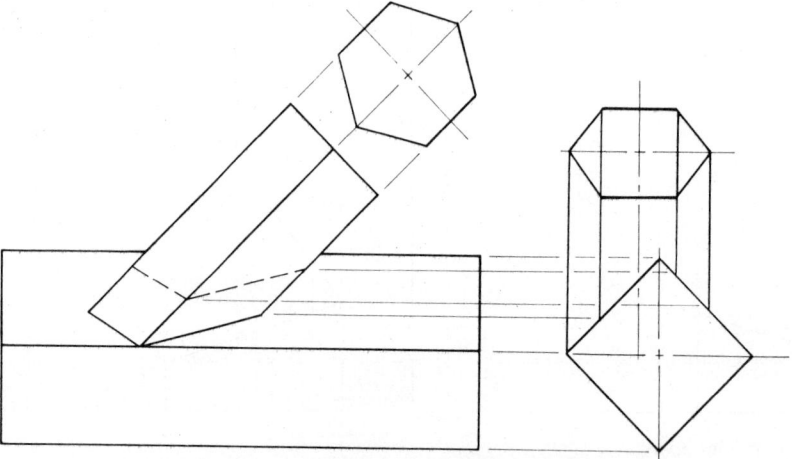

PARALLEL-LINE DEVELOPMENT

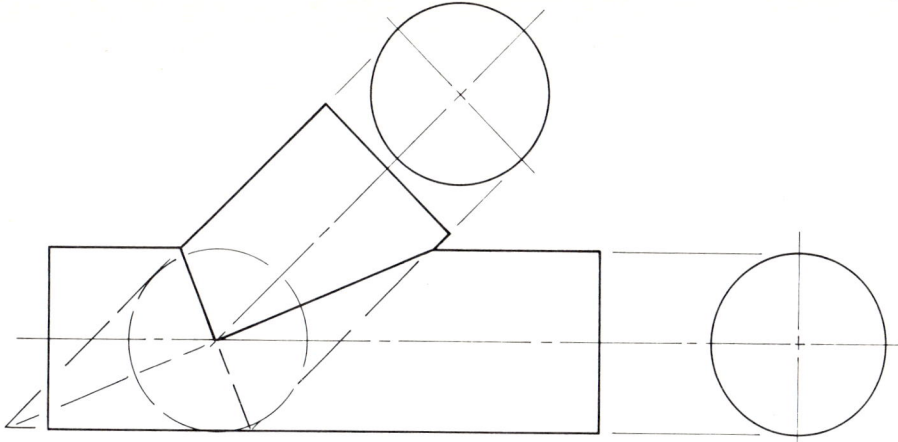

Fig. 9-38 Tangent-sphere method of locating miter lines.

intersecting axes; otherwise, the line will appear as a curve in all views. Figure 9-38 illustrates the intersection when pipes are alike in shape and size and have intersecting axes. Because cylinders and cones of circular shape have identical sections when drawn tangent to a sphere, the sphere method of locating miter lines is also shown in Figure 9-38. Notice that the lines are established by connecting the intersection points of the tangent elements. Figure 9-39 illustrates an intersection of identically shaped but different size pipes using intersecting axes.

Because of the many different shapes of pipe and the varied conditions in which joints may be designed, exceptions to the rules may occur. For example, the line of intersection of a square or rectangular pipe with a round pipe, as shown in Figure 9-40, would project as straight lines in the front view.

Fig. 9-39

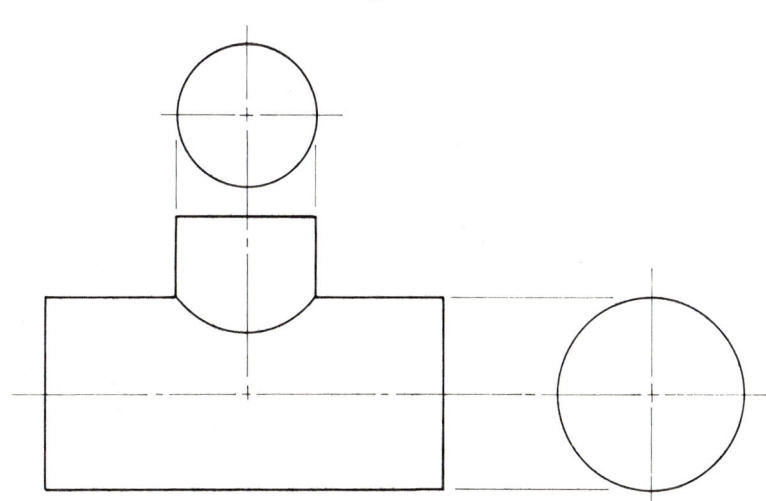

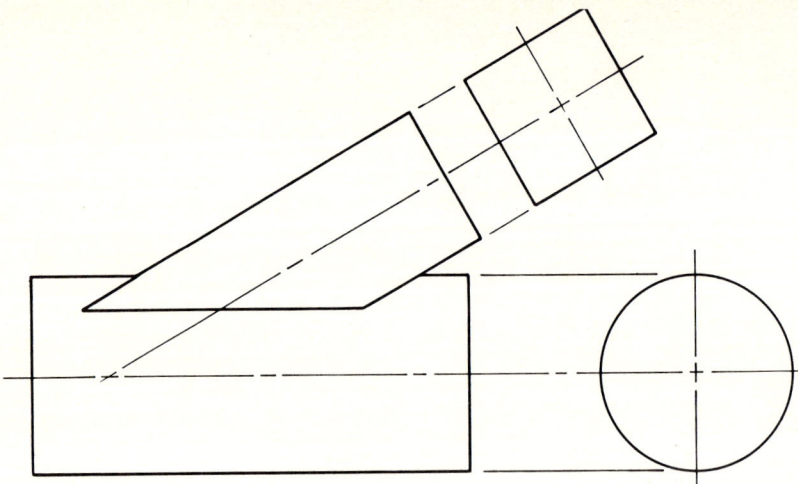

Fig. 9-40

9-19 To find the line of intersection

There are five methods for finding points on the line of intersection of surfaces. Two of these, being practical for prisms and cylinders in sheet metal, will be discussed at this time. The others will be discussed later when dealing with cones and pyramids. The two methods best suited to cylinders and prisms are (1) the edge view of one surface; (2) the cutting-plane method.

The Edge-view Method. In this method, views of the fitting are selected so that one pipe appears edgewise in a view. This view then makes apparent the points at which the edges or elements of the other pipe intersect the first. For example, in Figure 9-41 the round pipe appears edgewise in the side view. Edges 1, 2, 3, 4, and intermediate elements *A, B, C, D* of the square pipe intersect the round at 1', 2', 3', 4', etc., in the side view. The intersection points are projected to the front view to locate $1_F, 2_F, 3_F, 4_F, A_F,$

Fig. 9-41

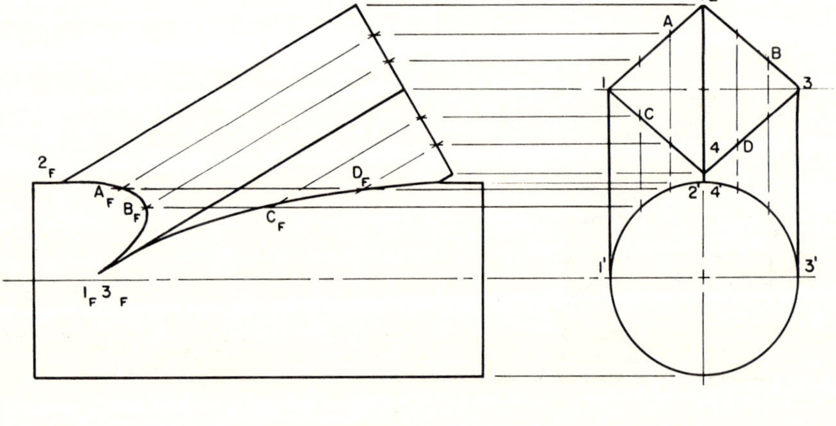

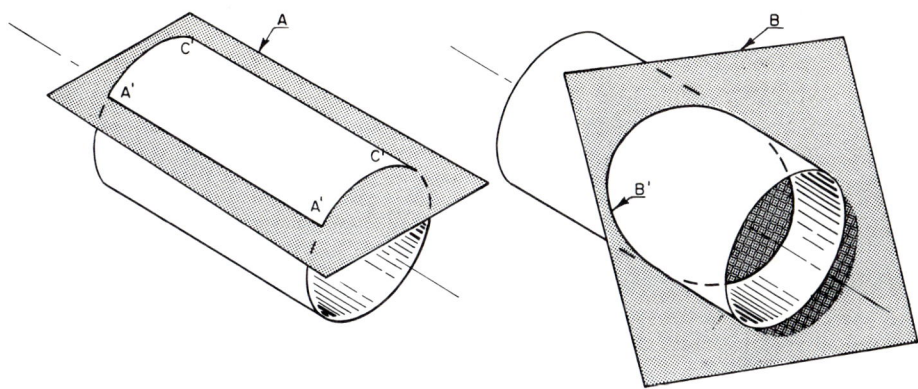

Fig. 9-42 Cylinders cut by planes.

B_F, C_F, and D_F. By connecting the points thus found, the line of intersection is established.*

The Cutting-plane Method. An arbitrary cutting plane, passed through two geometric surfaces, will trace a line on each surface. If the traces intersect, they intersect at one point on the line of intersection. When cutting cylinders, the line will be straight only when the cutting plane is parallel to the cylinder axis. The straight line is an element of the cylinder. In Figure 9-42, cutting plane A is parallel to the cylinder axis and produces elements $A'A'$ and $C'C'$. Cutting plane B is not parallel to the cylinder axis; hence it produces the curve B' (an ellipse). When prismatic surfaces are cut, the traces produced are always straight lines, but they are parallel to the prism edges only when the plane is parallel to the prism. The parallel trace may be considered an element of the prism. As illustrated, cutting plane A of Figure 9-43 being parallel to the prism produces element $A'A'$. Cutting plane B, not being parallel, results in the oblique lines

---

* The edge-view method is similar to the "piercing-point method" because it makes obvious the points at which elements pierce.

Fig. 9-43 Prisms cut by planes.

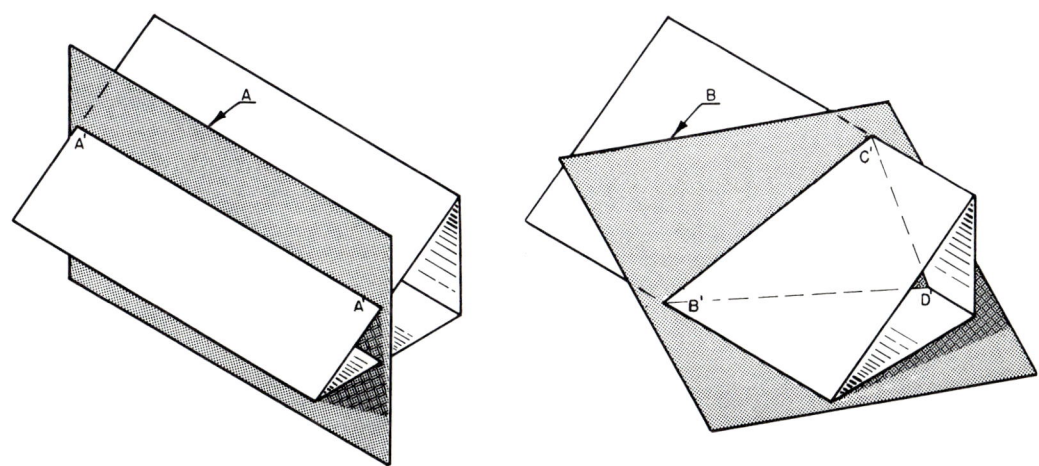

$B'C'$, $C'D'$, and $D'B'$. Because of the ease in locating and drawing elements on cylinders and prisms, only cutting planes which produce element traces are of value in locating points; *therefore, cutting planes parallel to the axes of both pipes or prisms are most valuable in finding the line of intersection.*

Figure 9-44 illustrates the use of seven cutting planes to produce elements on each cylinder. The elements produced by each cutting plane intersect at points on the line of intersection.

9-20 To develop the patterns for a 90-degree tee

Figure 9-45 consists of a front view and partial top and side views. Only partial top and side views are required because the line of intersection appears as straight miter lines

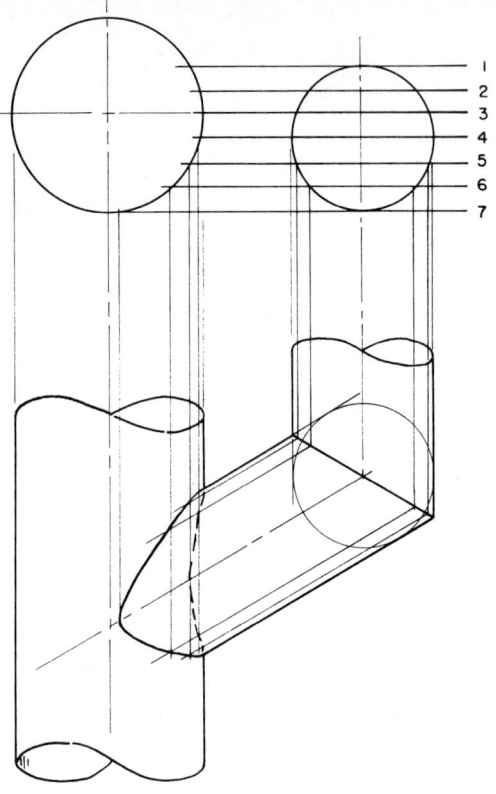

Fig. 9-44 The cutting planes produce elements which are used to locate points on the line of intersection.

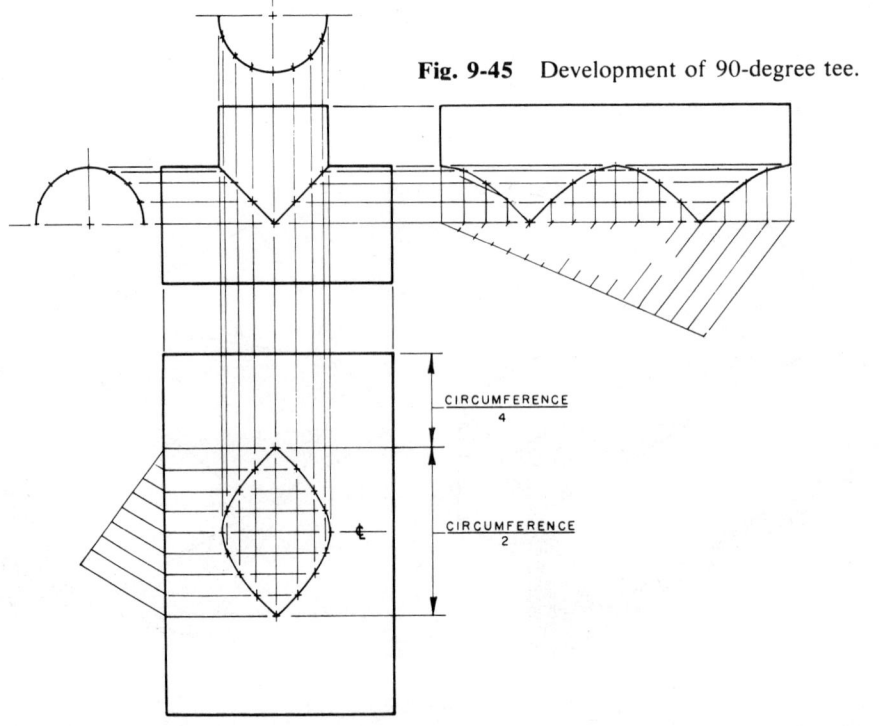

Fig. 9-45 Development of 90-degree tee.

PARALLEL-LINE DEVELOPMENT

Fig. 9-46

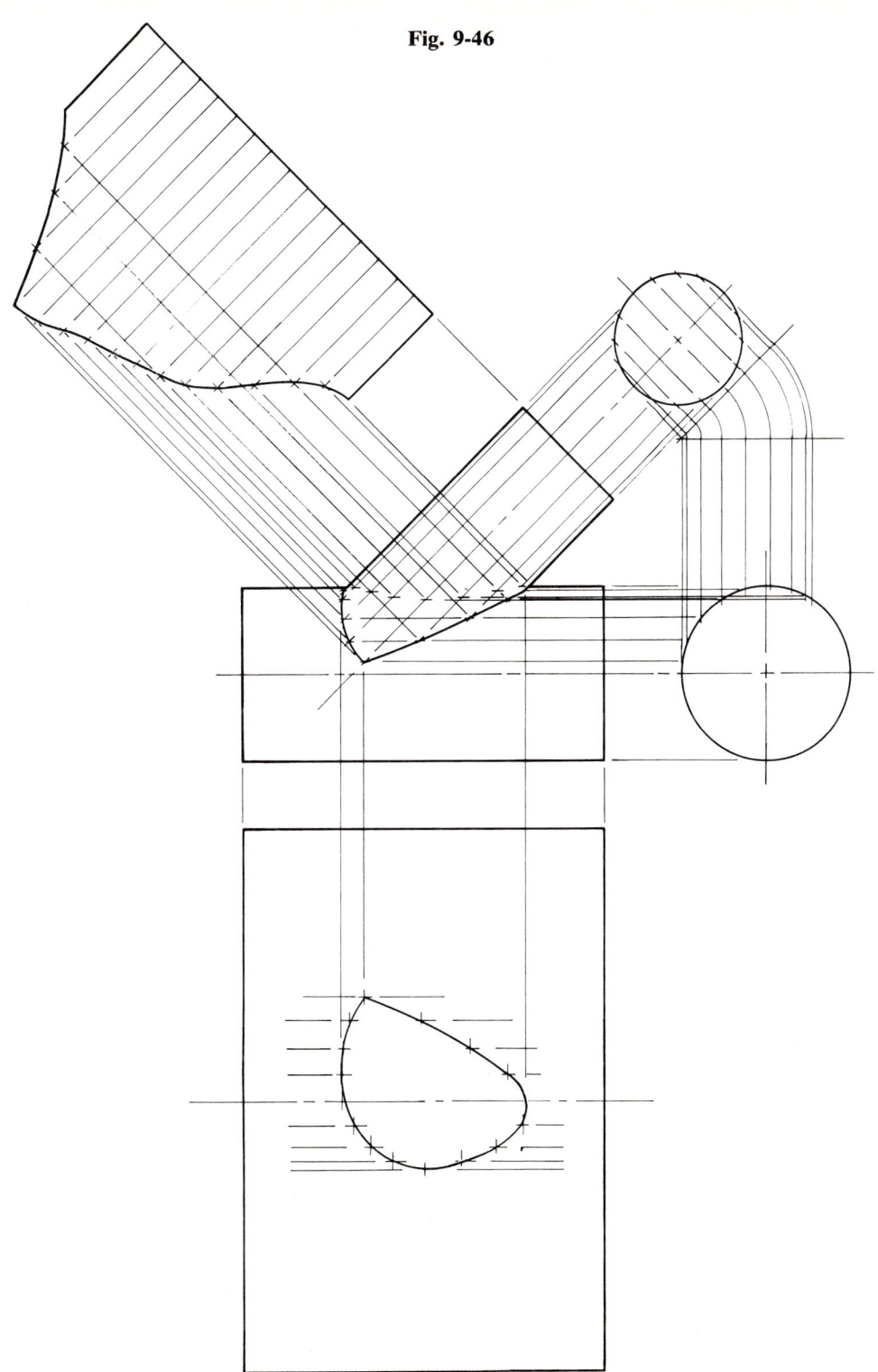

in the front view. The partial views are used to space the elements uniformly on each pipe. The over-all stretchout for each pipe is computed and this distance is divided into the required number of elements. The true length of each element is projected to the stretchout to develop the curve for the stem of the tee. The hole in the main pattern is developed in the same manner by spacing the elements uniformly in the center of the strip and by projecting the true lengths from the front view.

9-21 To develop the patterns for a pitched tee

Items to be specified for a fitting, as shown in Figure 9-46, are (1) the size and shape of the pipes; (2) the amount of off-center relationship; (3) the degree of pitch to be given the small pipe. Draw the front and side views and determine the line of intersection, using the cutting-plane method described in section 9-19. This is done by cutting the auxiliary view of the small pipe with a convenient number of planes as shown. Transfer the planes to the side view, using the front surface of each pipe as a reference. From the auxiliary view and from the intersecting points in the side view, elements extended to the front view cross at the points on the line of intersection, as shown in Figure 9-46. After the line of intersection is established, the development of each piece is made in the same manner as for previous problems.

9-22 To develop a pitched tee (varied pipe shapes and nonintersecting axes)

One version of the problem, consisting of a square pipe, pitched at 45 degrees, intersecting a round pipe, is shown in Figure

Fig. 9-47 Intersection development.

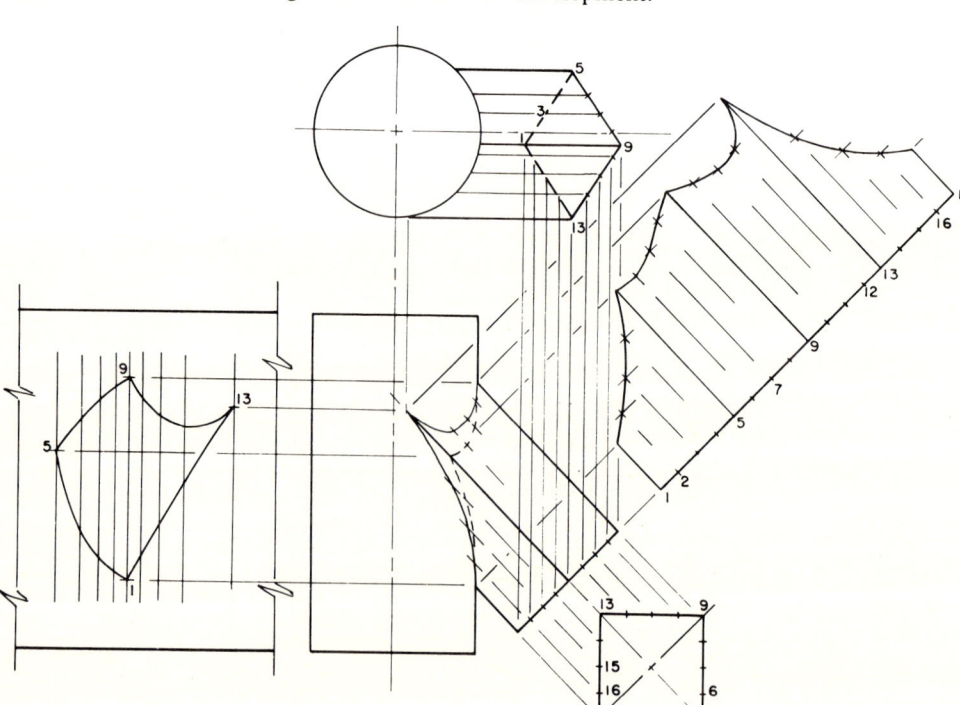

PARALLEL-LINE DEVELOPMENT

9-47. This problem is similar to the previous one, but certain features should be noted. (1) The line of intersection appears as a curve in the front view. (2) The curve is plotted, using the edge-view method (see Figure 9-41). (3) Total stretchout of the square pipe equals the perimeter of the square. (4) Total stretchout for the round pipe equals pipe diameter multiplied by π. (5) The hole in the main pattern is developed using elements spaced on the round pipe.

9-23 To develop a Y breeching

Figure 9-48 illustrates the construction and development of one type of Y elbow. Notice that the pipes involved are cylinders of equal diameter and intersecting axes; therefore, the line of intersection appears as straight miter lines in the front view. Because of symmetry, develop a pattern only for either pipe A or B, and also a pattern for pipe C. The development procedure for

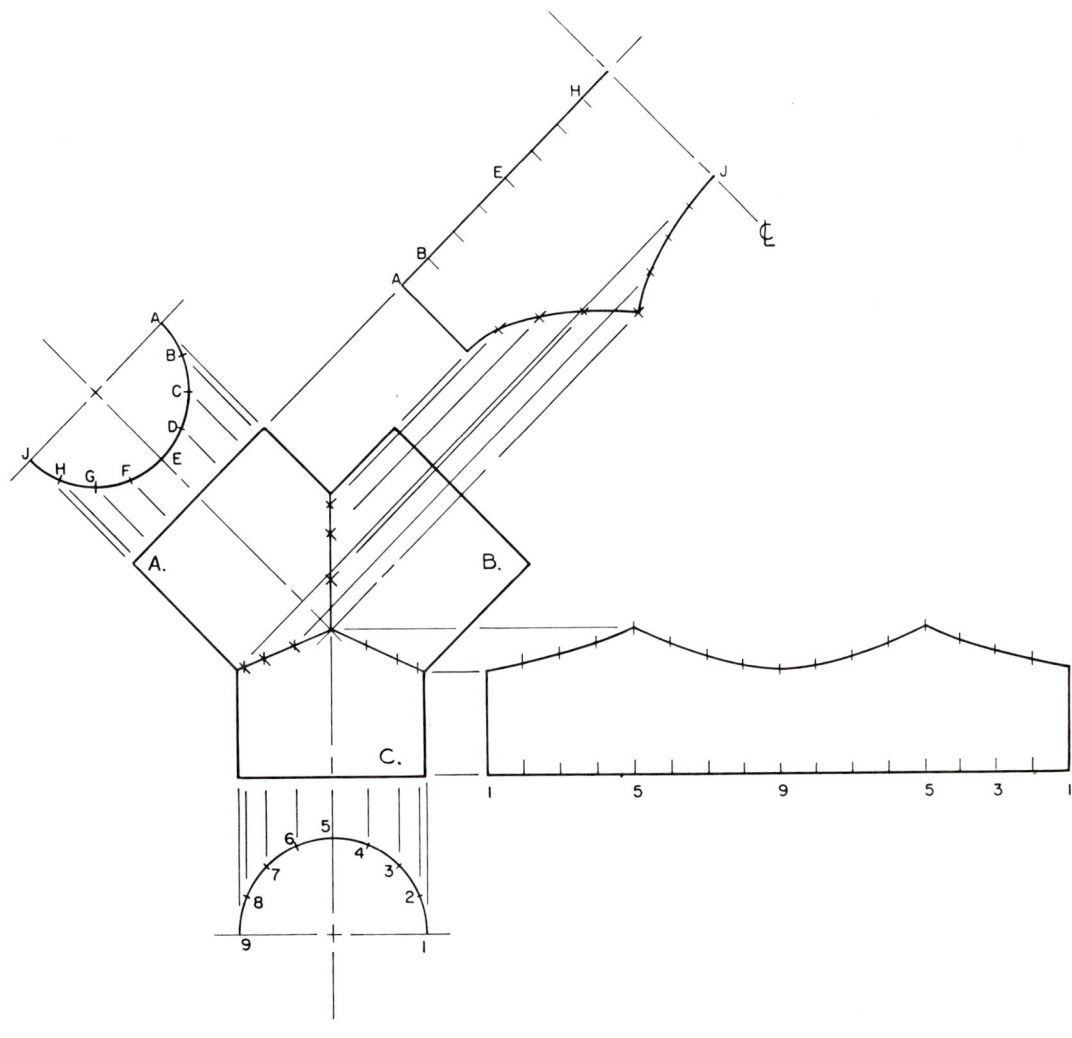

Fig. 9-48 Y branch.

each piece is the same as for the previous problems.

9-24 Transition, round-to-oblong

The surface of the fitting in Figure 9-49 is composed of elliptical semicylinders *A* and *B*, separated by flat triangular cheeks *C*. The top ends of the semicylinders join and appear as a circle in the top view. The right section of each of the semicylinders, however, is elliptical, as shown in the auxiliary view. The wedge-shaped cheeks are already true shape and size in the front view; therefore, they are transferred to their proper place in the development after the semicylinder is developed.

The top and front views are constructed from specifications. Three measurements are needed: (1) the diameter of the round pipe; (2) the center-to-center distance of the oblong pipe; (3) the run or height of the fitting.

To construct the auxiliary right section, one of the semicylinders is divided into a convenient number of points in the top view (they represent elements). Project the elements to the front view (where they are true length) and thence to the auxiliary view (where they appear as points). The points are located in the auxiliary view by transferring the distance from each element to center line, from the top view. For example, points 3 and 5 carry to the front view and

Fig. 9-49 Round-to-oblong transition.

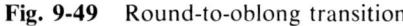

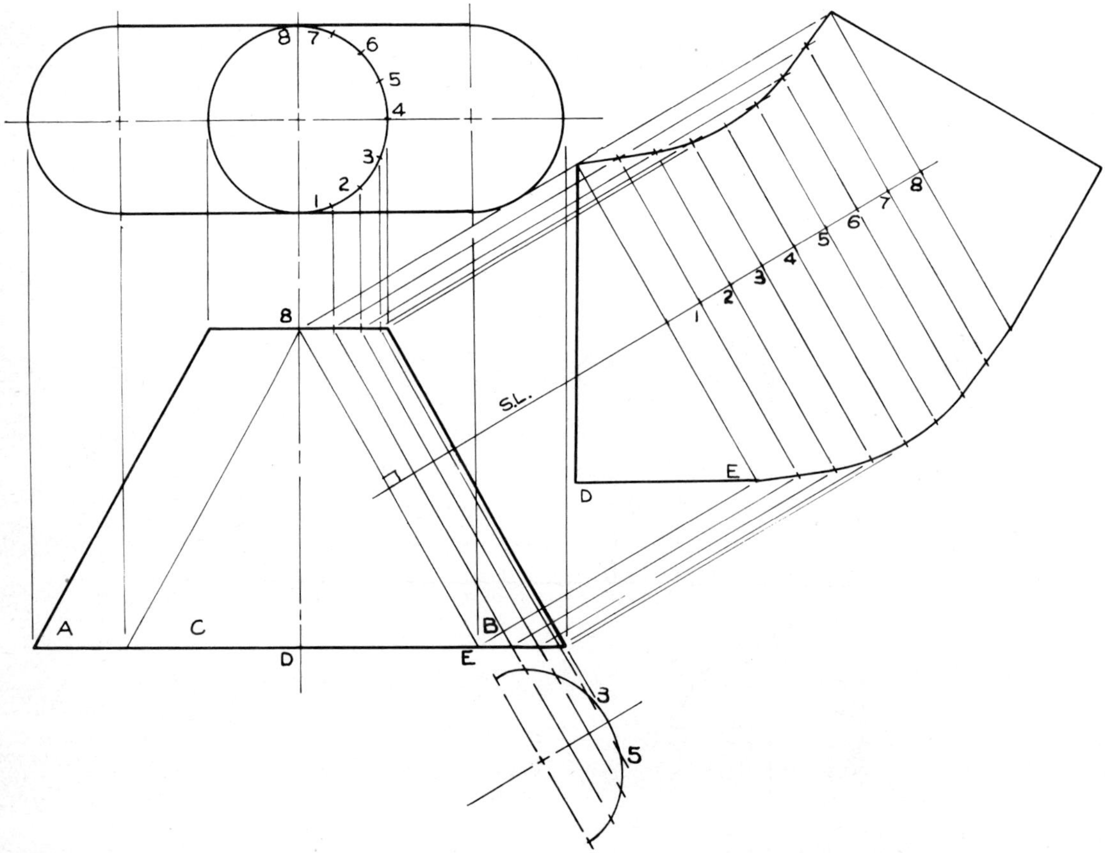

to the auxiliary view on a single projector, but are on opposite sides of the center line in the top view and thus are measured on opposite sides of center in the auxiliary view.

To develop one-half of the pattern: (1) Project a stretchout line (SL) perpendicular to the front-view elements. (2) Transfer the *girth from between elements in the auxiliary view* to the stretchout line. (3) Construct the elements perpendicular to the stretchout line and parallel to each other. (4) Project the true length of each element from the front view to the corresponding element in the stretchout. (5) Sketch and draw the curves thus plotted. (6) The triangular piece 8-*D*-*E* is added to both sides of the stretchout thus forming a half-pattern. This results in a pattern which is symmetrical, with the seams at the center of the flat area on the front and back of the fitting. The pattern is shown "outside up" in Figure 9-49 but regardless of which way the metal is formed, the fitting would make up the same. Allowances for seams are added and two pieces per fitting are necessary.

CORNICES AND GUTTERS

9-25 Definitions

Sheet metal, formed in decorative strips and mitered at corners, is used in several ways to improve appearance and control run-off on buildings. The metal strips are known as *moldings* and are used to create *cornices, gutters, panels, pediments, dormers, bays, finials,* etc.

Fig. 9-50 K gutter fabrication. (*Bethlehem Steel Co., Bethlehem, Pa.*)

The *cornice* of a building is used to create a pleasing juncture of wall and roof surfaces. On masonry, the cornice is usually of stone, but many buildings have cornices made of sheet metal. This is especially true when the cornice includes an *eave gutter*. On houses, the effect of a cornice is obtained by using a *molded eave gutter* which is sometimes called a "K" gutter (Figure 9-50). A cornice between the first and second floors is known as a *lintel cornice*.

Eave gutters are used at the roof edge to gather runoff while *roof gutters* are placed at any convenient location on the roof.

Cornices and gutters, as mitered fittings, are called *bevel, square-return,* or *butt miters*.

A bevel miter contains two parts that lie in the same plane and make angles other than 90 degrees with each other. A square-return miter is the special case involving the 90-degree relationship of the two parts. A butt miter is used at the beginning or end of the gutter or cornice, where it intersects some other surface of the building. Both parts of *face miters* lie in the same vertical plane and are used on panels, pediments, etc.

SHEET METAL DRAFTING

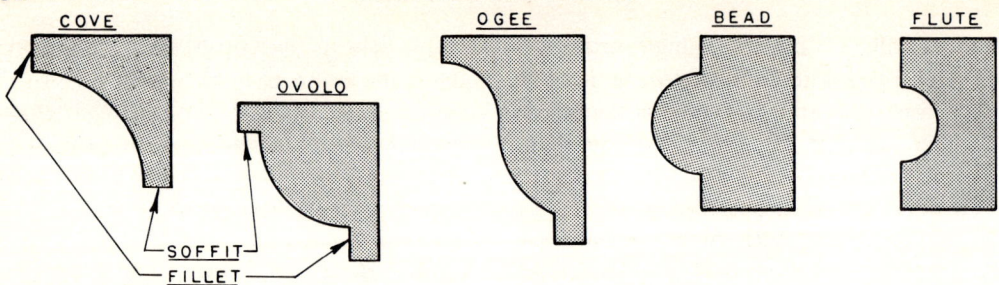

Fig. 9-51 Molding curves.

9-26 Classical forms

Sheet metal shapes for architectural purposes are based upon the classic architectural forms of ancient Greek and Roman structures. The basic forms are shown in Figure 9-51 and consist of: *cove, ovolo, ogee, bead,* and *flute.* Horizontal flats are known as *soffits,* vertical flats as *fillets.* Thus the classic forms, individually and in combinations, are used to provide effective treatment of otherwise bare surfaces and to create a pleasing play of light and shadow on the building.

9-27 Ogee eave gutter

Figure 9-52 illustrates an ogee beveled eave gutter. The necessary views and development are made as shown in Figure

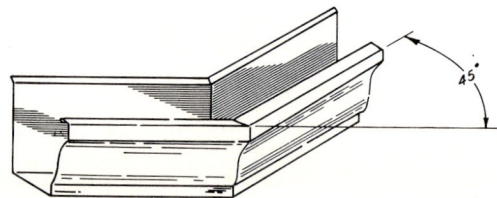

Fig. 9-52 Ogee eave gutter on a 45-degree miter.

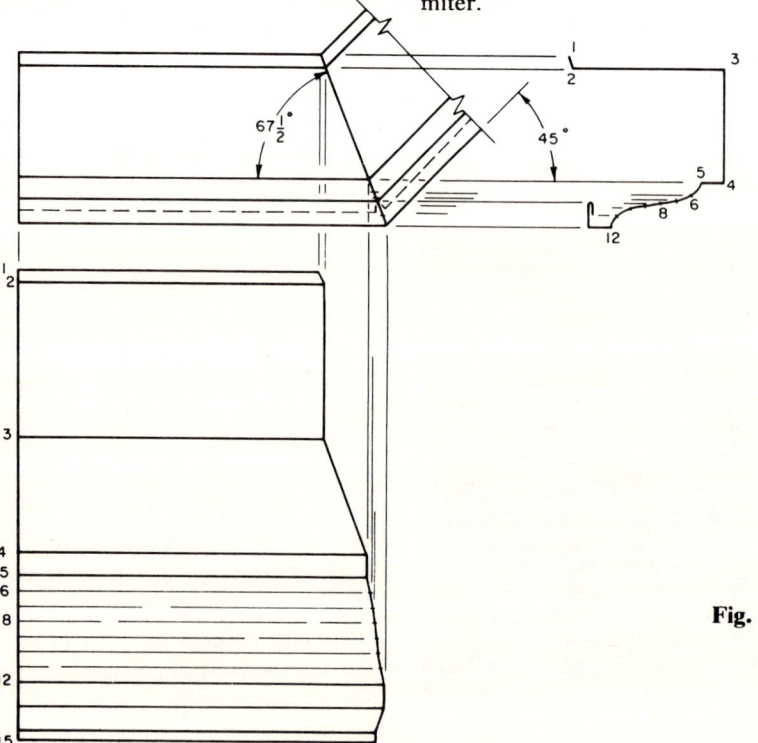

Fig. 9-53

9-53. A plan, or top view, is needed since that view shows the miter plane edgewise. (In the top view, the miter line is the bisector of the angle of the gutter.) A side view, adjacent to the top view and showing only the contour of the front piece, is used to step off the true length of the stretchout. The true length of each element is projected downward onto the stretchout. (Spacings 1-2, 2-3, 3-4, etc., are transferred from the side view, as shown.) The points thus determined are joined to produce the cut required for each arm of the fitting. Two pieces are made from the pattern and are bent in opposite directions so as to produce a right and left arm. Both pieces can be shaped in the brake simultaneously, if they are placed with the square ends butting.

9-28 To develop a square-return cornice

A typical square-return cornice is shown pictorially in Figure 9-54. The fitting can be developed from a single view because both parts (A and B, Figure 9-55) are the same shape and size and are at 90 degrees to each other. Thus, in this special case, the view may be read as a front view of one arm A, and as an edge view of the other arm B. The development is made from the front view (Figure 9-55).

To make the stretchout, the true lengths of the elements of part A are projected downward. The elements are spaced their

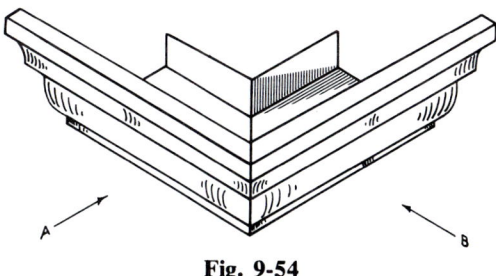

Fig. 9-54

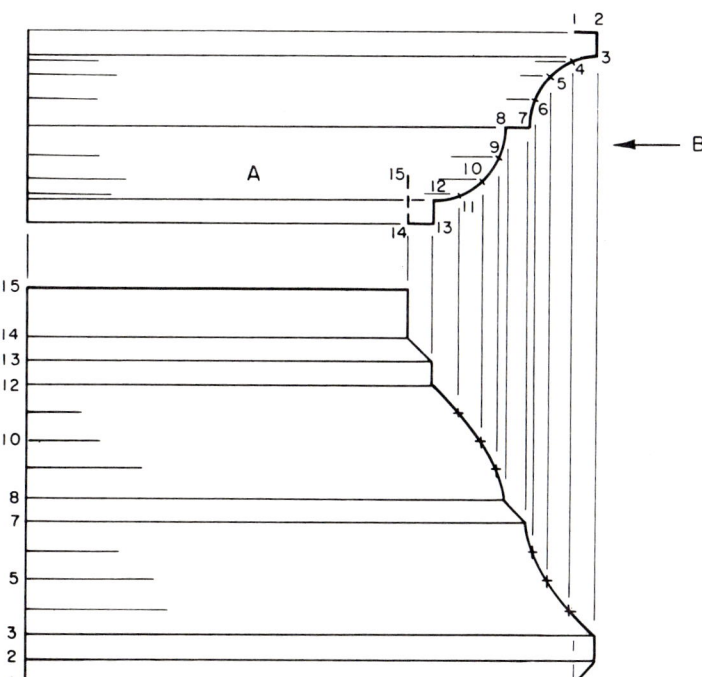

Fig. 9-55

true distance apart, as stepped off from the edge view of part *B*. The points thus determined are joined to produce the cut for each arm of the fitting. The procedure is similar to the previous problem except that only one view is needed.

9-29 Square-return cornice with reduced miter

A square-return cornice with one arm reduced in width is sometimes desired in order to maintain clearances, allow more light in restricted areas, or whenever a full cornice would be unsatisfactory. The square-return cornice in Figure 9-56 shows the width *W* of arm *B* as narrower than the width *W'* of arm *A*. The height *H* of each arm is the same and the arms are of matching shape. The reduction results in a fitting in which the miter line is *not* the bisector of the fitting angle.

The procedure for constructing the necessary views of the cornice is as follows: (1) Determine the width *W'* and the length *L'* of arm *A* in the top view. (2) Determine the amount of reduced width *W* and length *L* of arm *B*. (3) Complete the outline of both arms in the top view. This establishes the miter line *XY*. (4) The shape of *A* is designed and drawn edgewise in a side view projected from the top view of *A*. The projected shape produces visible and hidden lines in the top view. (5) The visible and hidden edges of arm *B* are drawn to match those of arm *A* (intersecting on the miter line). (6) The lines of arm *B* are projected to a front view and the corners and curves of the front edge view of *B* are located by transferring heights from the side view of *A*

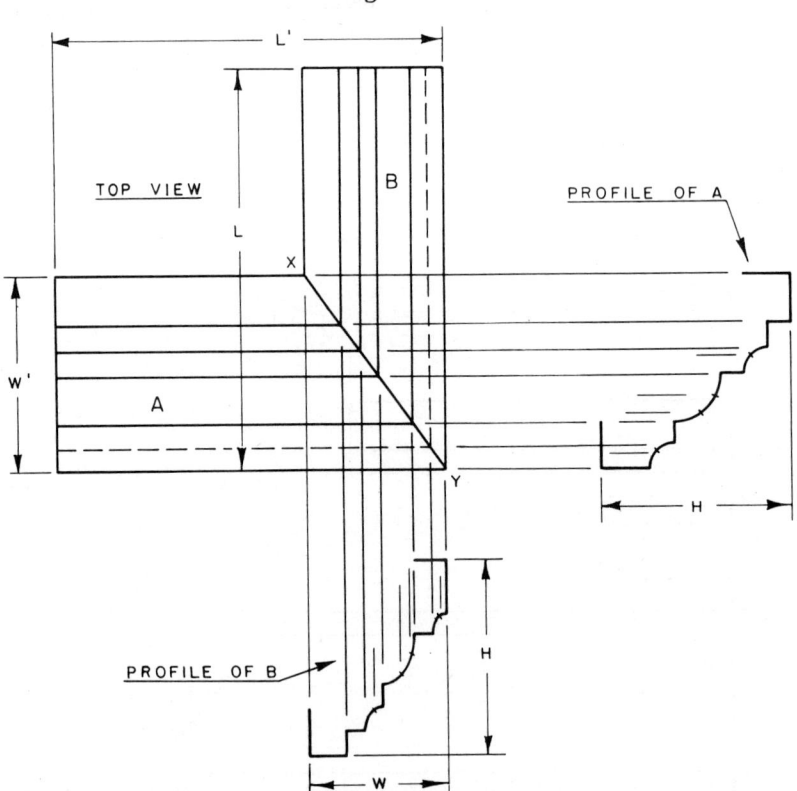

Fig. 9-56

PARALLEL-LINE DEVELOPMENT

135

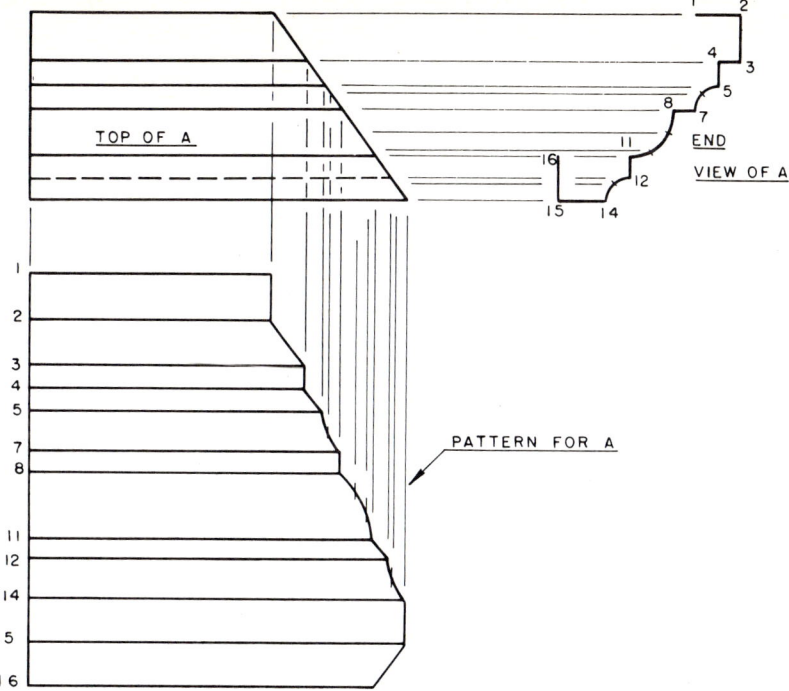

Fig. 9-57

Fig. 9-58

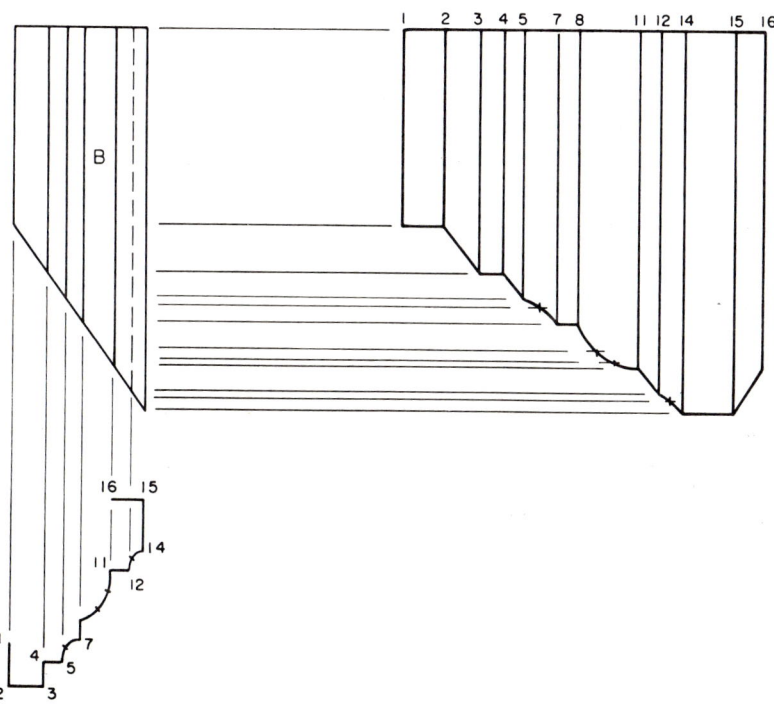

as shown in Figure 9-56 (just as any front view could be constructed when top and side views are given adjacent to each other). Notice that in the case of curves several points are necessary in order to plot an accurate shape. (7) The development of a pattern for each arm is made by projecting the true lengths and by transferring the stretchout distances from the edge view involved, as shown in Figure 9-57 for arm *A* and 9-58 for arm *B*. The developments are shown in separate illustrations for clearness; in practice, however, only one drawing would be sufficient to develop patterns for both arms. (The development is similar to that of the previous problems.)

EXERCISES

The following problems should be drawn full size when possible. When full size is not possible, as some problems are taken directly from industry, select a suitable scale in order to adapt them to the approved classroom materials.

1. Make a pattern for the body of a scoop, as shown in Figure 9-59. The diameter of the cylindrical body is to be 3¼ in [83 mm]; place the seam at the top.
2. Make a pattern for the square body of a scoop, as shown in Figure 9-59. The square body is to be 3 in [76 mm]; place the seam in the middle of the top flat.
3. Make a scoop pattern, as shown in Figure 9-60. Cylinder diameter is 3¼ in [83 mm]. Make the seam as short as possible.
4. Make a rectangular scoop pattern, as shown in Figure 9-60. The rectangular body is 3½ in [89 mm] wide and 2 in [51 mm] high. Place the seam in the middle of the top flat.
5. Refer to Figure 7-32. Given the top and front views of the center line of a cylindrical tube 12 in [305 mm] in diameter, which intersects a floor (horizontal plan) at *A*, and a wall (vertical plane) at *B*. Draw only those views necessary to develop a pattern for the tube and develop it.
6. Refer to Figure 7-32. Given the top and front views of the center line of a duct 12 in [305 mm] square, which intersects vertical and horizontal planes as indicated. The duct is positioned so that one diagonal of a *right section* is parallel to the horizontal (H) plane. Draw only those views necessary to develop a pattern for the duct.
7. Develop a fishtail pattern for the following elbow: three-piece; 90-degree; 3½-in [89-mm] diameter; 3-in [76-mm] throat radius.
8. Develop a pattern for a four-piece 105-degree elbow 3¼ in [83 mm] in diameter. Use a 1⅝-in [41-mm] throat radius and place seams alternately on throat and heel.

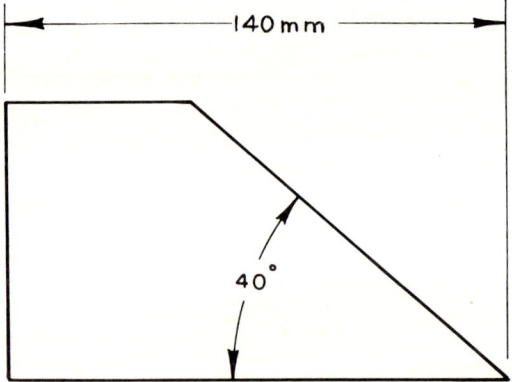

Fig. 9-59

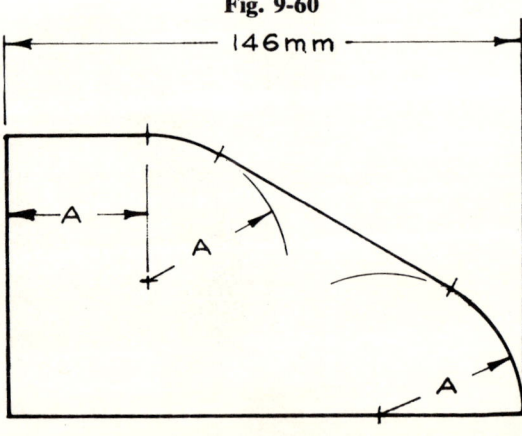

Fig. 9-60

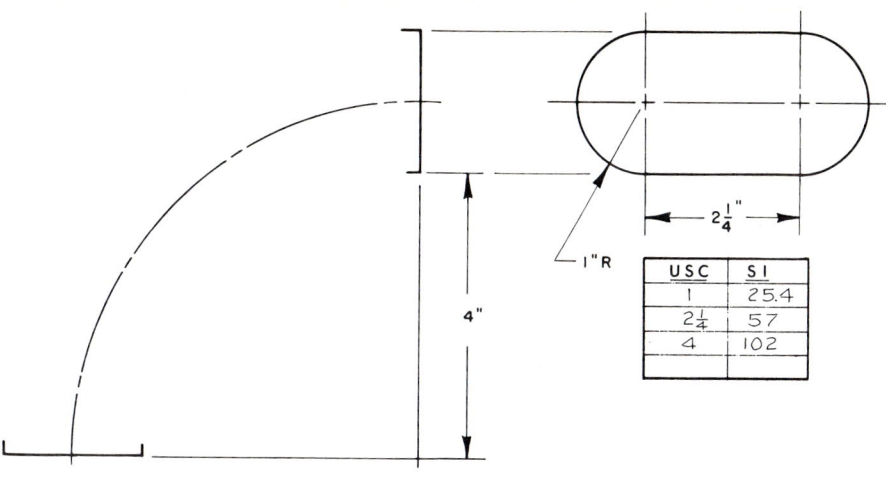

Fig. 9-61

USC	SI
1	25.4
2¼	57
4	102

9. Develop a pattern for a four-piece 90-degree elbow, using an oblong pipe placed "on the flat," as shown in Figure 9-61. All the seams are to be in the throat.

10. Develop patterns for the elbow shown in Figure 9-62.

11. Develop heel, cheek, and throat patterns for a curved 90-degree duct elbow. The duct size is to be 12 in [305 mm] square. The throat radius is to equal the duct size.

12. Develop patterns for a three-piece 90-degree duct elbow. The duct size is to be 10 by 14 in [254 by 356 mm], the throat radius is to be 14 in [356 mm], and the 14-in [356-mm] measurement on the horizontal duct is to be on the flat.

13. Develop heel, cheek, and throat patterns for the 90-degree transition elbow to be used in the drier installation in Figure 9-31. Elbow specifications are as follows: the throat radius equals 18 in [457 mm]; the heel radius equals 42 in [1067 mm]. The elbow openings join the 24- by 36-in [610- by 914-mm] reverse elbow to the 18- by 36-in [610- by 914-mm] curved offset.

14. Develop heel, cheek, and throat patterns

Fig. 9-62 Three-piece 45-degree elbow. (*Naylor Pipe Co., Chicago, Ill.*)

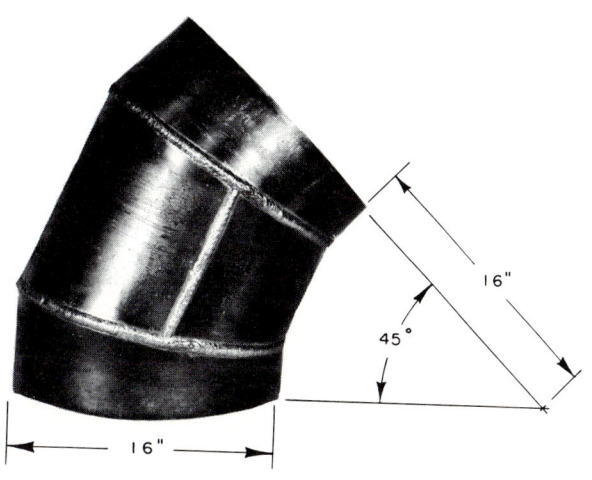

for a 75-degree curved transition elbow, which is to be used to conncet a vertical stack, 16 in [406 mm] square, to a rectangular duct 8 by 16 in [203 by 406 mm].

15. Develop the patterns necessary for the compound curved duct elbow in Figure 9-63.
16. Develop patterns for a round pipe offset. Specifications are as follows: the offset is 3 in [76 mm]; the pipe diameter is 2½ in [64 mm]; the total run of entire fitting is 6½ in [165 mm].
17. Develop patterns for a square duct offset. Specifications as follows: the duct size is 2½ in [64 mm]; the offset is 3 in [76 mm]; the run of the center section of the fitting is 4¾ in [121 mm]. Make each end section equal to one-half the center section.
18. Develop a pattern for an oblong duct offset. Place the oblong duct on the flat of the offset. Specifications as follows: the oblong size is 2 by 3 in [51 by 76 mm]; the offset is 2½ in [64 mm]; the total run of fitting is 7 in [178 mm].
19. Develop patterns for the curved duct offset to be used in the drier installation in Figure 9-31. Specifications are as follows: the duct size is 18 by 36 in [457 by 914 mm]; the offset is 30 in [762 mm]; total run of fitting is 72 in [1 829 mm].
20. Develop patterns for the compound offset shown in Figure 9-64.
21. Develop patterns for the 45-degree lateral shown in Figure 9-65. Use dimensions as assigned.
22. Use formula to compute minimum run for a curved duct elbow using assigned duct size and throat radius.

Fig. 9-63

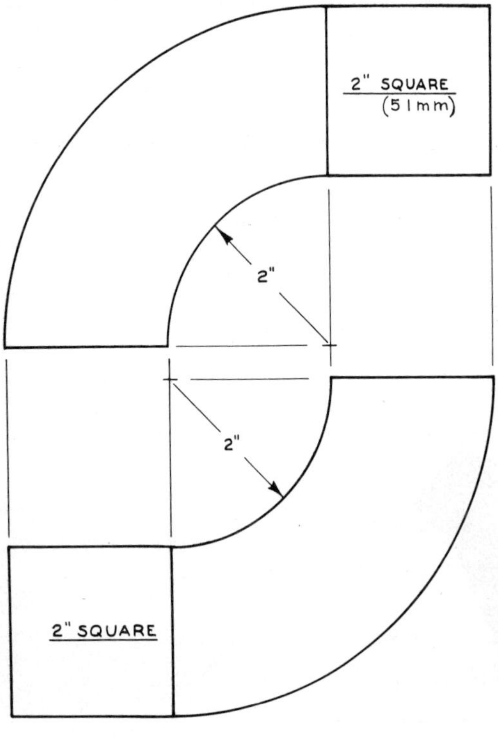

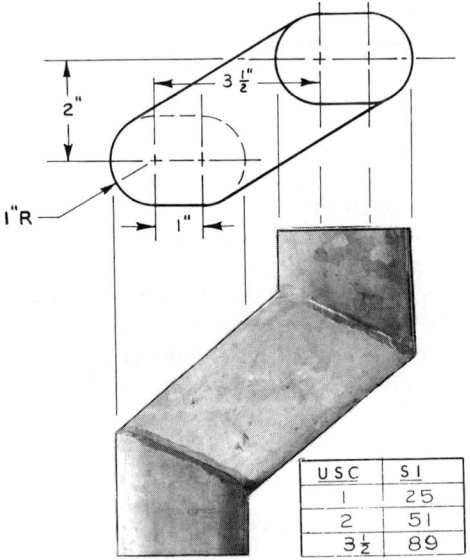

USC	SI
1	25
2	51
3½	89

Fig. 9-64

Fig. 9-65 (*Naylor Pipe Co., Chicago, Ill.*)

PARALLEL-LINE DEVELOPMENT

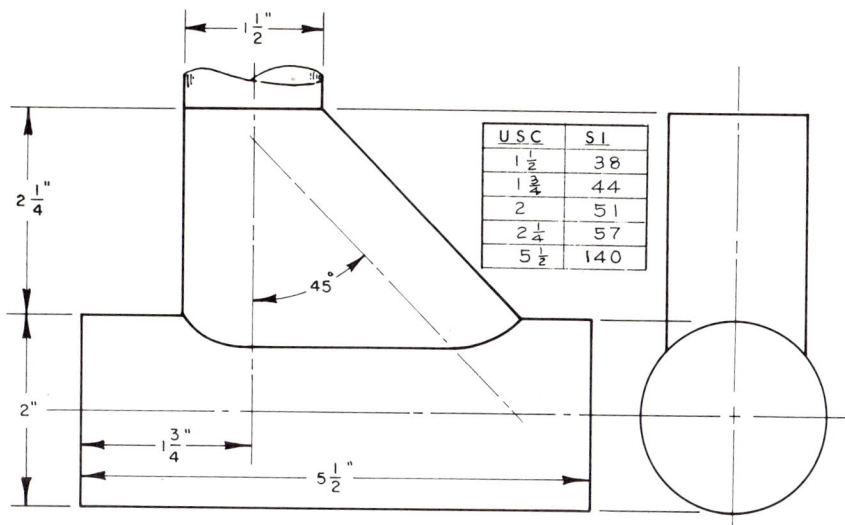

Fig. 9-66 Shoe tee.

23. Develop patterns for the shoe tee (sometimes called boot tee) shown in Figure 9-66.
24. Develop patterns for the 60-degree tee shown in Figure 9-67.
25. Develop patterns for the 60-degree square-to-octagon tee shown in Figure 9-68.
26. Lay out and develop patterns for the Y breeching indicated in Figure 9-69.
27. Lay out and develop a pattern for the three-

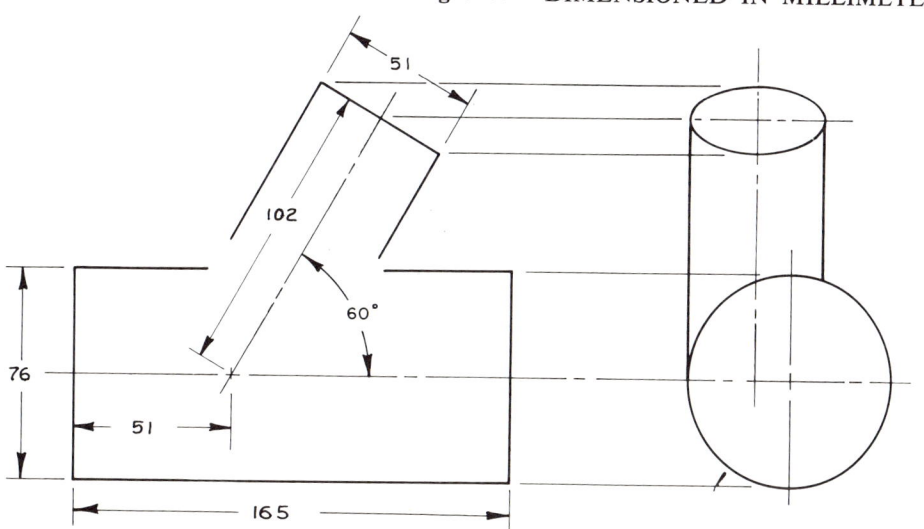

Fig. 9-67 DIMENSIONED IN MILLIMETERS

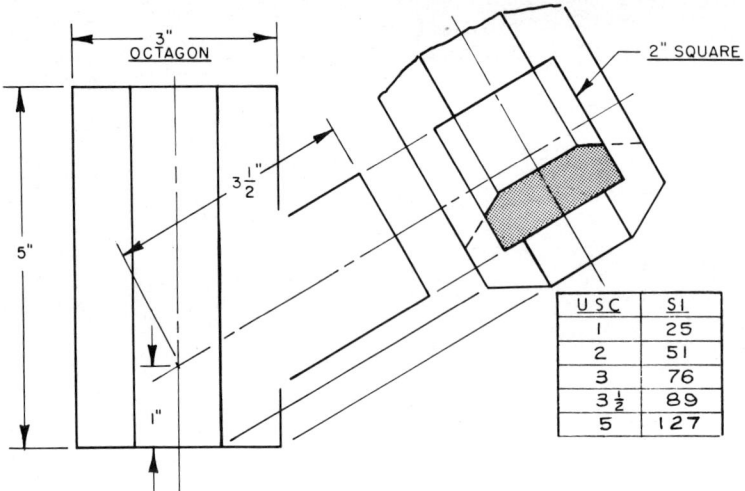

Fig. 9-68

Fig. 9-69

DIMENSIONED IN INCHES AND MILLIMETERS

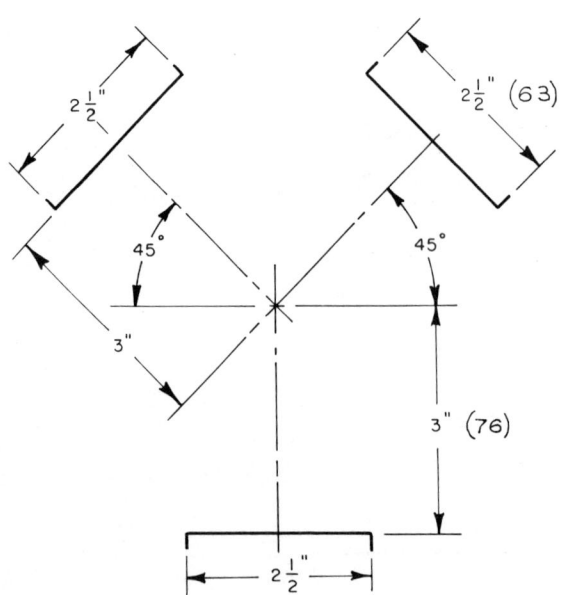

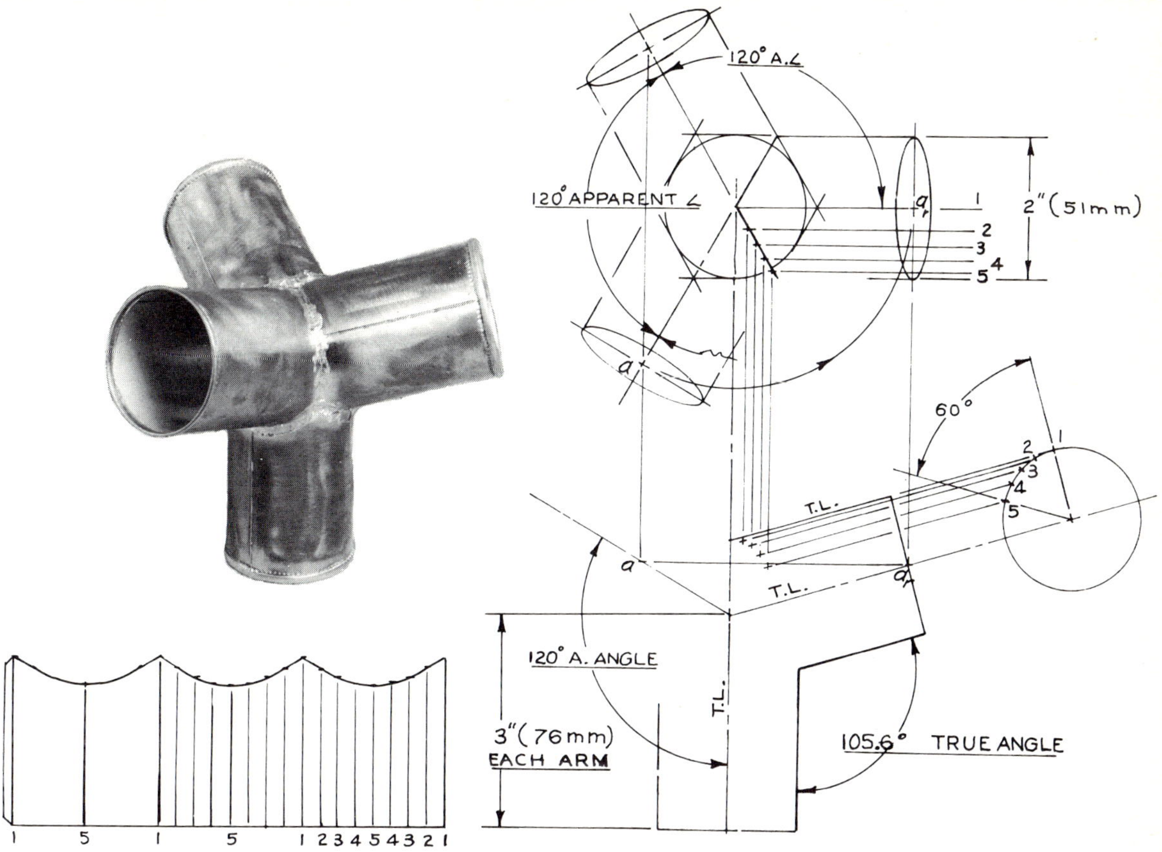

PARALLEL-LINE DEVELOPMENT

Fig. 9-70 Three-branch Y cluster.

Fig. 9-71

DIMENSIONED IN MILLIMETERS

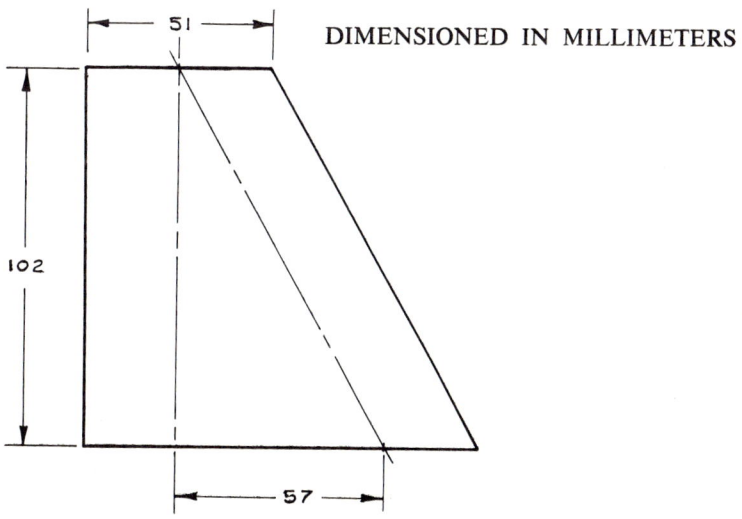

branch clustered Y breeching of Figure 9-70.

If all four branches are to be at the same angle and the pattern the same for each intersection, then the true angle for any two intersecting center lines must be established. This is accomplished, as shown in Figure 9-70, by drawing the top view of the three intersecting center lines, using *an apparent angle* of 120 degrees. The true angle is found in the front view by rotating arbitrary point *A*. The true angle for this problem, if measured, is 105.6 degrees. Now an auxiliary view of the circular end can be divided so that one-sixth the girth (60 degrees) contains uniformly spaced elements (five). The five elements are transferred to the top view to intersect the mitreline and then are projected to the front view to establish their true length.

To lay out a complete pattern, repeat the five elements and the resulting curve six times as a stretchout. Thus one pattern will serve for each of the four arms.

28. Lay out and develop patterns for the round-to-oblong transition which is shown in Figure 9-71.

chapter **10**

RADIAL-LINE DEVELOPMENT

Just as parallel-line development is used to create patterns for all manner of parallel forms, *radial-line development* is used for objects having straight lines (elements) radiating from an apex; these are commonly referred to as *radial forms*. Cones and pyramids, or portions of them, are good examples of radial forms. Sheet metal objects assuming the above shapes might be hoppers, separators, funnels, measures, ventilating heads, shanty caps, roof collars, etc.

Figure 10-1 shows the surface of a cone being "unrolled" on a flat plane. Note that the apex of the cone remains at one point and that the form unrolls in a circular path about that point as a center. Figure 10-2 shows the surfaces of a pyramid being "unfolded" on a flat plane. Again notice that the pyramid apex also remains fixed while the surfaces of the pyramid unfold in a

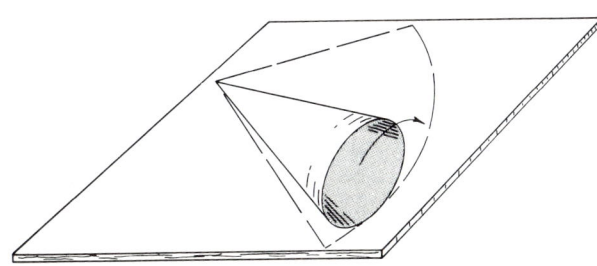

Fig. 10-1 Developable conical surface.

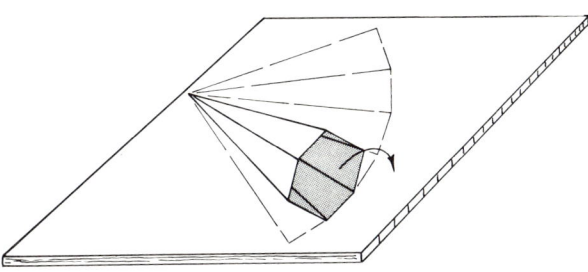

Fig. 10-2 Developable pyramidal surface.

circular manner about the apex as a center. Thus the patterns for radial forms of sheet metal are developable when the following two specifications have been found: (1) the true length of the elements; (2) the circumference or perimeter of the base.

THE CONE

10-1 Necessary views

Basically the right circular cone as a geometric shape is completely described when front and top views are shown, as in Figure 10-3. The front view furnishes the true length of all elements (slant height R). The top view shows the shape and the true circumference of the base C (Figure 10-3).

10-2 The stretchout arc

The pattern for the cone surface may be drawn by using the slant height of the cone as the radius of the *stretchout arc R* (Figure 10-4). Draw the stretchout arc more than long enough to contain a distance equal to the circumference of the cone base. The base circumference may be transferred to the stretchout arc by dividing the base into a sufficient number of equal spaces (16 in Figure 10-4) and by transferring the chordal distances, which are true length in the top view to the stretchout arc as shown.

An alternative and perhaps more accurate method in the case of a complete right circular cone or cone frustum, would be to find the circumference of the base mathematically

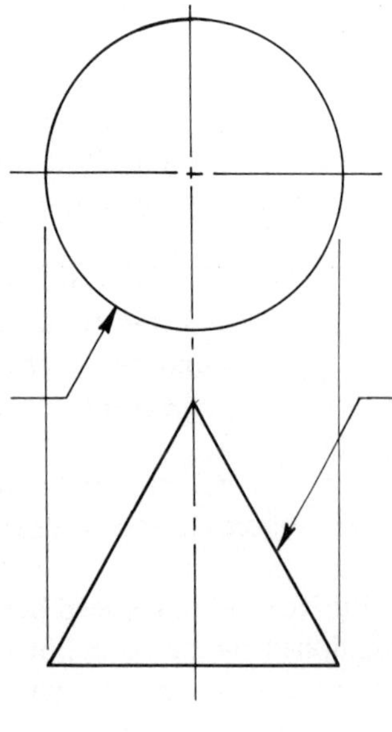

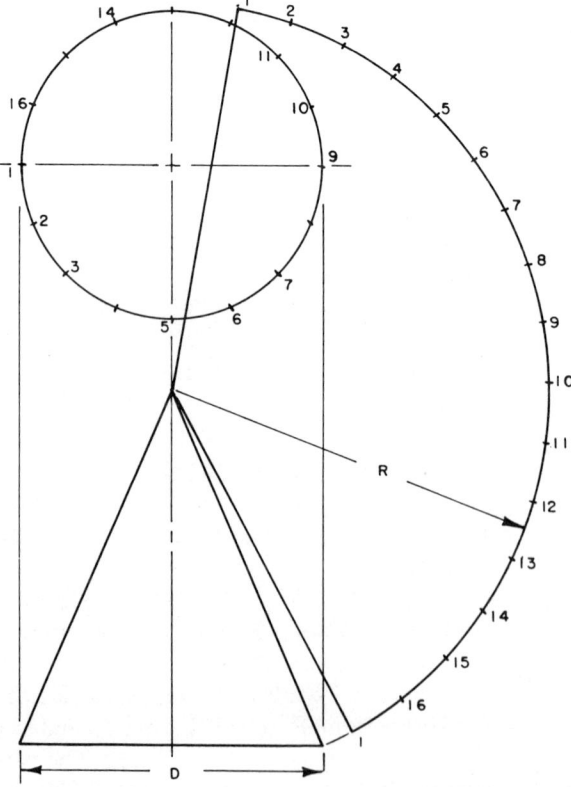

Fig. 10-4 Right-circular cone development.

Fig. 10-3

RADIAL-LINE DEVELOPMENT

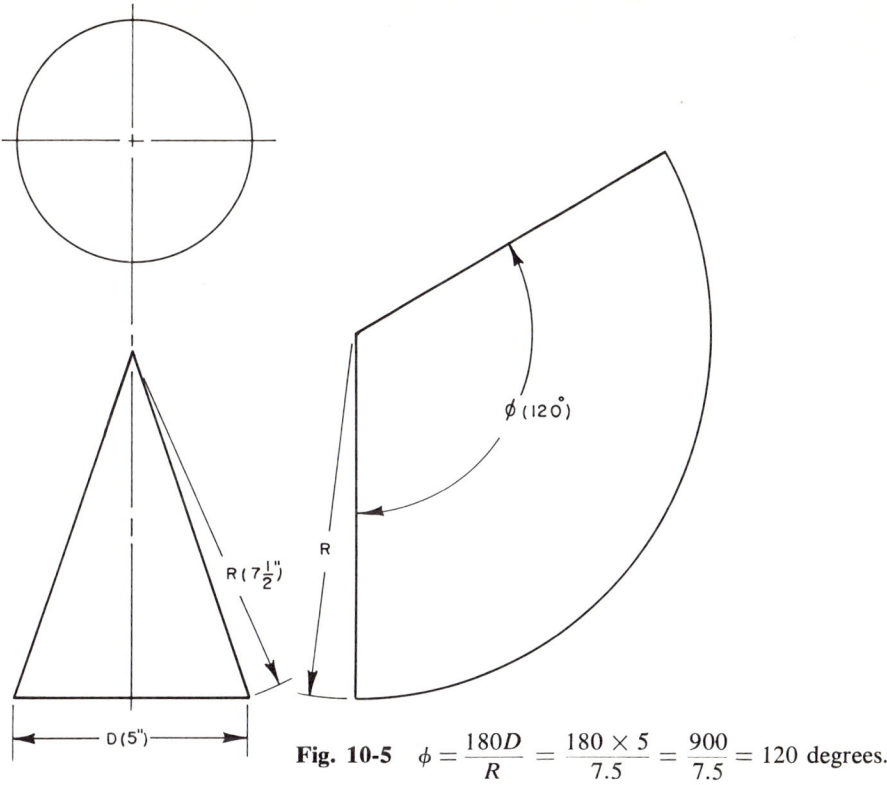

Fig. 10-5 $\phi = \dfrac{180D}{R} = \dfrac{180 \times 5}{7.5} = \dfrac{900}{7.5} = 120$ degrees.

(πD), or with the circumference rule of the sheet metal shop. With the circumference determined, set the dividers at, say, ½ inch for small cones, 1 inch or more for cones of long radius, and step off the required number of divider spaces to establish the base circumference on the stretchout arc.

The stretchout arc can also be computed by determining the enclosed angle ϕ (Figure 10-5). This method might serve as a check on the two previously mentioned methods. For example, develop the pattern for the cone of Figure 10-5. D (diameter of base) is to equal 5 inches; R (slant height) is to equal 7½ inches. By means of the formula $\phi = 180\,D/R$, the enclosed angle is found to be equal to 120 degrees. SI units can also be used in the formula, thus 127 mm (D) and 191 mm (R) also result in 120 degrees. The pattern can be drawn using a sweep of 120 degrees and a radius equal to the cone elements.

The sheet metal designer should be aware of the fact that a cone whose front view is an equilateral triangle will have a sweep of 180 degrees on the stretchout arc; therefore, when a cone or cone frustum is required in a proposed design, if the design will permit, a cone of equilateral appearance should be considered. For example, the funnel in Figure 10-6 is designed with a front view which appears as an equilateral triangle. Since, in this case, both R and D are of the same value, the formula can be resolved, as shown, by cancellation.

For the sheet metal student familiar with the natural functions of angles, the sweep of the stretchout arc may be computed, as shown in Figure 10-7, the sweep being equal to 360 multiplied by the cosine of the angle between the elements and the base. In ex-

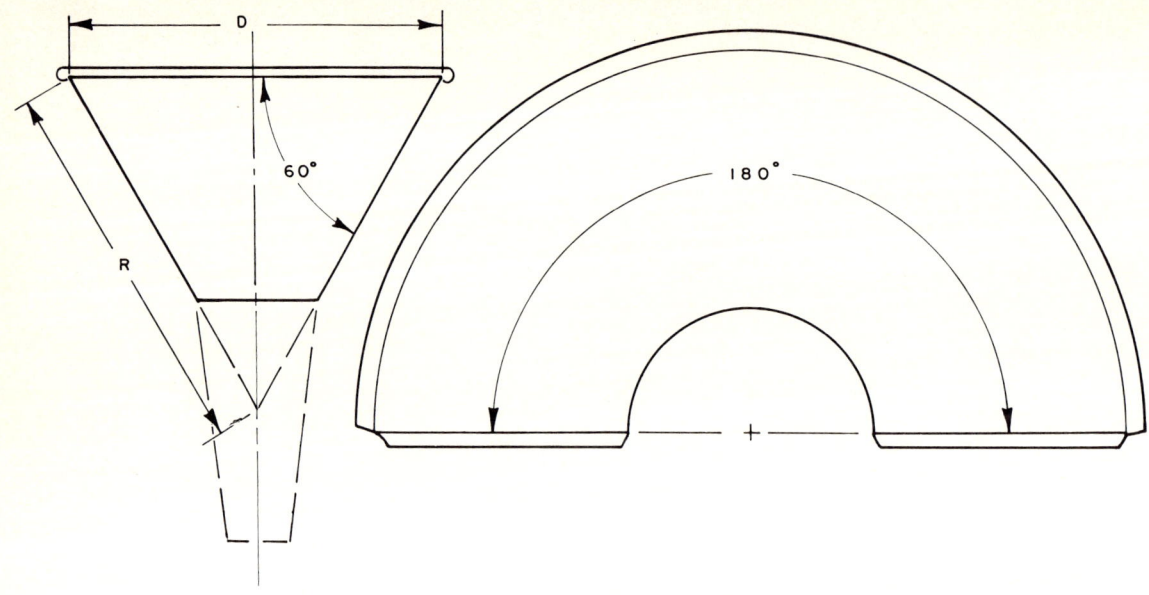

Fig. 10-6 $\phi = \dfrac{180D}{R} = \dfrac{180\cancel{D}}{\cancel{R}} = 180$ degrees.

Fig. 10-7 $\phi = 360 \cos \theta$. Example 1: ($\theta = 60$ degrees; $\cos \theta = 0.5$) $360 \times 0.5 = 180$ degrees.
Example 2: ($\theta = 75$ degrees; $\cos \theta = 0.25882$) $360 \times 0.25882 = 93$ degrees.

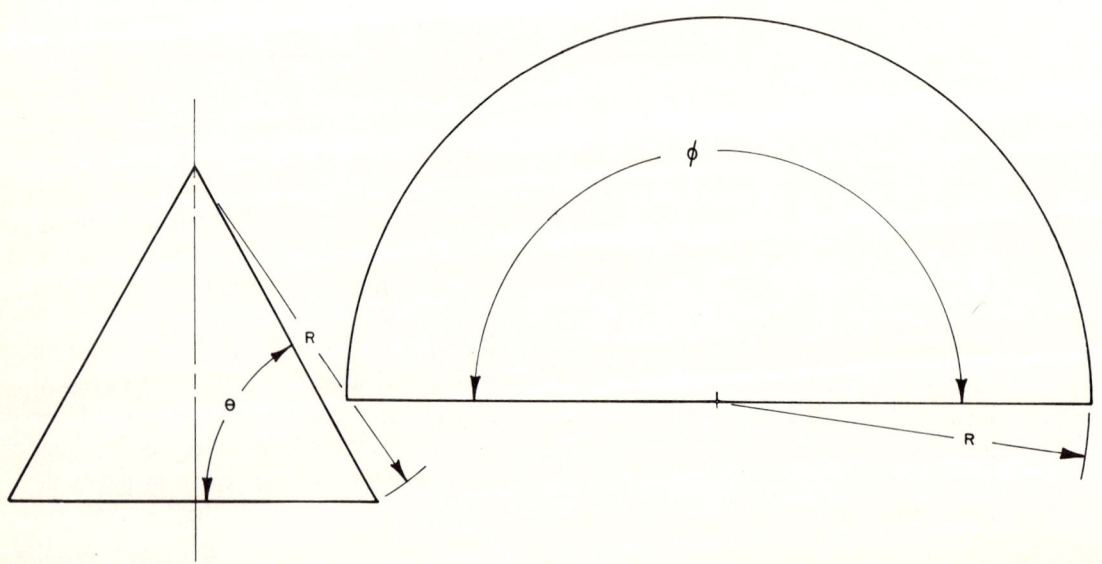

RADIAL-LINE DEVELOPMENT

Fig. 10-8 Cone development sweep angle plotted for base angle.

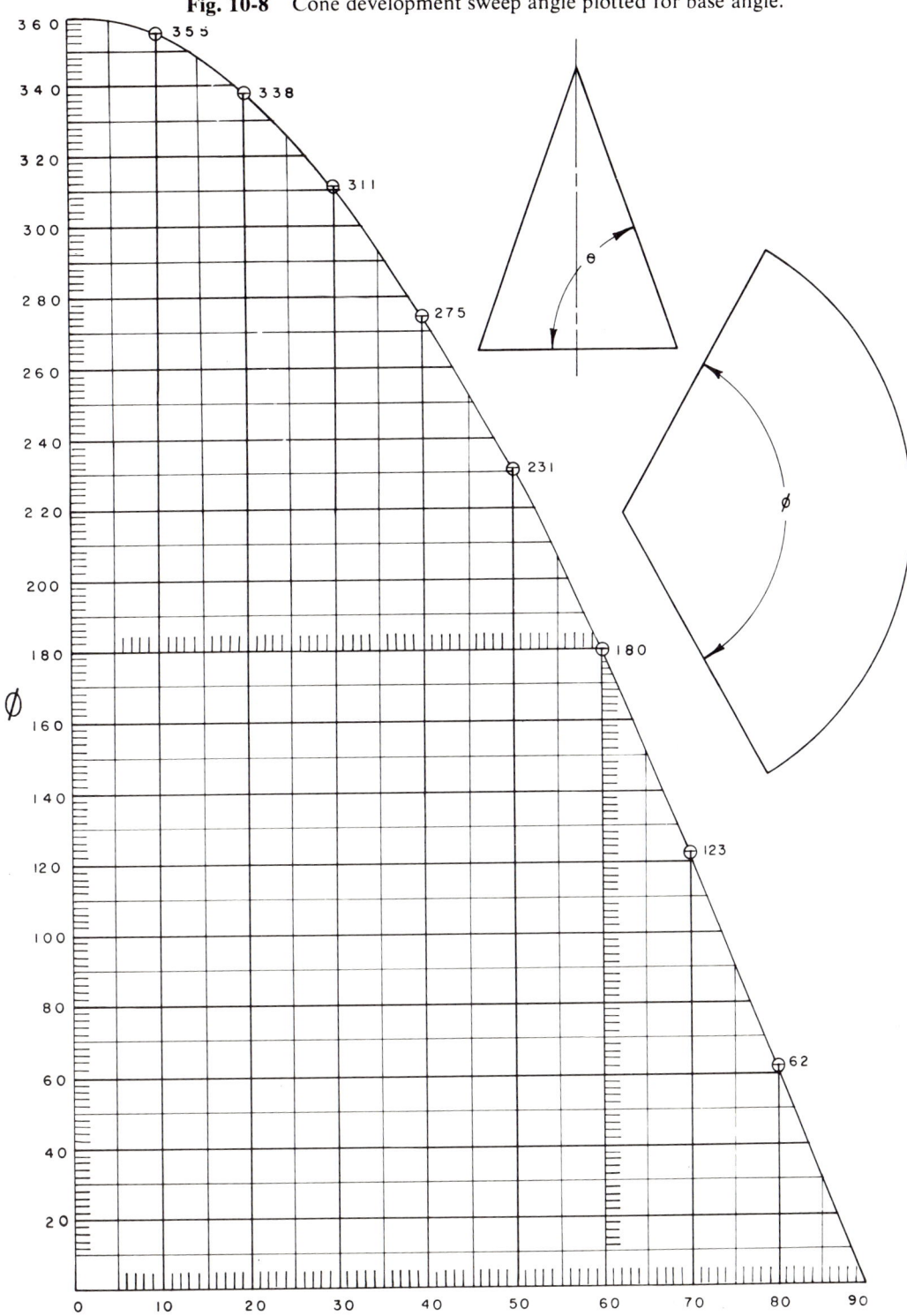

ample 1 of Figure 10-7, the cosine of 60 degrees is 0.5; therefore, multiplication gives 180 degrees as sweep angle. In example 2, the cosine of 75 degrees is 0.25882; the multiplication indicates that the sweep angle for a 75-degree cone should therefore be 93 degrees.

The graph of Figure 10-8 was constructed by means of the formula. This graph may be used to determine the sweep of the stretchout arc *for any cone* when the angle θ is known. To use it, find the given θ value horizontally, project upward to intersect the curve; then follow the horizontal line to read the sweep angle from the vertical calibrations. For example, a cone with θ angle of 33 degrees would indicate, by using the graph, a sweep of 302 degrees.

10-3 To develop the pattern for a round pitched cover

A round pitched sheet metal cover will serve as a good example of the development of the pattern for a "low altitude" or "flat" cone. By reference to the graph of Figure 10-8, notice that cones of low angle require a large sweep on the pattern. In other words, the pattern very closely approaches a flat disc in shape. In this case, it is more practical to determine the amount of metal to be removed from the disk than to lay out the sweep of the disk. The *sector* to be removed, measured in degrees, can be found by adapting the formula previously used to compute the sweep. In the example of Figure 10-9, where $D = 37\frac{1}{4}$ inches, $R = 20\frac{1}{8}$ inches,

Fig. 10-9

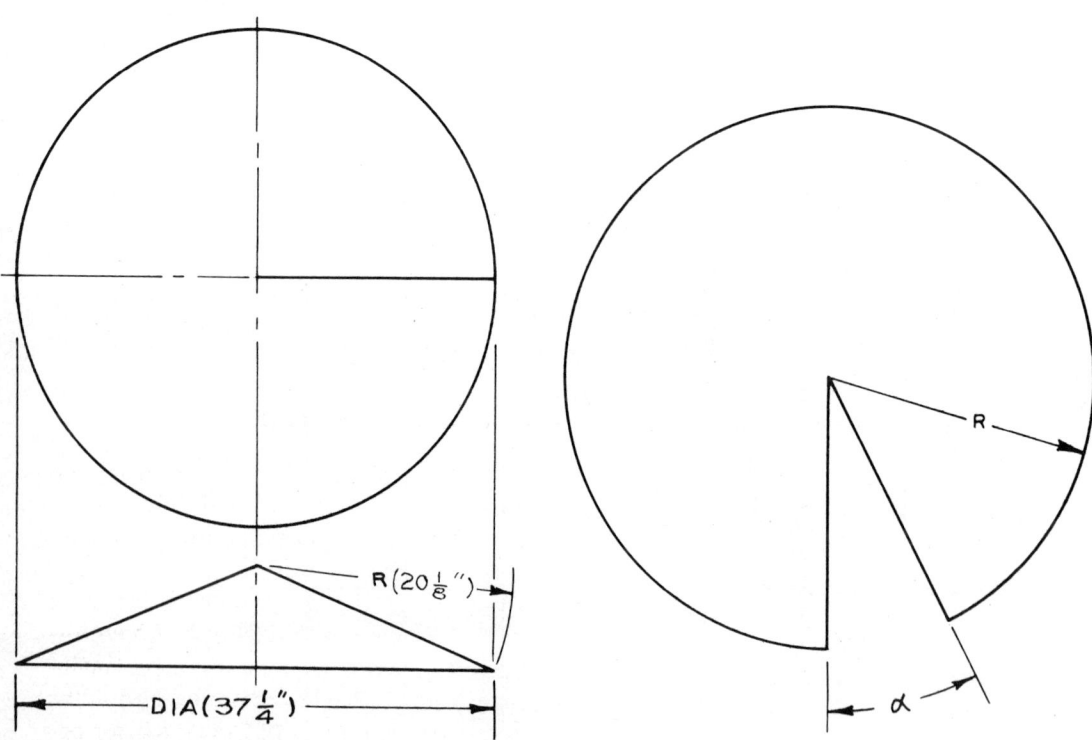

RADIAL-LINE DEVELOPMENT

149

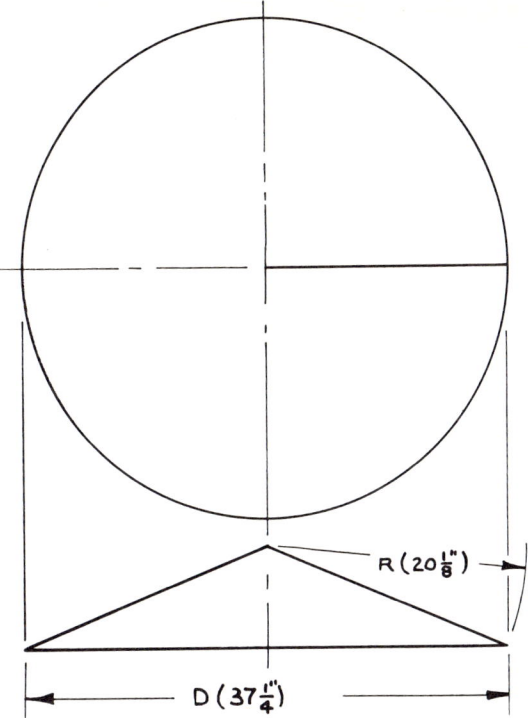

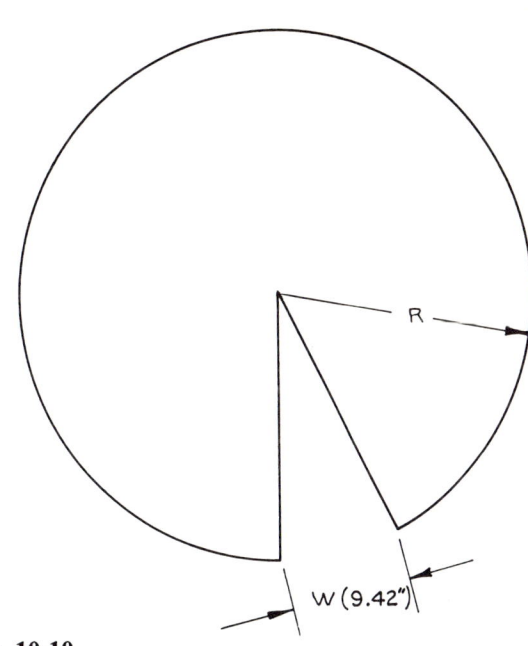

Fig. 10-10

Fig. 10-11 Cyclone collector. (*Quinn Bros. Inc., Philadelphia, Pa.*)

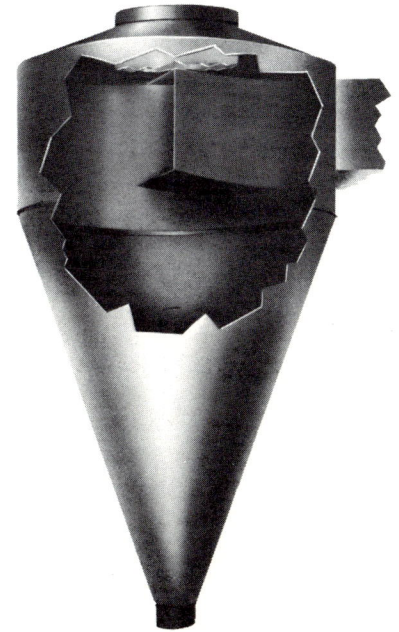

use the formula $\alpha = 360 - 180\, D/R$; therefore, $360 - 333 = 27$ degrees.

Using the same cone as an example in which the *width* of the *sector* to be removed is computed, refer to Figure 10-10. Use the formula Width $W = \pi(2R - D)$. By substitution and multiplication, the distance W is determined as equal to $9\frac{7}{16}$ inches. Using SI units, the answer is as follows: Width $W = \pi(2 \times 511 - 946)$; therefore the distance W is equal to 239 mm.

10-4 Cone frustums

Because they are used in so many industries for the recovery and collection of dust, lint, wood shavings, sawdust, grains, cereals, livestock feeds, etc.; cyclone collectors are a familiar sight in most industrial areas. They are popular because of their low cost, low power requirements, low maintenance, and high efficiency. As shown in the illustration (Figure 10-11), the collector operates upon the principle of centrifugal force.

The air carrying the particles is fed into the collector in such a manner as to create a "cyclone." This causes the particles to fly to the outside and to be conducted down the surface of the large cone frustum while the air is exhausted from the relatively calm eye or center of the cyclone. Intake and exhaust volumes and velocities can be regulated to control the action of the collector relative to various weights and sizes of particles.

The body of the collector, as shown in Figure 10-11, is composed of a cylinder (with intake duct and baffle through the side), a cap and outlet (cone frustum), a long taper as the lower portion of body (cone frustum), and connection flanges at the top and bottom.

To develop the patterns for the tapered body and cap (cone frustums), draw the front view as shown in Figure 10-12. The specified dimensions would likely be those indicated by the letters A, B, C, D, E, F, and G.

Since the dimension B represents the altitude of the cone frustum only, the entire cone is constructed in order to obtain the slant height R, as shown in Figure 10-13. With R as radius, the stretchout arc is drawn, the length of the arc computed, and then stepped off (or the sweep angle calculated instead). Since the tapered body is a cone frustum, the radius R' is necessary in order to complete the pattern, as shown in Figure 10-13. Just a reminder, at this time, that the drawings may be constructed to a reduced scale, one-half or one-fourth size; yet the patterns may be developed, or the layout on metal made full size (section 5-12).

If the layout is large, as is so often the case with collector bodies, it may be advantageous to calculate all required measurements R, R', and the sweep angle ϕ, from the given specifications D, B, C. Using the calculations, the pattern can then be constructed directly.

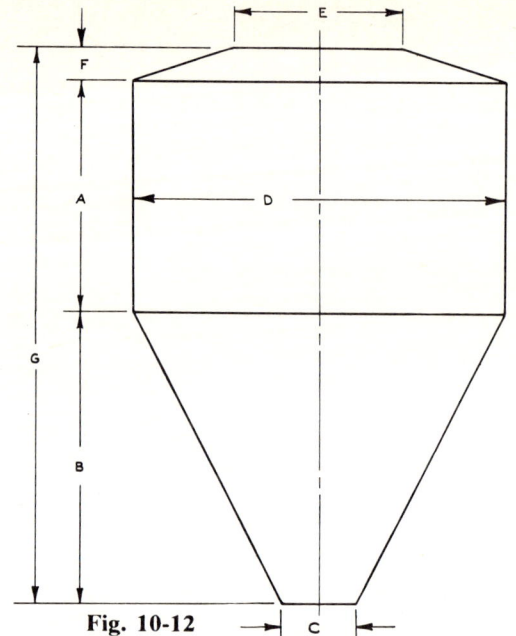

Fig. 10-12

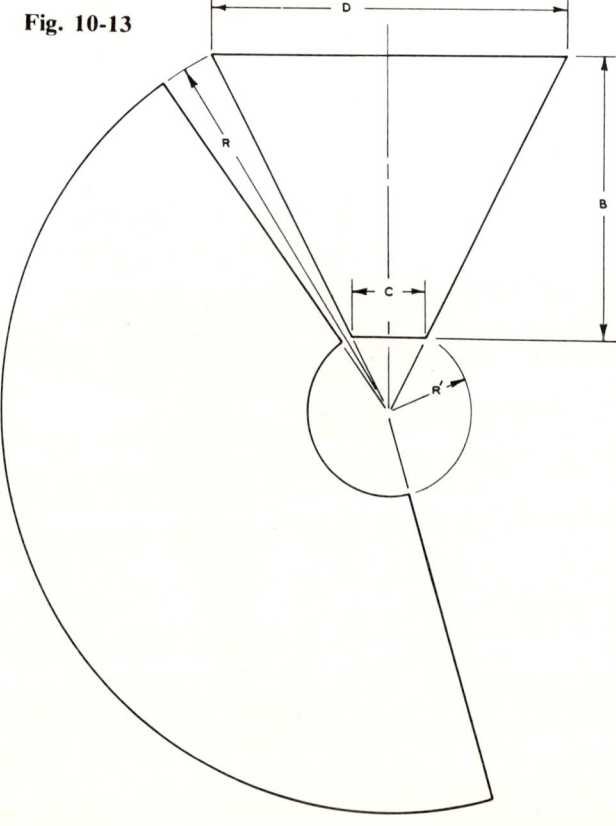

Fig. 10-13

RADIAL-LINE DEVELOPMENT

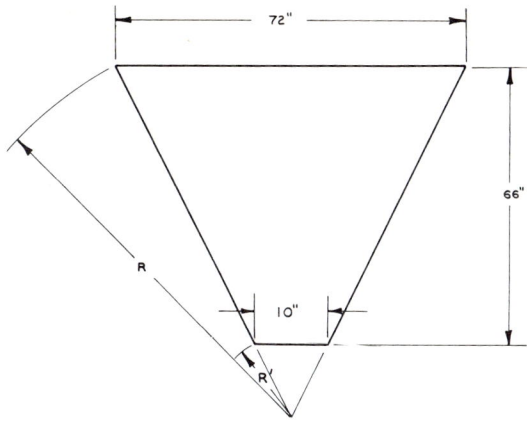

Fig. 10-14a. $R = \sqrt{\left(\dfrac{DB}{D-C}\right)^2 + \left(\dfrac{D}{2}\right)^2}$

$R = \sqrt{\left(\dfrac{72 \times 66}{62}\right)^2 + 36^2}$

$R = \sqrt{7{,}170} = 84.6$ inches

b. $\dfrac{R'}{C} = \dfrac{R}{D}$ ∴ $\dfrac{R'}{10} = \dfrac{84.6}{72}$

∴ $72R' = 846$ or

$R' = \dfrac{846}{72}$ or $R' = 11.8$ inches

c. $\phi = \dfrac{180 \times 72}{84.6} = \dfrac{12{,}960}{84.6}$

$= 153$ degrees

Fig. 10-15 $R = \sqrt{\left(\dfrac{72 \times 6}{72 - 36}\right)^2 + \left(\dfrac{72}{2}\right)^2}$

$= \sqrt{144 + 1{,}296}$

$= \sqrt{1{,}440}$

$= 38$ inches $\dfrac{R'}{36} = \dfrac{38}{72}$

∴ $72R' = 1{,}368$ or $R' = 19$ inches

$\phi = \dfrac{180 \times 72}{38} = 341°$

For example, in Figure 10-14, D equals 72 inches, B equals 66 inches, and C equals 10 inches. To find the slant height R, use the formula

$$R = \sqrt{\left(\dfrac{DB}{D-C}\right)^2 + \left(\dfrac{D}{2}\right)^2}$$

By substituting the known values into the formula, the equation is reduced as shown in Figure 10-14A until $R = \sqrt{7{,}170}$. The answer is then found by extracting the square root. R is found to equal $84\tfrac{5}{8}$ inches.

To find the slant height R', establish a proportion, as shown in Figure 10-14B. Putting the proportion into words would be to say that "The small slant height is to the small base as the large slant height is to the large base." Solve the proportion by cross multiplication. R' is found to equal $11\tfrac{3}{4}$ inches.

To find the sweep angle ϕ, use the formula $\phi = 180\,D/R$, as shown in Figure 10-14C. The sweep angle is found to equal 153 degrees.

The collector cap is shown in Figure 10-15. The computations, using the specified dimensions as shown, determine the values for R, R', and ϕ. When the given data and desired results are to be in meters or millimeters, the formulas can be used equally well using the SI units. The pattern is completed, as shown, using the computed values.

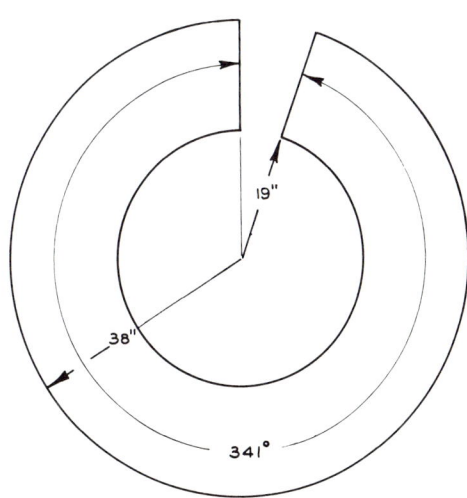

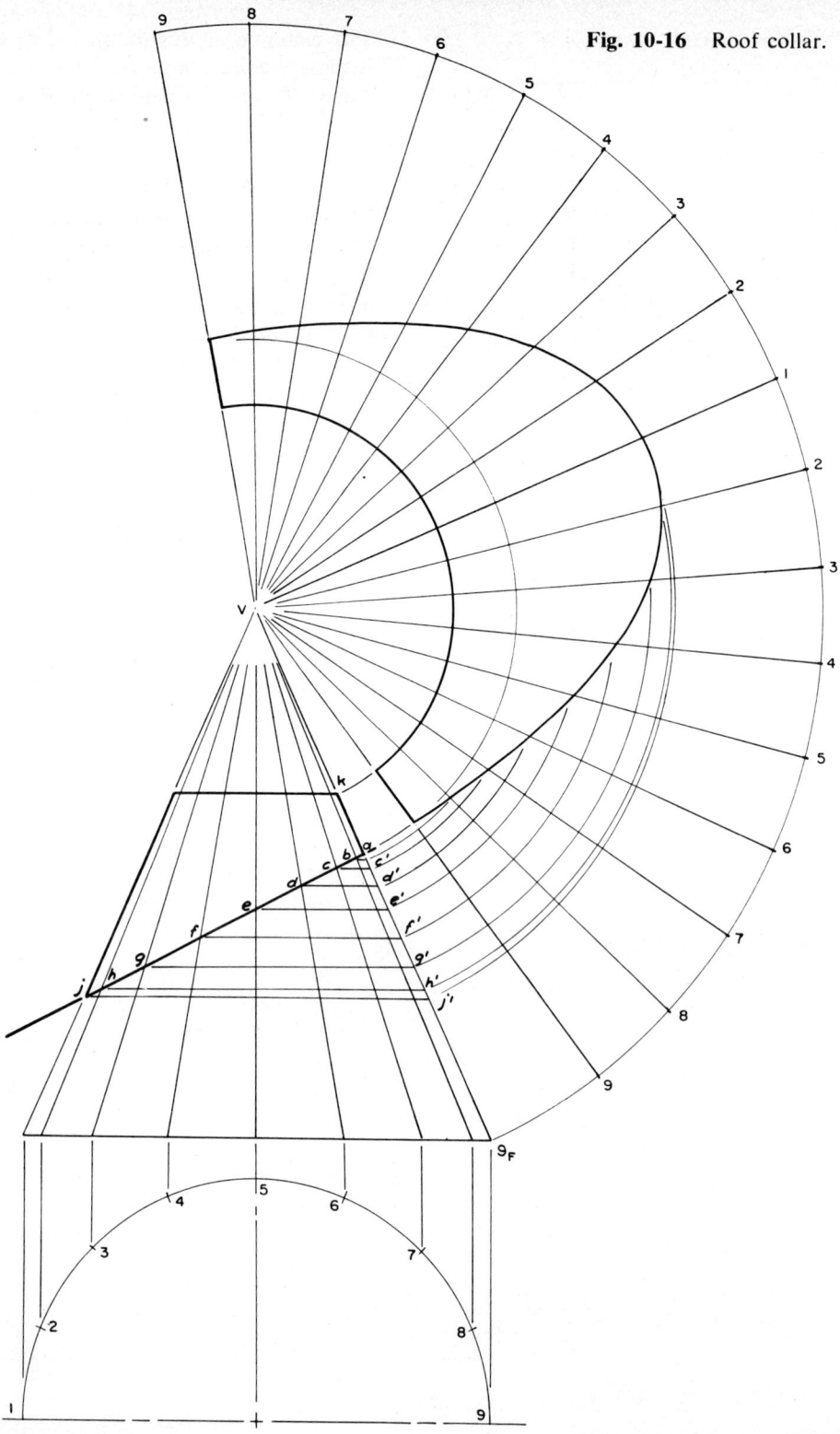

Fig. 10-16 Roof collar.

10-5 Truncated right circular cone

The roof collar of Figure 10-16 represents a right circular cone truncated by the plane of a roof. This is a typical example of many situations in sheet metal where only part of the cone construction on the drawing actually represents the required metallic piece. Besides roof collars; conical elbows, measure pouring lips, tapering tee intersections, and breechings are other examples of truncated cones.

To develop the pattern of the roof collar, draw the front view of the collar and its cone as was done in Figure 10-16. Then construct a bottom or a half-bottom view of the base and apex only. Divide the bottom view into a suitable number of elements (1 through 9 and back to 1, in the figure). Project the elements to the front view and draw the elements piercing the roof plane. The piercing points thus located are at a, b, c, d, e, etc. Because they are contour elements, the elements Va and Vj are true length in the front view. All others are foreshortened in the front view. The foreshortened elements must be rotated about the cone axis until they appear as true length elements at b', c', d', e', etc. (Chapter 8).

The pattern can now be developed as for a complete right circular cone using the slant heights $V\text{-}9_F$ and Vk as radii. Step off the elements on the pattern, using the distances 1 to 2, 2 to 3, 3 to 4, etc. Beginning with the shortest element Va as a seam element, sweep the true distances from the front view to the pattern, as shown, to establish points on the pattern curve. Complete the curve using the techniques described in section 3-14. Make the necessary allowances for seams and flanges and the pattern is complete.

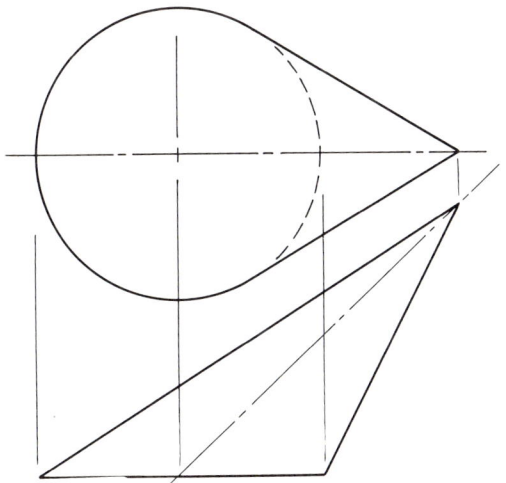

Fig. 10-17

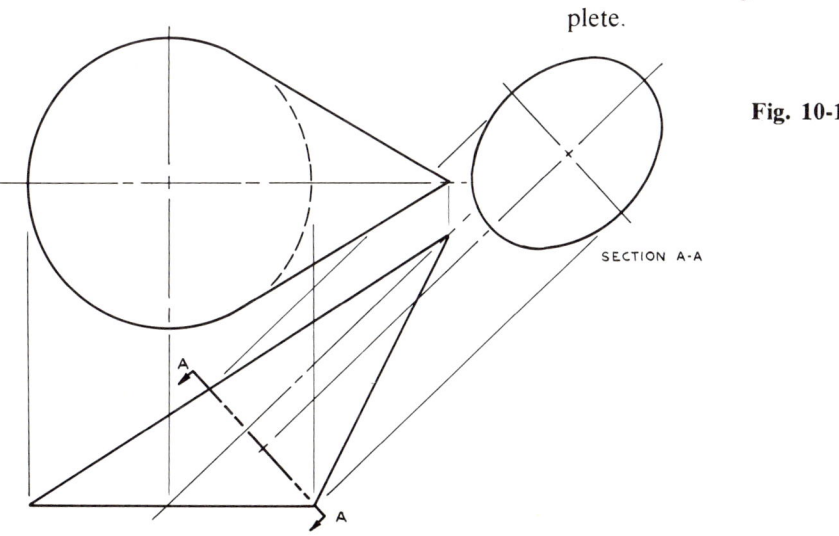

Fig. 10-18

SECTION A-A

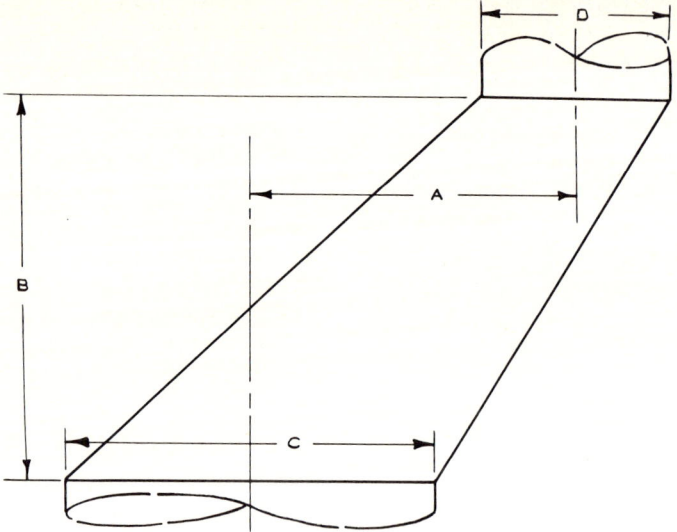

Fig. 10-19 Tapered offset.

10-6 Oblique cone with circular base

The top and front views of an oblique cone are shown in Figure 10-17. Any oblique cone with circular base is actually a *right elliptical cone* because any right section of the cone is an ellipse (see Figure 10-18).

While the base of the oblique cone may be a circle and the base circumference therefore equal to πD, the true length of each of the elements from base to apex varies for all points on the base. The development problem thus created is a bit more complicated than for the right circular cones previously mentioned.

In sheet metal, the *tapered offset* is a typical application of the circular based elliptic cone. The tapered offset is used as shown in Figure 10-19 to connect offset pipes of different diameters. The specifications for the fitting are indicated by the measurements A, B, C, and D.

To develop the pattern: (1) Draw a front view from specifications. (2) Extend the cone to its apex, as in Figure 10-20. (3) Draw a half-bottom view of cone and apex only. (4) Divide the base into a sufficient number of equal spaces (eight in Figure 10-20) and number the elements thus established (1 through 9 and back to 1). (5) Using the apex as center, rotate the elements *in the bottom view* until each is parallel to the front plane of projection. (6) Project the rotated elements to the front view. (7) Draw the rotated elements (1' through 9') in the front view where each is now true length. (8) Using the apex as a center, sweep the true lengths into the pattern development area as shown by the arrows in Figure 10-20. (9) Because the cone is oblique the base circumference will be a curve on the pattern. To establish the curve, set the compass equal to the true distance between elements (as taken from the bottom view). (10) Starting at any place on the arc from 1', strike the compass setting to cross arc 2'; using this point as center, strike the setting to cross arc 3'; using this point as center, strike the setting to cross arc 4';

RADIAL-LINE DEVELOPMENT

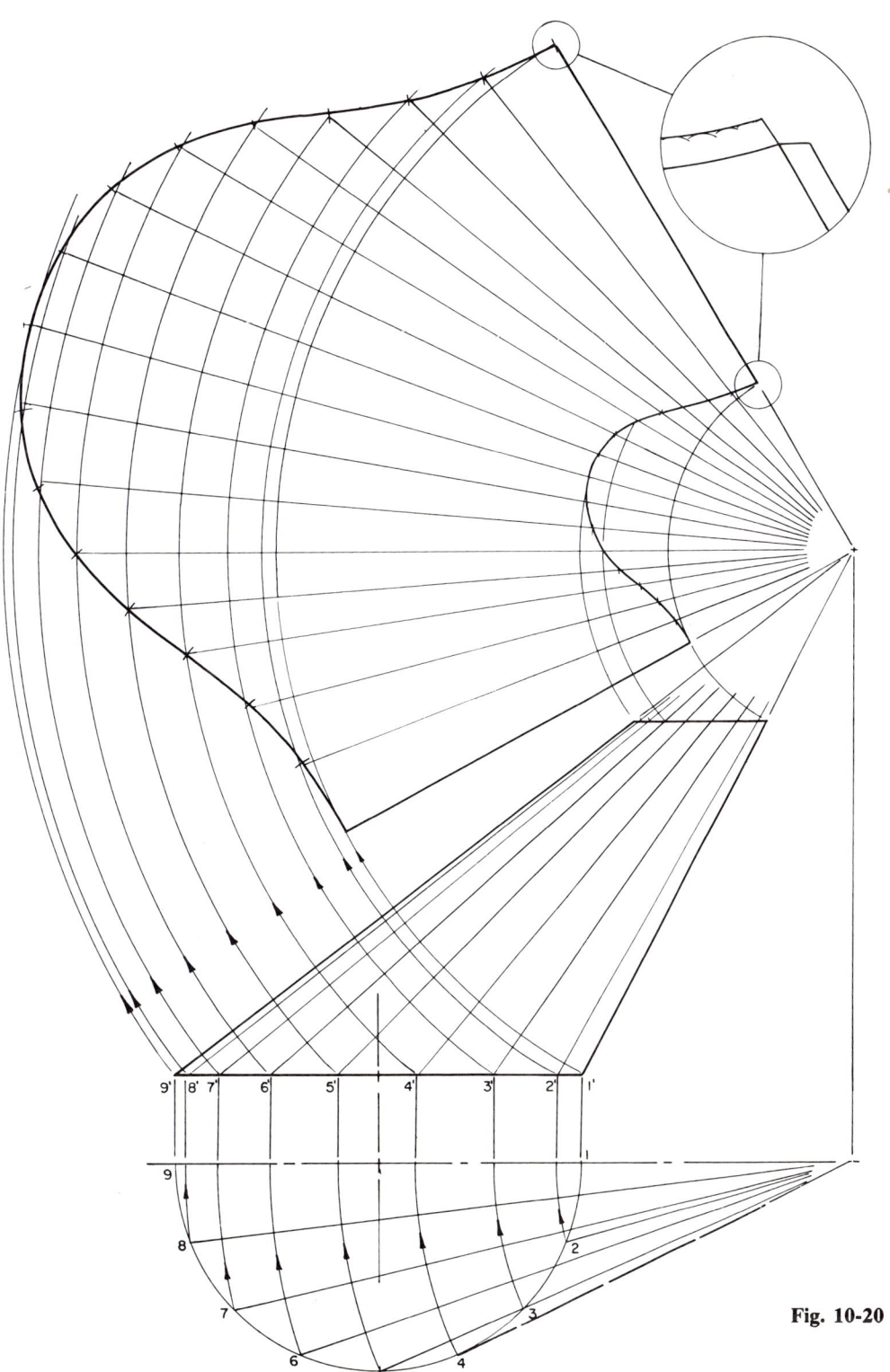

Fig. 10-20

repeat this procedure until points for all elements have been determined. (This procedure is, in effect, triangulation as described in Chapter 11.) (11) Draw the curve. (12) Draw elements on the pattern from the apex to each point on the curve. (13) Again using the apex as center, sweep the top diameter and element intersection points of the front view to their respective elements on the pattern. This establishes points for the short curve and the curve may be drawn. In order to avoid making the pattern construction difficult to follow, typical allowances for seams are illustrated in the removed detail in Figure 10-20. In practice of course, the seams are added to the pattern as developed above.

10-7 Conical reducing elbow

A five-piece 90-degree conical reducing elbow is shown in Figure 10-21. Since the purpose of the elbow is to produce, not only a change in direction, but also a *uniform* change in diameter, the well-designed reducing elbow should consist of sections taken from right circular cones.

The angular spacing of a conical reducing elbow of any number of pieces and of any desired angle of bend is calculated in much the same manner as for cylindrical elbows (see section 9-7) but with certain differences, because the elbow is taken from cones rather than a cylinder.

To illustrate the construction and development of the cone sections required to make a conical elbow to specifications, consider the following items:

    a. Large diameter to be 24 in [610 mm]
    b. Small diameter to be 12 in [305 mm]
    c. Elbow to be four-piece 75 degree
    d. Center-line radius to be 36 in [914 mm]

To review the above specifications as a designer, items *a* and *b* have probably been established by the situation for which the elbow is needed. Item *c* is based upon the angle of bend and the angular smoothness desired in the elbow. Item *d*, the center-line radius, as well as item *c*, affects the length of cones required. The center-line radius may in some cases be controlled by installation limitations. A good rule of thumb might otherwise be to let the radius equal from one to two times the large pipe diameter.

To construct the front view from specifications: (1) Draw a base line and vertical center line. On the base line, measure the large diameter and the center-line radius to establish *the center of the elbow, V* (Figure 10-22*A*). (2) From *V* draw the angle of bend (75 degrees), again establish the center-line radius and the small-pipe diameter (Figure 10-22*B*). *Notice that the center line of the small pipe is perpendicular to the end of the elbow.* (3) Calculate the miter *angle,* $\alpha = A/(2N-2)$ where *A* equals the angle of the elbow (75 degrees) and *N* equals the number

Fig. 10-21 Conical reducing elbow. (*Naylor Pipe Co., Chicago, Ill.*)

RADIAL-LINE DEVELOPMENT

of pieces. Thus, in the example, α equals $12\frac{1}{2}$ degrees. (4) Using the α value draw a line from V to establish the center-line distance 1-2 (Figure 10-22B). Transfer that 1-2 distance to the center line at the small end of the elbow ($1_T - 2_T$, Figure 10-22C). (5) Using V as center, an arc from point 2 at the large end will also pass through point 2_T at the small end of the elbow (Figure 10-22C). Use the distance 1-2 to step off along the arc, thus locating points 3-4-5 (Figure 10-22C). (6) Starting at the base line, the center lines of the cones forming the elbow sections intersect at points 2, 4, and 2_T (Figure 10-22C). (7) The contour elements of the required cones are found by placing spheres (imaginatively) at the intersection points.

The size of the spheres is calculated as follows: Large diameter − small diameter = diametral difference to be distributed uniformly through the elbow. Thus, in the example, 24 in − 12 in = 12 in. The difference is divided by the number of uniform spaces within the angle of bend ($2N - 2$), where N equals the number of pieces. Thus (2×4) $- 2 = 6$ spaces. (8) The difference (12 in) thus is divided up into the following diametral reductions: 22 in at point 2; 18 in at point 4; and 14 in at point 2_T. *Notice that the reductions for center sections are twice those for end sections*, just as for cylindrical elbows (section 9-7). (9) The contour elements of the cones intersect. The intersections locate the *miter lines* as at x and y of Figure 10-22C. The front view of the elbow can now be completed. The joints between sections of the elbows are alike in size and shape. The theory of the "tangent spheres" method of locating

Fig. 10-22

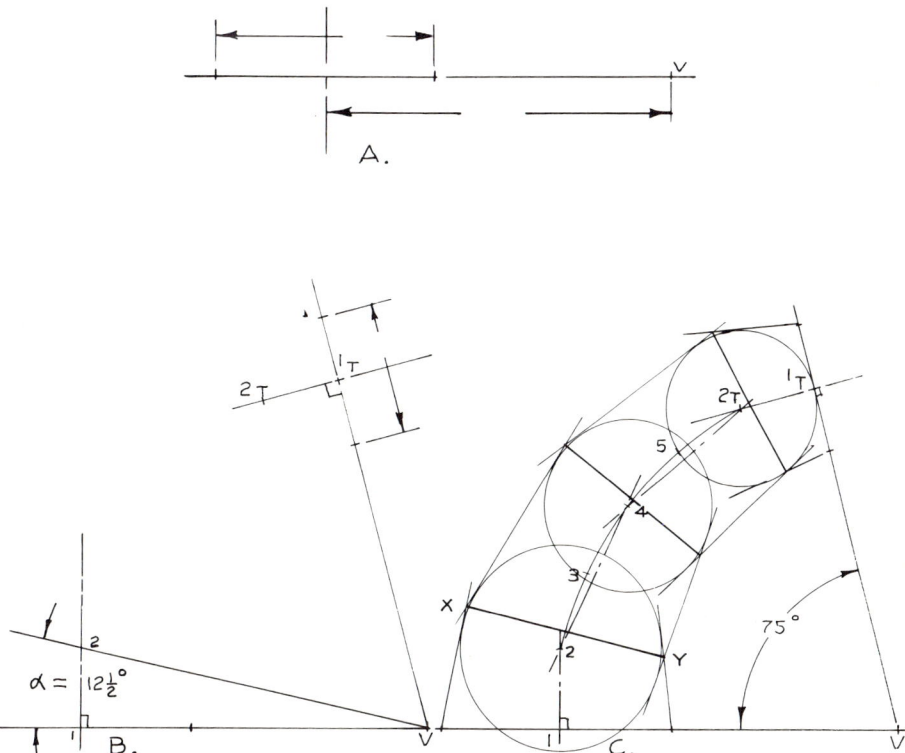

the miter lines guarantees that they will fit. This theory was first explained in section 9-18 where three criteria are listed explaining when and only when an intersection will appear as a miter line and not as a curve. The intersecting cones of Figure 10-22C result in miter lines between sections because they are of *equal size, identical shape,* and *have intersecting axes* at the center of each "ball."

To make the patterns for the elbow sections, the right circular cone for each section is developed. That portion of the cone representing the elbow section generates curves on the entire pattern and can be used as the section pattern. See Figure 10-23 where the large diameter cone and the cone of a center section have been developed. To save space, a half-pattern for the center section is shown. The remaining elbow sections are developed in the same manner. Notice that the curves can only be plotted by using true-length intersection points determined in the front view. The completed elbow which meets the specifications is shown in Figure 10-24.

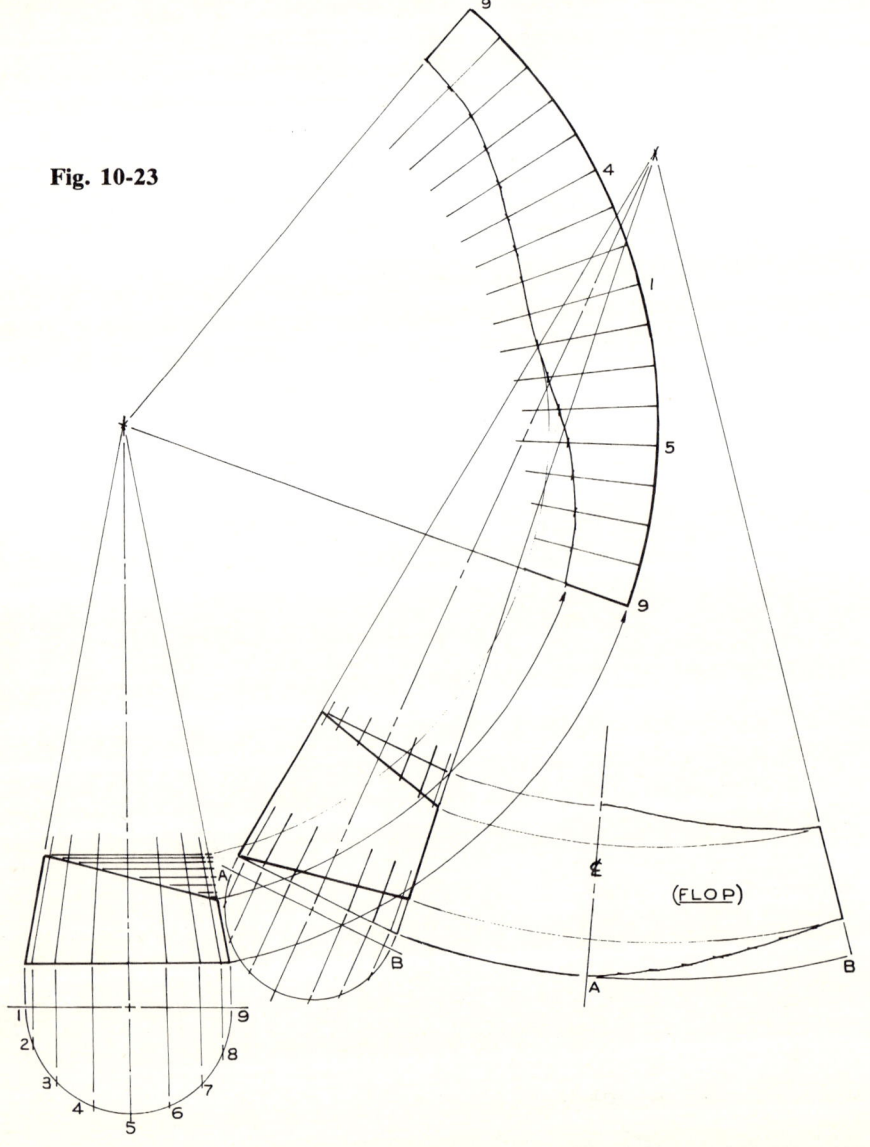

Fig. 10-23

RADIAL-LINE DEVELOPMENT

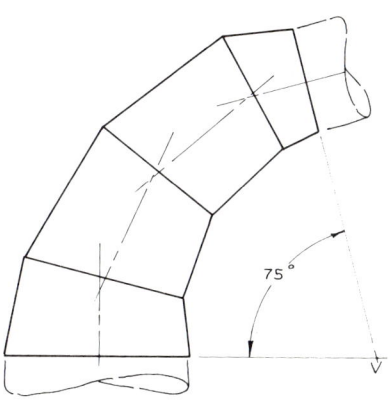

Fig. 10-24

The elbow as shown in Figure 10-24 need not be drawn. It is shown in the text only to clarify the way in which pieces cut from a cone pattern are fabricated to form the elbow.

Conical elbows to fit special situations may be designed by means of tangent spheres. If the amount of taper or flare within the elbow is to vary, then separate right circular cones are needed for each elbow section.

As an example, in drawing the front view of a conical elbow made up from different cones (and to be used as a ventilating cowl), consider the specifications indicated in Figure 10-25. The three-piece 90-degree reducing elbow desired tapers from a 36- to a 12-in [914- to a 305-mm] diameter in steps of 24 and 16 in [610 and 406 mm], respectively. Figure 10-25A shows the center lines of the intersecting cones as established by the specifications. Figure 10-25B shows the 24- and 16-in [610- and 406-mm] spheres drawn with centers at the intersection of the cone center lines. In Figure 10-25C, the contour elements have been added tangent to the spheres. To complete the view, the miter lines are then drawn from the intersection of contour elements on one side of the sphere to the intersection point of contour elements on the other side of the sphere (see Figure 10-25D). To develop the pattern for each section of the elbow, as shown in Figure 10-26, draw a base line for each cone and add a half-base view with a sufficient number of elements. Develop the pattern for each cone just as for any right circular truncated cone (section 10-5, Figure 10-16). Notice that pattern development is required for three different cones and that an arbitrary base for each cone may be selected which will represent the longest required element on each cone (Figure 10-26).

Fig. 10-25 DIMENSIONED IN MILLIMETERS

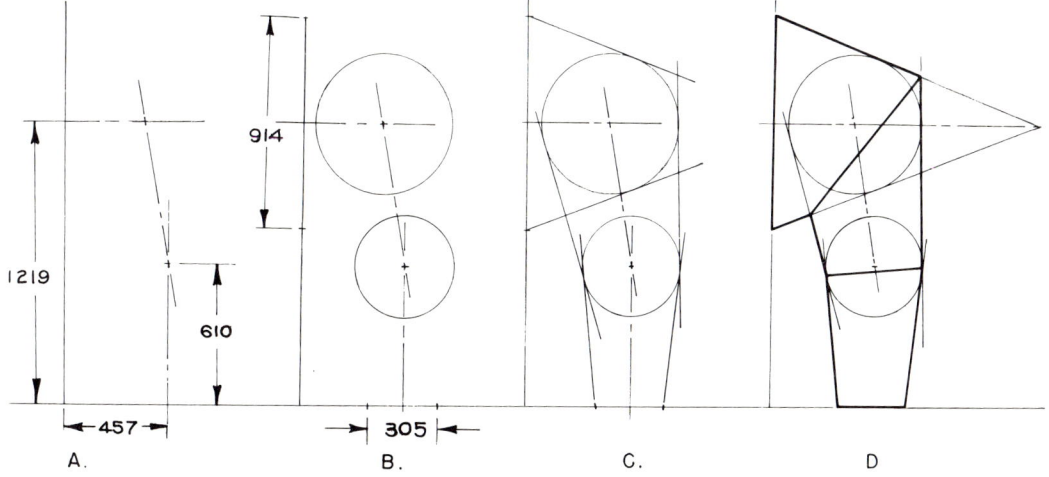

A.   B.   C.   D.

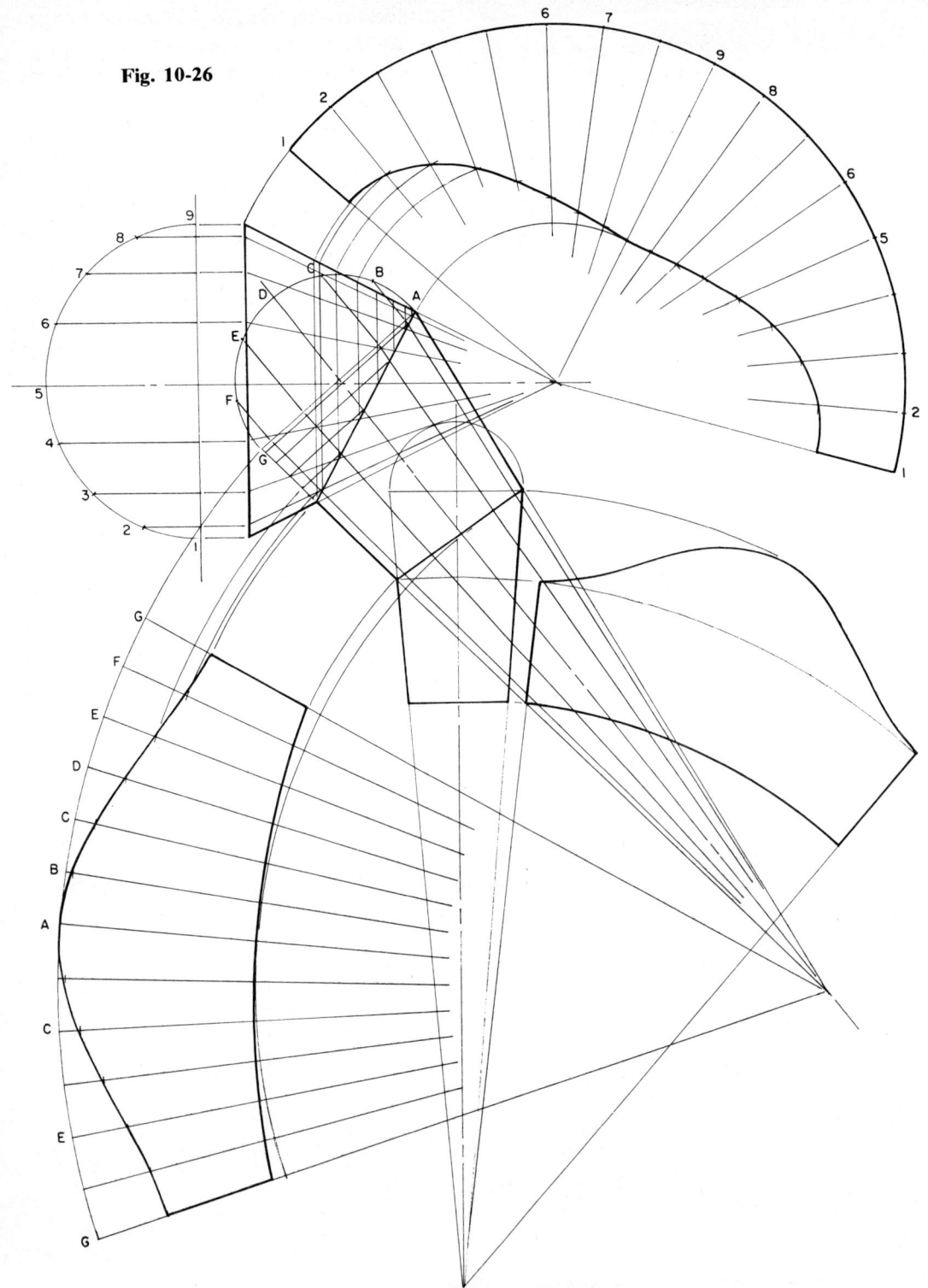

Fig. 10-26

10-8 Intersections of cones and planes

In section 5-15, Conic sections, definitions are given for the various curves created by intersecting planes and cones. It would be well for the student to review these definitions at this time. Plane surfaces and cones frequently intersect on sheet metal drawings, hence the draftsman is often required to find the true size and shape of the intersections. This may also be needed in order to develop patterns or to calculate waste, flashings, weight, etc. Furthermore, a knowledge of curve plotting will aid in developing patterns for combinations of intersecting geometric shapes as assigned in the following chapters.

a. Triangular Section. Any plane cutting the base of a right circular cone and passing through its apex will cut an isosceles triangular section (see Figure 10-27). The base line of the isosceles triangle is equal to the span of the cut across the base, the altitude of the triangle is equal to the slant height of the cut (base to apex), and the sides of the isosceles section are elements of the cone. In order to cut a triangular section (in other words, sides which are straight lines), the cutting plane *must pass through the apex of the cone;* all other planes cut curves. Study Figure 10-27 carefully to establish the above principles. (Cutting planes will be used to establish lines of intersection in problems to be discussed later.)

b. Ellipse. If a conical collar is to pass through an inclined roof, then the opening

Fig. 10-27 Cutting plane passing through base and apex of a cone.

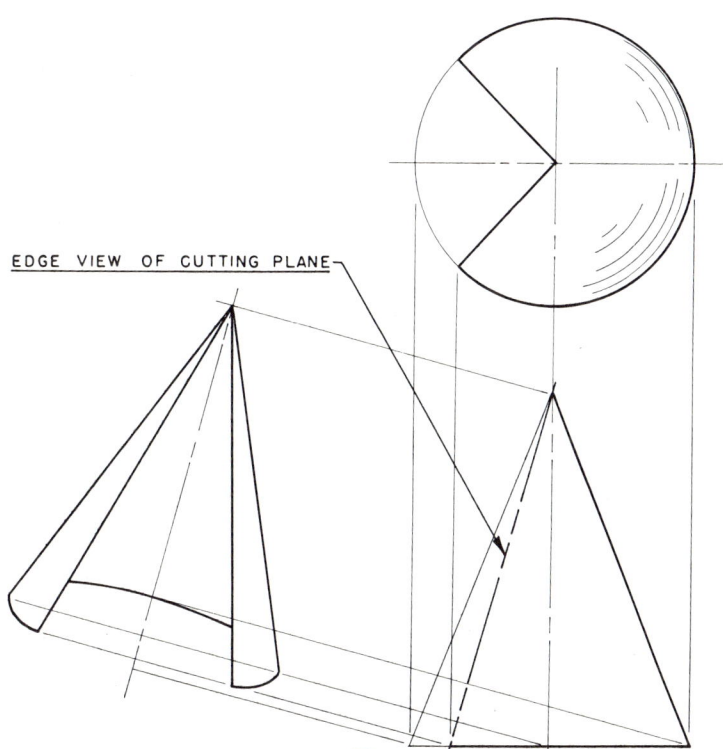

required would be elliptical. As an example of the development of the opening refer to Figure 10-28. As in the example, draw the front and bottom views of the required cone. Next, cut the cone with the required plane (a ¼ pitch roof in Figure 10-28). Thus, points 1 and 2 for projectors, and the line of sight 3 are established (the line of sight is perpendicular to the roof plane). The distance 1-2 represents the major diameter of the required elliptical opening. The minor diameter is perpendicular to the major therefore, the minor diameter appears as a point, 4 in the front view, midway between points 1 and 2. To find the true length of the minor diameter; cut the cone with an arbitrary plane through point 4 and the apex. The arbitrary cutting plane produces elements 5 and 6 on the cone. Elements 5 and 6, in the bottom view, cross the projector from point 4 at the piercing points X and Y. The distance XY represents the true minor diameter of the ellipse and may be transferred to the auxiliary view by reference. The ellipse, as shown in Figure 10-28, may then be completed by trammel (section 5-14a). Notice that the trammel method may be used in this case because it produces a *true ellipse* and a true ellipse is required in order to have the hole and cone fit accurately. The trammel method represents a short-cut that is entirely satisfactory in problems of this type. The ellipse could be completely plotted, point by point, in the bottom and auxiliary views as shown for a different roof collar and roof pitch in Figure 10-29. The procedure is the same as previously described except that the bot-

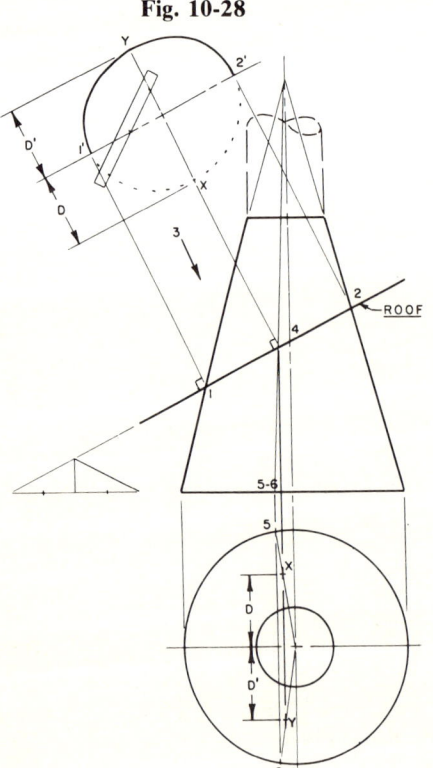

Fig. 10-28

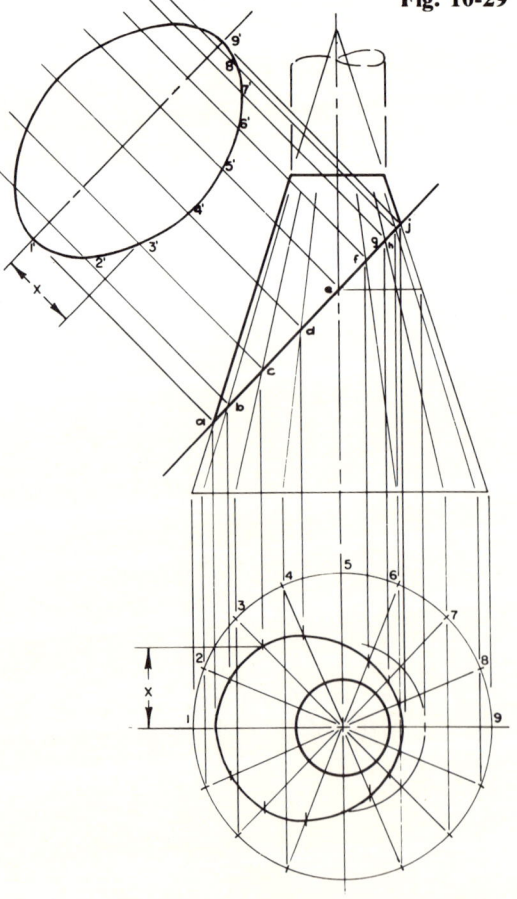

Fig. 10-29

RADIAL-LINE DEVELOPMENT

tom view of the cone base is divided uniformly into a sufficient number of elements. These are projected and drawn in the front view; thus establishing piercing points *a, b, c, d, e,* etc. Projectors from the piercing points back to the elements in the bottom view establish the elliptical curve as seen from the bottom. Projectors from the piercing points to the auxiliary view and transfer distances taken from the bottom view establish points in the auxiliary view for the true size and shape of the elliptical hole.

c. Parabola. As defined in section 5-15, when a cone is cut by a plane which makes the same angle to its axis as the cone elements, a parabolic curve results. Such a situation occurs in the problem of Figure 10-30. A conical air scoop is shown intersecting a flat, horizontal, motor hood. A pattern for the air scoop is desired and the

Fig. 10-30 Conical air scoop.

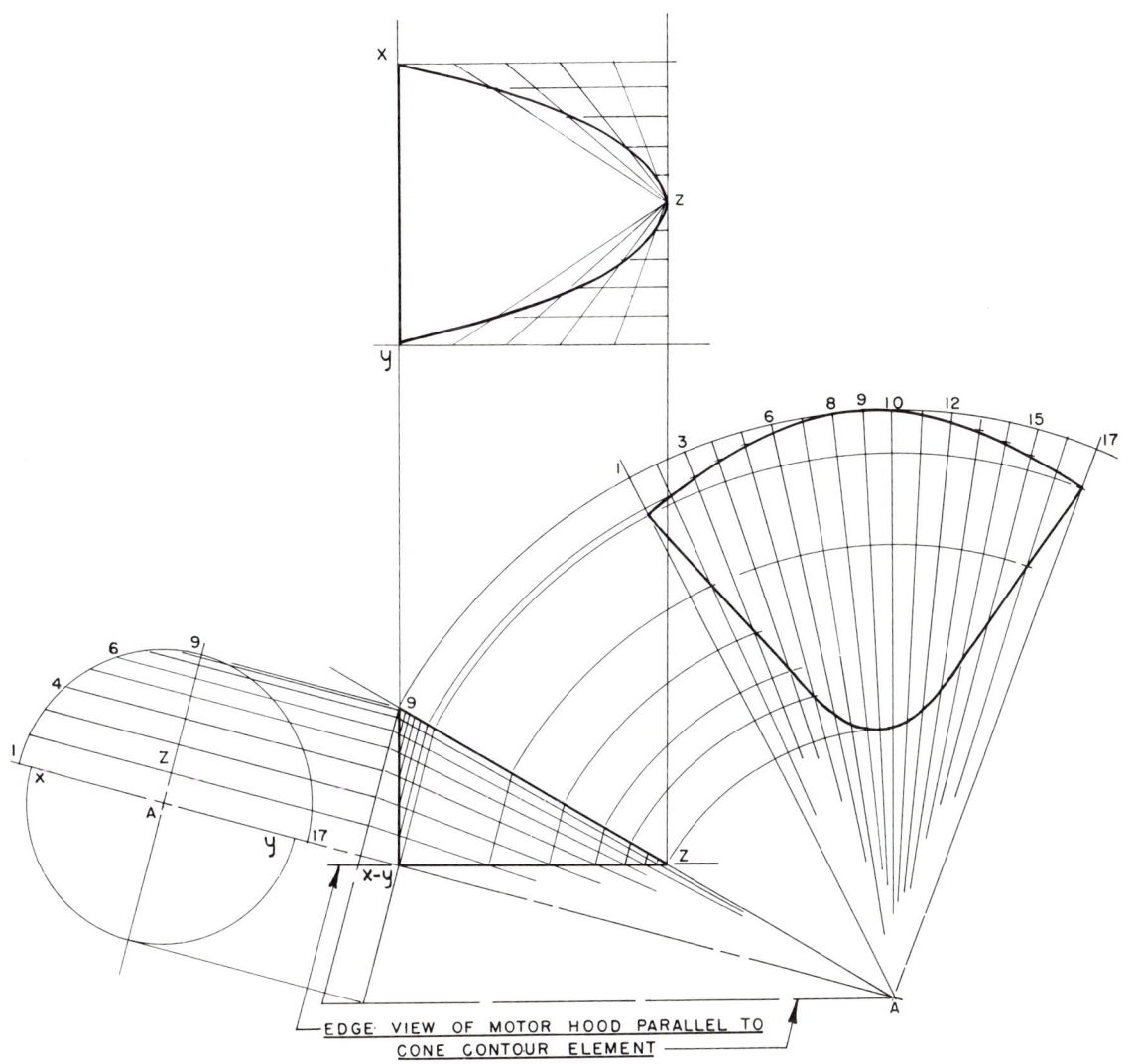

true size and shape of the opening in the motor hood required.

From specifications, draw the front view of the basic right circular cone and "heavy in" that which is to be used as the air scoop. (Notice that the base of the scoop xyz is parallel to a contour element of the cone; hence makes the same angle with cone axis.) Draw an auxiliary view which represents one-half the cone base (all that is needed for the scoop design shown). Divide the half-base into equally spaced points for seventeen elements. Project and draw the elements in the front view, thus establishing piercing points for both curves of the scoop. Rotate the piercing points about the cone axis to establish the true length for each element as shown on the contour element 9-Z. Using the spaces between elements, as taken from the base, draw a radial stretchout as for any right circular cone. Sweep the true-lengths onto the stretchout to establish points for the curves of the air scoop pattern.

The parabolic opening in the hood may be drawn as a partial top view. Two dimensions are required, namely, the altitude of the parabola (as projected from the front view) and the base xy (as transferred from the auxiliary view). The parabola may then be drawn. Either the tangent or the parallelogram method represents a short cut in this problem, because, except for x and y, the plotting of the piercing points may be omitted in the auxiliary view. Note the location of x and y as seen in the auxiliary view. Also note that a right section, taken in the front view so as to contain x and y, is smaller in diameter than the cone base. This diameter is projected to the auxiliary view where x and y are easily located by projection onto the circular view of the section taken.

Because of its reflective characteristics, the parabola is an important curve in sheet metal design. Since the focus is the single point from which sound, light, electromagnetic, or infrared waves may be projected in a parallel manner, or concentrated when being collected by a reflector, the parabola is useful in designing searchlights, radio telescopes, radar, public address systems, auditoriums, band shells, solar collectors, etc.

In most cases the altitude and base (sometimes called rise and span) of the parabola are known. The design may then be completed using one of the methods previously described. Should the focus of a specified parabola be desired (in order to properly locate a microphone, loudspeaker, antenna, filament, collector, etc.) the point can be found as shown in Figure 10-31. (1) Measure

Fig. 10-31 Locating the focal point of a parabola.

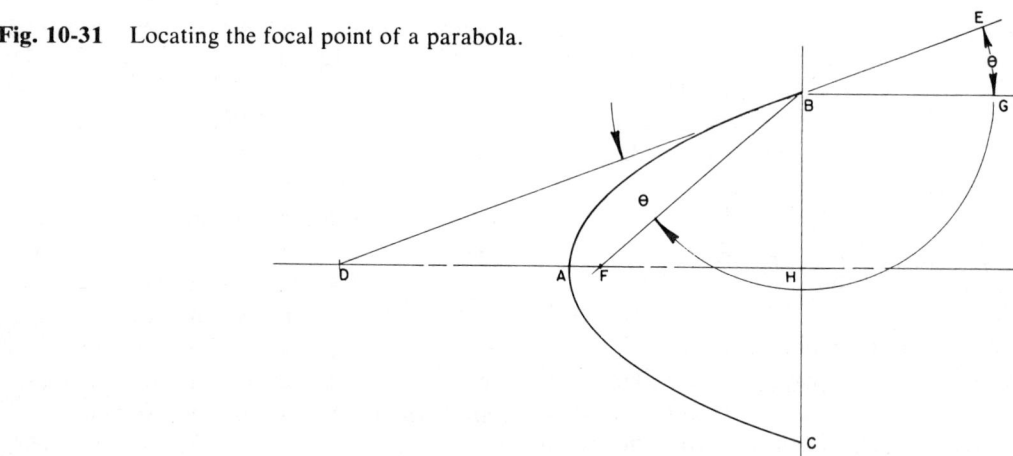

on the axis, behind the parabola, a distance *AD* equal to the altitude *AH*. (2) Draw a line *DE* through a base point *B*. (3) Draw a line *BG* parallel to the parabola axis *DH*. (4) With *B* as vertex, construct angle *DBF* equal to angle *EBG*. (This may be done with compass, as shown.) (5) Point *F* is the focus of the parabola. Other points on the curve could be used in the same manner to further check on the location of *F*.

d. Hyperbola. To generate a hyperbolic curve, cut a right circular cone with a plane parallel to the cone axis, or at a smaller angle to the cone axis than are the cone elements. The latter situation was used in Figure 10-32 to design a hyperbola as a possible cross-sectional shape for an autobody tail fender. (1) The top and front views of a high, slender, right-circular cone are drawn with the diameter of the base *BC* equal to the required span of the hyperbola. (2) An arbitrary cutting plane is taken through *BC* and tilted so as to break through the cone at *A*, the desired rise of the hyperbola. (3) The edge view of the curve plane is divided into seven equal spaces and the points numbered as shown. (4) Pass a series of horizontal cutting planes through the numbered points and project the radii of the circles thus determined to the top view. (5) Project the numbered points to the top view and strike each projector with its related radius. (6) This determines the half span of each projector in the auxiliary view as shown by sample dimension *D*. (7) When the remaining points have been transferred from top view to auxiliary view, the curve may be drawn.

10-9 Cone intersections

When a sheet metal fitting is to be developed, the draftsman must first analyze the fitting in terms of basic geometric shapes and the resulting lines of intersection (section 9-18). The intersection of two cones, as a

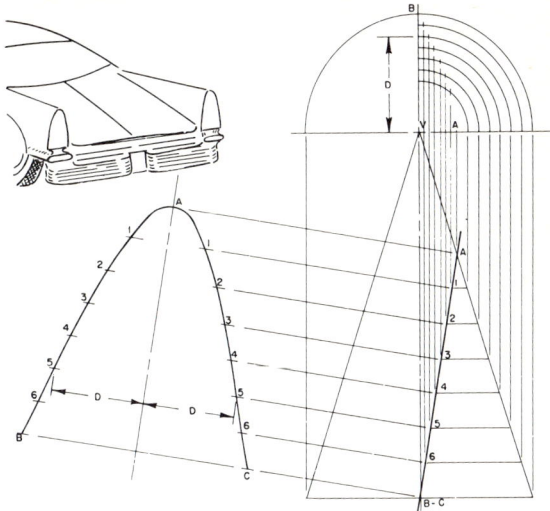

Fig. 10-32 Constructing a hyperbola.

fitting, may be analyzed by considering the following three questions:

 a. Are the cones right circular or oblique?
 b. Do the axes of the cones intersect?
 c. Are the bases in parallel planes?

Only after the above questions have been answered, can the best method for finding the line of intersection be selected.

There are five methods for finding points on the line of intersection of surfaces (section 9-19). Some are best suited to cylinders and prisms, while others are adaptable only to cones and pyramids. The three methods, suitable for *cone-to-cone intersections,* are:

 a. The cutting plane which cuts circular sections from each cone. This method is usually the best for cones whose *circular* bases lie in parallel planes (problems of the type shown in Figure 10-33).

 b. The cutting plane which cuts elements on each cone (problems of the type shown in Figures 10-36 to 10-39, inclusive).

 c. Cutting spheres which cut circular sections from each cone. This method is restricted to right circular cones with intersecting axes (Figures 10-40 and 10-41).

To illustrate the methods, four examples

Fig. 10-33 Cone intersections.

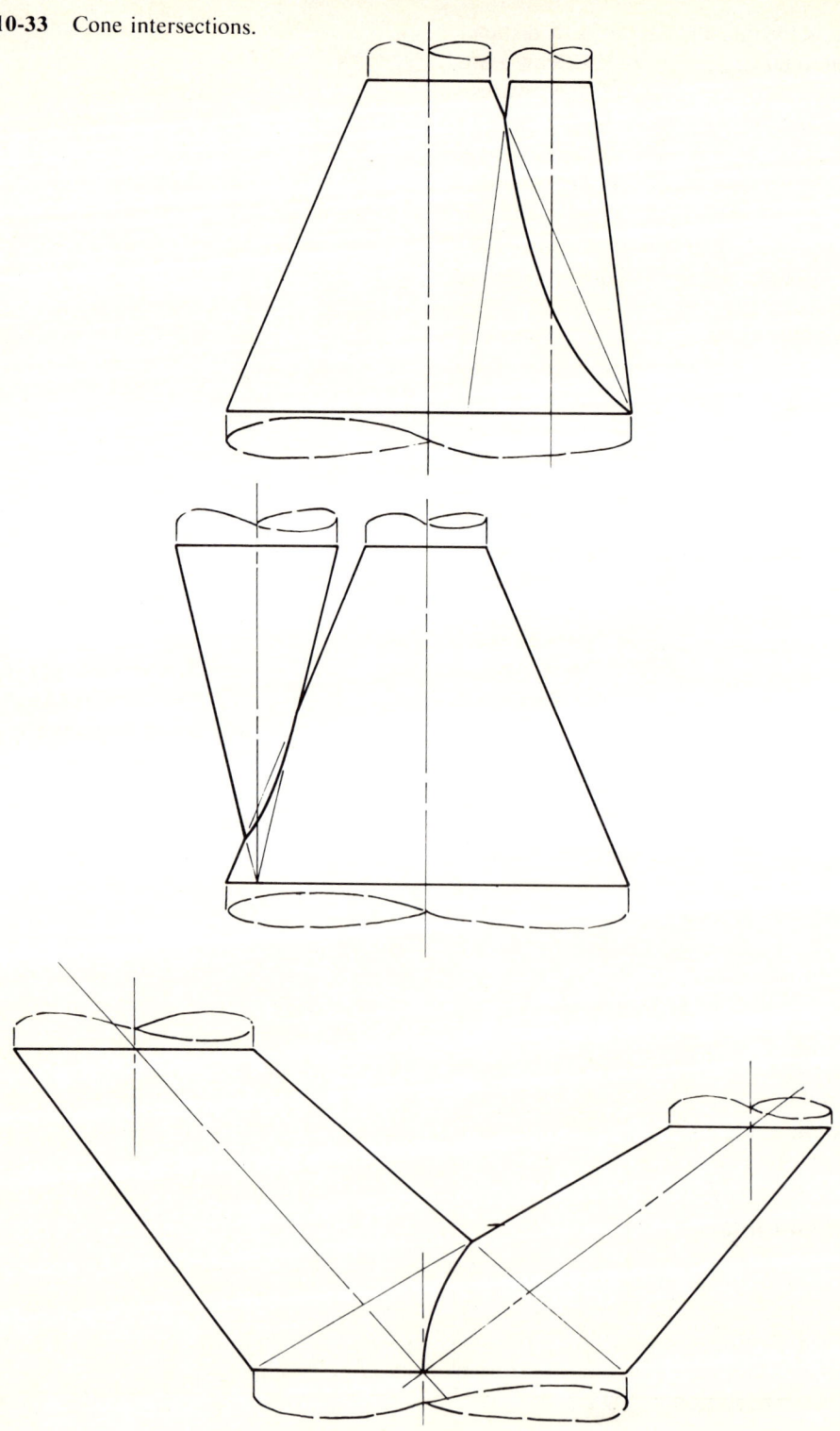

involving the most common applications will be shown. From these examples, the student can prepare to solve a variety of cone-to-cone intersections.

Example 1. If the circular sections cut from two cones by an arbitrary cutting plane touch or overlap each other, the points of contact are points on the line of intersection of the two cones. For example, in order to complete the patterns for the fitting indicated, the line of intersection is desired between the cones in Figure 10-34A. Cutting plane 1 is taken at the highest possible point of intersection; therefore, the resulting circular sections shown in the top view are tangent. Cutting plane 7 is taken at the lowest possible point of intersection. Notice that again the resulting circular sections, shown in the top view, are tangent. Figure 10-34B shows the additional cutting planes 2 through 6 taken uniformly in the front view. The resulting circular sections are identified in the top view. Notice that the circular sections for any one cutting plane overlap; therefore, the points of contact are points on the line of intersection. They are projected to the front view to their respective planes to complete the curve.

After the true intersection has been established, the patterns for each cone are developed, as shown in Figure 10-35. Develop the pattern for the main cone frustum using any of the methods previously described in sections 10-2 to 10-4. For the hole, draw a center line midway between the ends of the pattern. Strike arcs which repre-

Fig. 10-34

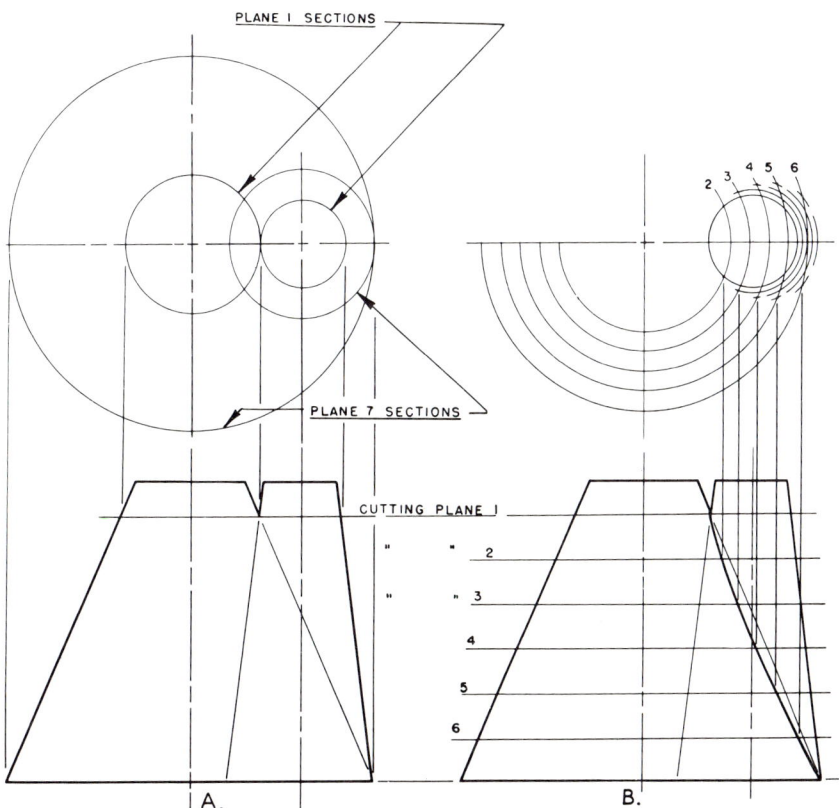

SHEET METAL DRAFTING

Fig. 10-35

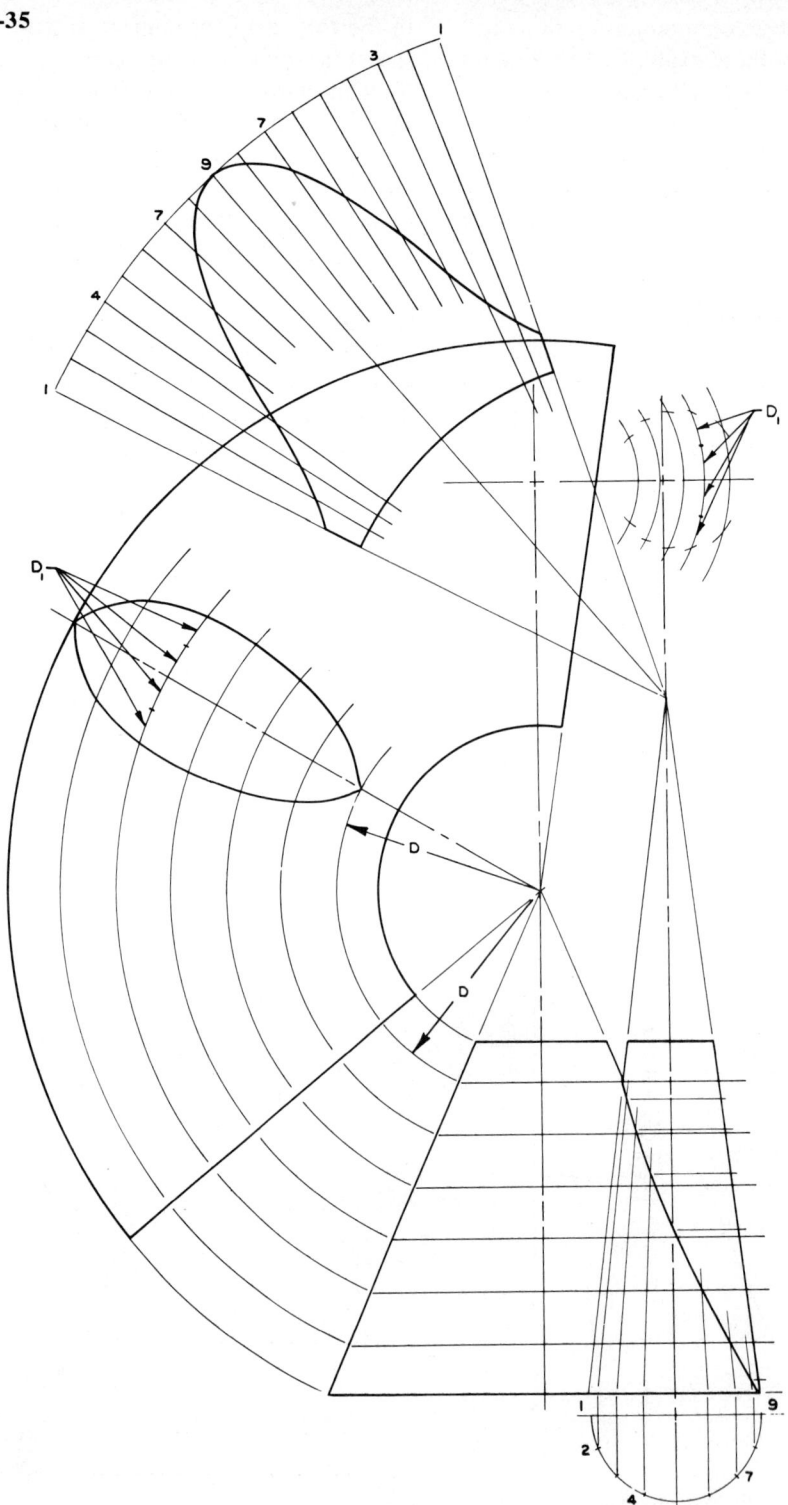

sent the slant height of the cutting planes (hence the true slant height of the points on the curve) as shown by the sample dimension D in Figure 10-35. By stepping off with dividers, measure the *curved* distance each side of the center for each set of points in the top view. Transfer this *curved* distance by stepping off the same measurement on the respective arcs in the pattern. See sample dimension D_1. When the points for all cutting planes have been transferred, the curve for the shape of the hole is completed.

To develop the pattern for the branch cone, draw a half-bottom view of the base as shown. Divide the base into a suitable number of points for elements (1 through 9 and back to 1 in the example). Project the elements to the front view and draw them piercing the curve. The foreshortened elements and their piercing points are rotated about the cone axis until they appear true length in the front view. Using the slant height of the full cone, strike an arc and step off on this arc the distances 1 to 2, 2 to 3, 3 to 4, etc. Beginning with the shortest element, transfer the true lengths from the front view to their respective elements on the pattern. Points for the pattern curve are thus determined and the curve may be drawn.

Example 2. The cutting-plane method is shown pictorially in Figure 10-36. *In order that a cutting plane will cut elements on two given cones, the plane must pass through both cone apexes* (section 10-8, Figure 10-27); therefore, to establish the cutting planes, the first line to be drawn is a line through both apexes (A and A_1 of Figure 10-36). Line AA_1 pierces the horizontal base plane at point H. The cutting plane also produces a *trace line HB* on the horizontal base. The trace line cuts the base circle at points 1 and 2; therefore, elements on each cone from 1 and 2 intersect at point 3—one point on the line of intersection. In Figure 10-37, the top and front views of the construction are shown. To further illustrate the location of

Fig. 10-36 Cutting plane passing through the apexes of two cones.

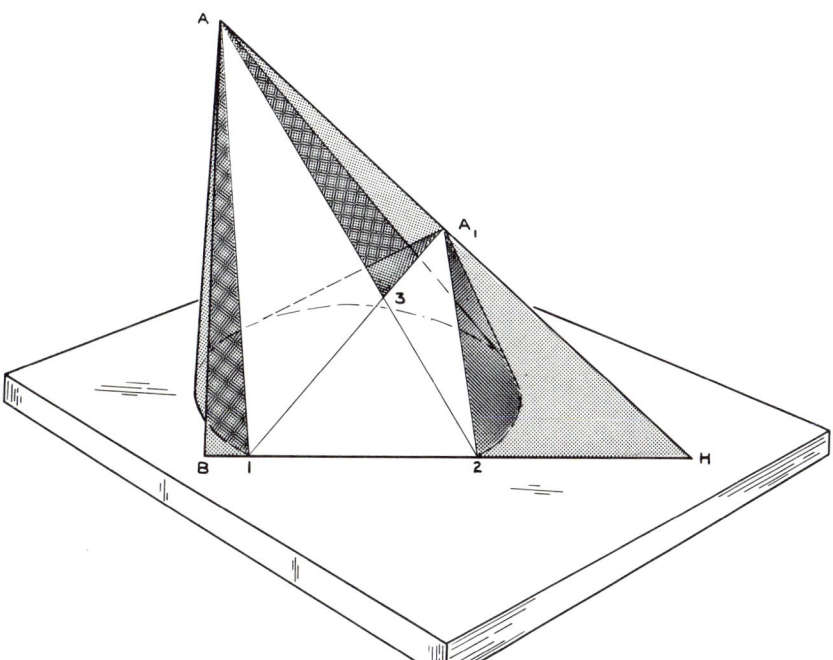

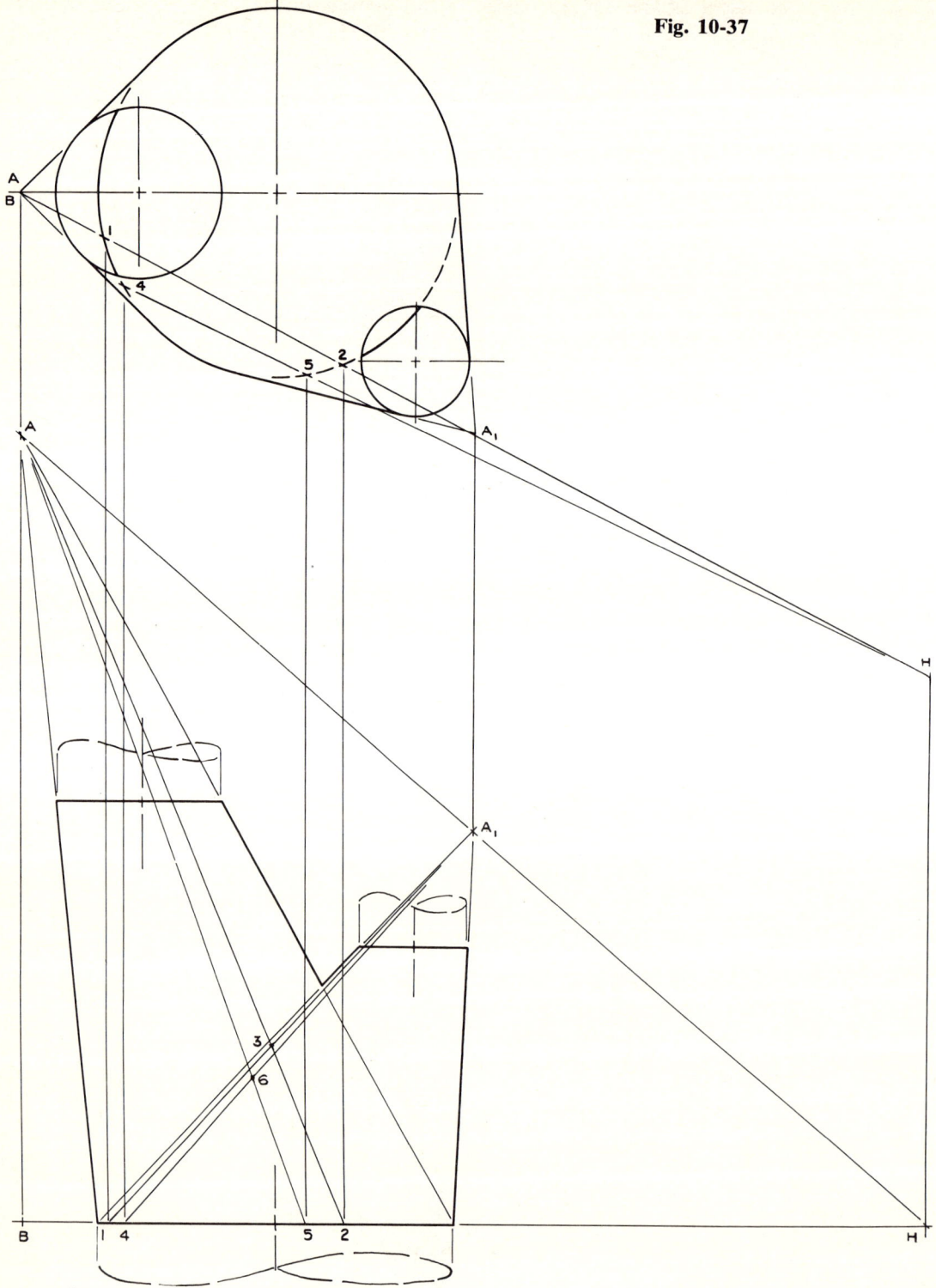

Fig. 10-37

RADIAL-LINE DEVELOPMENT

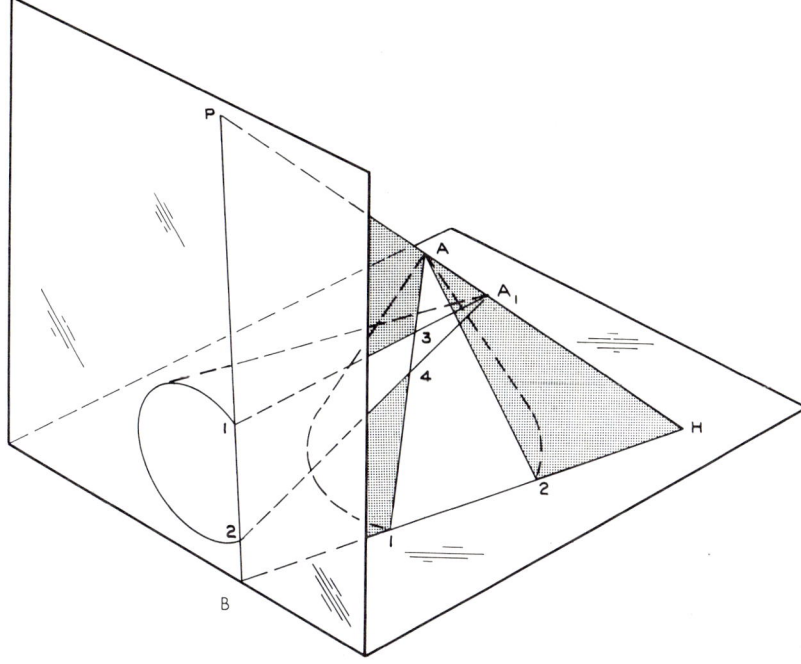

Fig. 10-38

elements and their intersection as a point, an additional plane with its trace line 4-5-H is drawn and the resulting elements intersect at point 6. In the same manner continue cutting the base with plane traces, using line AA_1 as a hinge line, to locate a sufficient number of points on the curve. After the line of intersection is found, the development of each cone may proceed as for previous similar shapes.

Example 3. The cutting-plane method is used to find the line of intersection between two right circular cones whose axes are nonparallel and nonintersecting, as shown pictorially in Figure 10-38. Again, as in the previous problem, the hinge line is a line through both apexes (A and A_1). Line AA_1 pierces the horizontal plane of one cone base at point H and pierces the profile plane of the other cone base at point P. Any cutting plane containing line AA_1 passes through points H and P and produces horizontal and profile traces from those points.

The traces join at the H and P plane intersection. (For example, see point B of Figure 10-38.) The plane PBH cuts the cone bases at points 1 and 2 (Figure 10-38). The elements from points 1 and 2 intersect in points 3 and 4 on the desired line of intersection.

In Figure 10-39, the front view of the two cones is drawn; then partial top and side views showing only the base circles and center lines for each cone are constructed. Draw the line AA_1 in the top and front views to locate points P and H. Include in the side view, by projection and reference, the piercing point P. In the top view a trace line, *drawn arbitrarily* from point H, intersects the profile base plane at B (B is transferred to the side view as shown by sample dimension D_1); therefore, the cutting plane PBH results in elements 1 and 2 on each cone. The intersecting elements locate points 3 and 4 on the desired curve. Another cutting plane PHC results in elements 5 and 6 on each cone; they, in turn, intersect and locate points

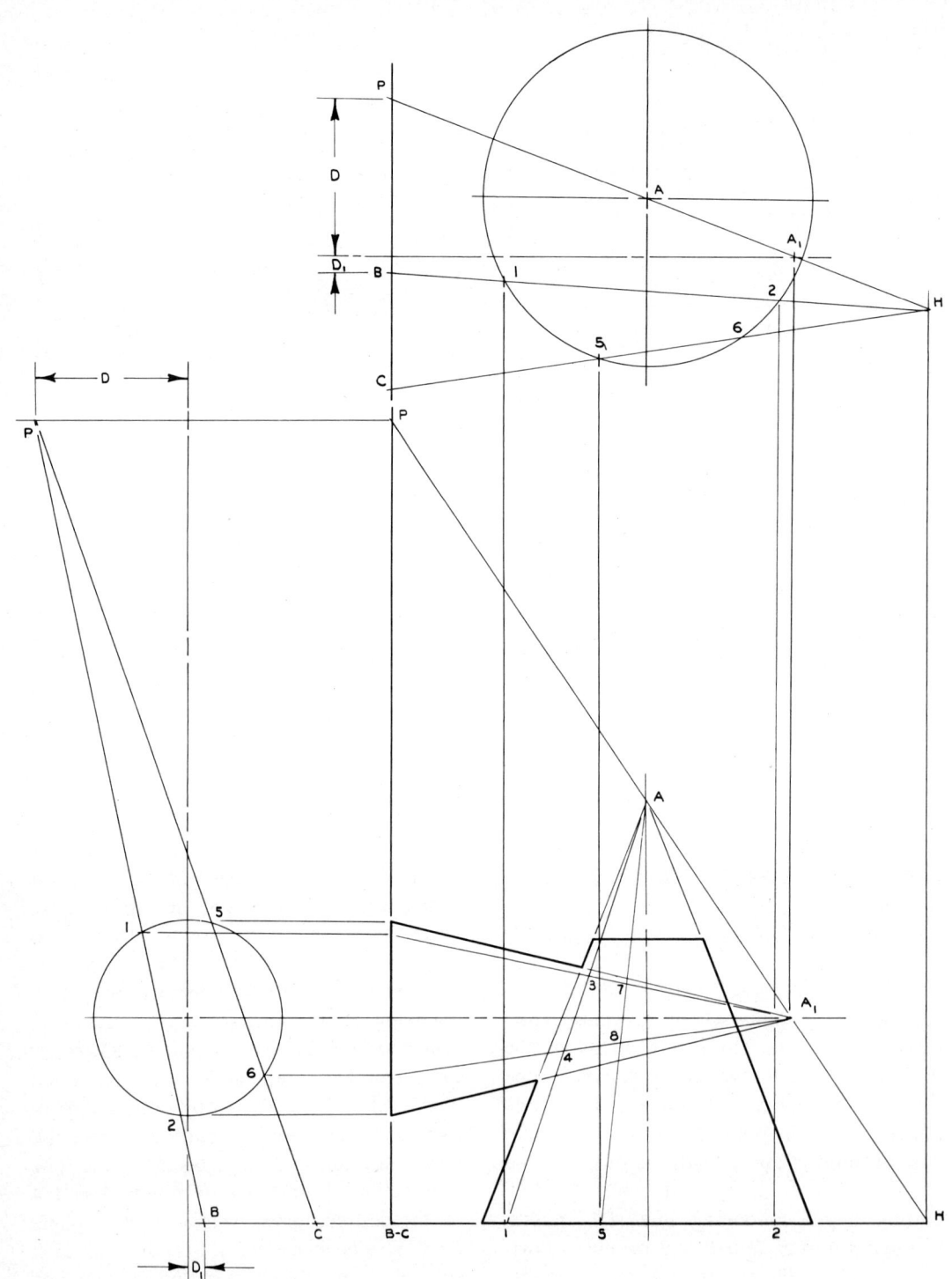

Fig. 10-39

7 and 8 on the desired curve. In the same manner, continue cutting the bases until a sufficient number of points for the curve have been secured.

Example 4. As stated previously, the cutting-sphere method is limited to right circular cones whose axes intersect. In Figure 10-40, a right circular cone is cut by a sphere and the line of intersection is a circle. *When two cones, as shown, are cut by the sphere, a circle on each cone results.* If the circles from each cone intersect, the point of intersection is a point on the line of intersection between the cones.

To make the drawing of the fitting, draw a front view of the two cones so that the axes are true length as shown in Figure 10-41. Pass cutting sphere 1 through both cones. This results in a line perpendicular to the axis of each cone. The line represents the edge view of the circular intersection of sphere and cone. If the edge views intersect as shown by the circled point in Figure 10-41, the intersection is one point on the line of intersection of the two cones. Cutting spheres 2 and 3 illustrate the finding of additional points on the line of intersection.

After the line of intersection has been determined, the pattern for each cone and the opening in the main cone pattern may be developed in the same manner as explained for example 1 (Figure 10-35).

10-10 Approximation of double-curved surfaces

The most common double-curved surfaces for which occasionally a pattern is needed are the sphere and paraboloid illustrated in Figure 10-42. Because the sphere and paraboloid are double-curved surfaces, they cannot, theoretically speaking, be developed. However, for all practical purposes, a very close approximation may be made by dividing the surface into areas and assuming each area to be a part of a cylinder, or of a cone.

The "polycylindric" or "gore" method is illustrated in Figure 10-43. Two views of the required hemisphere are drawn. The bottom view is divided into a number of sections or

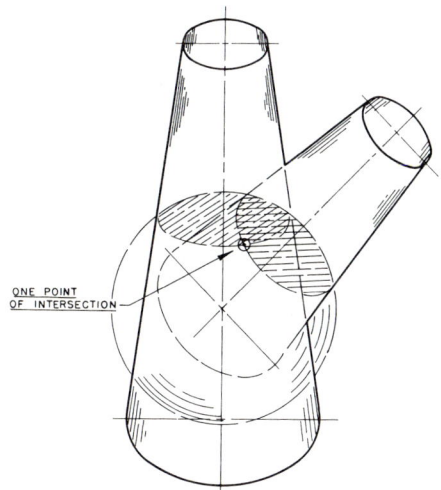

Fig. 10-40 The cutting-sphere method.

Fig. 10-41

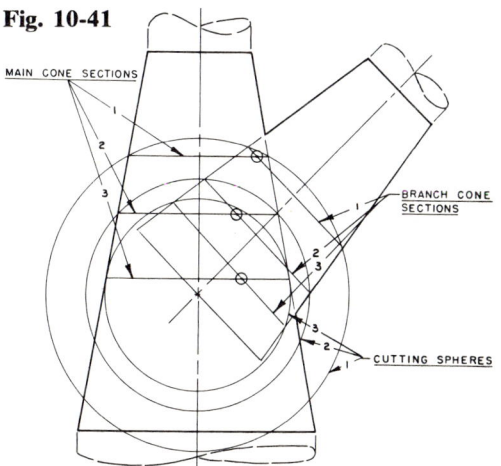

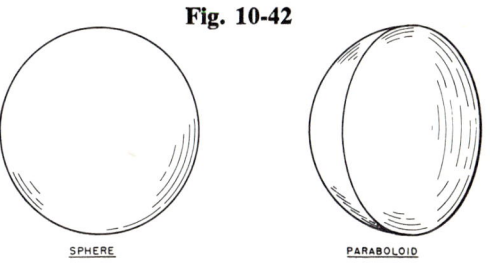

Fig. 10-42

Fig. 10-43 Development of a sphere by the polycylindric method (*Reynolds Metals Co., Louisville, Ky.*)

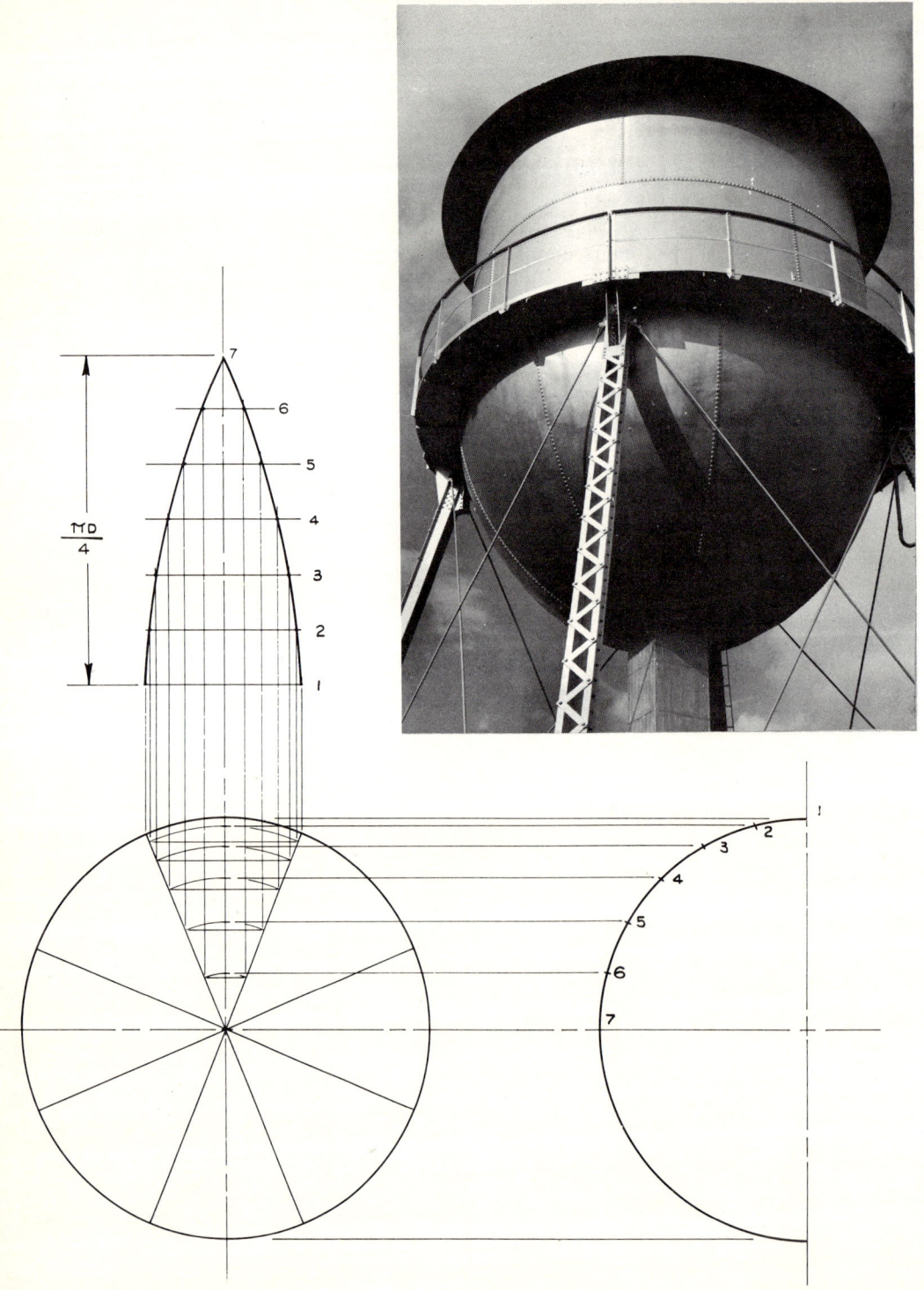

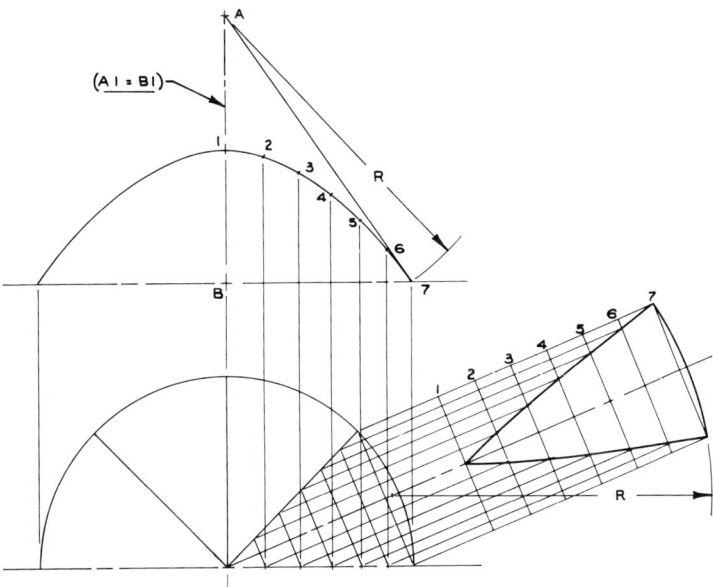

Fig. 10-44 Development of a parabolic dome.

gores (8 in Figure 10-43). The front (contour) view of one gore is uniformly divided into a sufficient number of points to represent elements, *as if for a cylinder* (1 through 7 in Figure 10-43). The elements, as points in the front view, are projected to the bottom view. Notice that, in the bottom view, the elements are positioned so that *only the ends* are actually on the surface of the original hemisphere. In effect, the octagonal dome is *enclosed* within the spherical dome. If the element ends were not rotated but remained with their projectors in the bottom view, *only the center* of the elements would actually be in contact with the original hemisphere and the spherical dome would be enclosed within the octagonal dome.

The length of the gore is computed and is shown to be equal to $\frac{\pi D}{4}$. The elements on the stretchout may be spaced by using the distances 1-2, 2-3, 3-4, etc., from the front view or the parallel-line method described in section 5-4. The true length of each element is projected to the stretchout from the bottom view as shown to complete the pattern for one sector of the hemisphere. If a complete sphere is desired, the length of the gore is equal to $\pi D/2$ and the spacing of the elements in the front view will encompass a half-cylinder. The same method, applied to a parabolic dome, is shown in Figure 10-44.

The "polyconic" or "zone" method is illustrated in Figure 10-45. As the names suggest, the method involves cones and zones. The sphere is divided into a suitable number of horizontal zones and each zone is developed as a cone frustum. In Figure 10-45A, the circle of the sphere has been divided into twelve equal parts, thereby establishing six zones whose slanted sides are equal.*

---

* Six zones were selected for the illustration in order to clarify construction. In a practical problem, however, eight or ten zones are needed to give a good spherical approximation. Also note that the polygon is within the circle of the sphere. This results in a development slightly smaller than the sphere; however, the metal sheet or plate is usually stretched or hammered to a spherical curvature, thereby compensating for the original undersize approximation.

Fig. 10-45 Development of a sphere by the polyconic method.

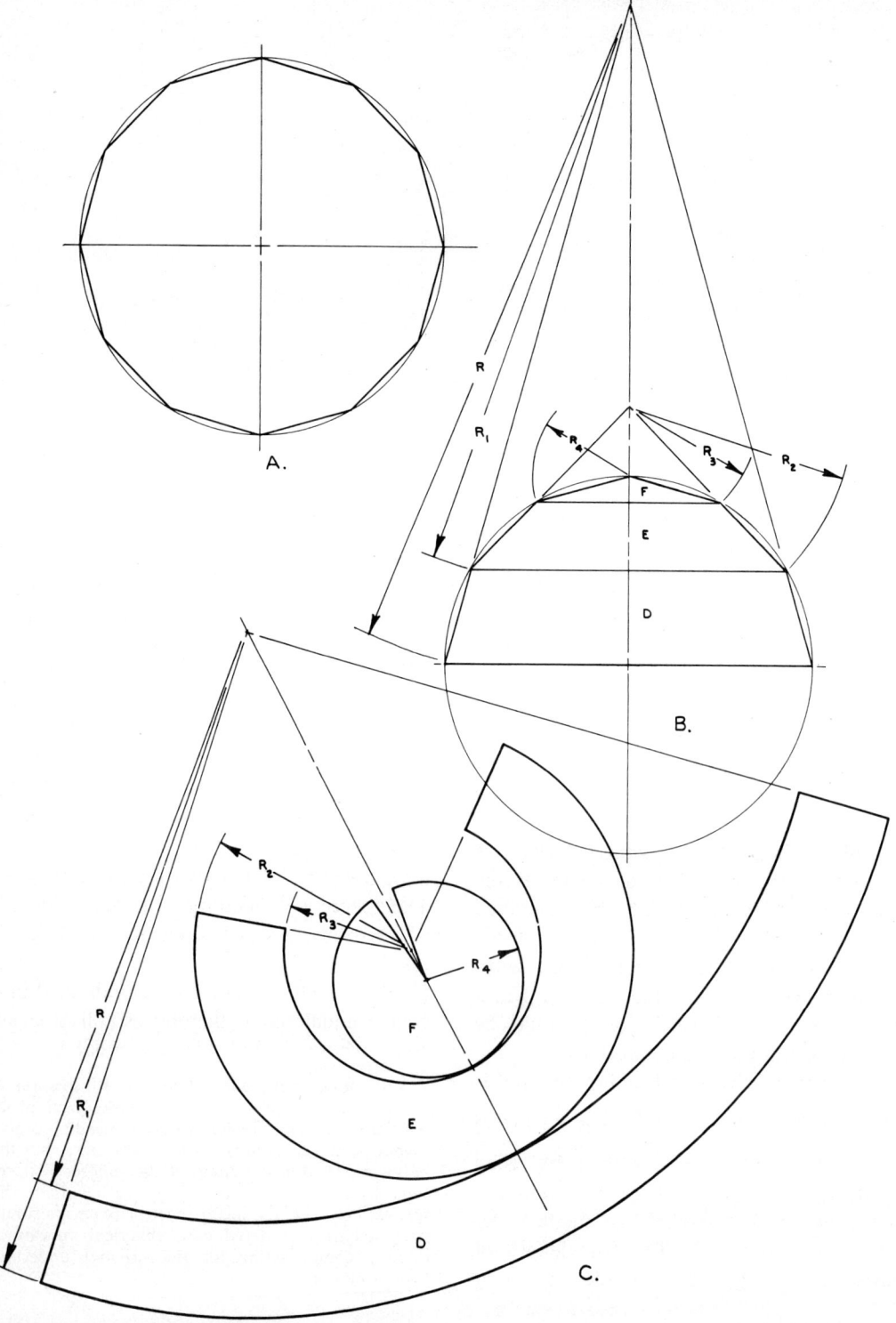

RADIAL-LINE DEVELOPMENT

Figure 10-45B shows the cones and their respective slant heights constructed for zones D, E, and F. The pattern for each zone is developed using the methods previously described as for any right circular cone or cone frustum. The development of each zone is shown in Figure 10-45C.

The method whereby the angular sweep of the pattern is determined from the cosine of the angle between base and elements as described in section 10-2 may again be used to rapidly determine the sweep of the pattern for the zones of a sphere. The computations remain fixed for *all spheres of a given number of zones, regardless of the sphere size.* For example, the six zones in Figure 10-45B concern three cones whose angles between base and elements θ are 75, 45, and 15 degrees; therefore, the sweep values are computed as 93, 254, and 347 degrees. The radii of the sweeps are as shown on the drawing (R values). The larger the sphere, the longer the R values.

An eight-zone sphere concerns four cones whose base angles are 78.75, 56.25, 33.75, and 11.25 degrees. The sweep values are, respectively: 70, 200, 299, and 353 degrees.

A ten-zone sphere concerns five cones whose angles are 81, 63, 45, 27, and 9 degrees. The corresponding sweep angles are 56, 163, 254, 321, and 355 degrees.

To use the preceding information as shown for a six-zone sphere in Figure 10-46, it is only necessary to draw the desired sphere as a circle. Divide a quarter of the circle into "flats" for three zones. Extend the flats to intersect the vertical center line, thus establishing the R values. The R values are then used to sweep the pattern the specified angle.

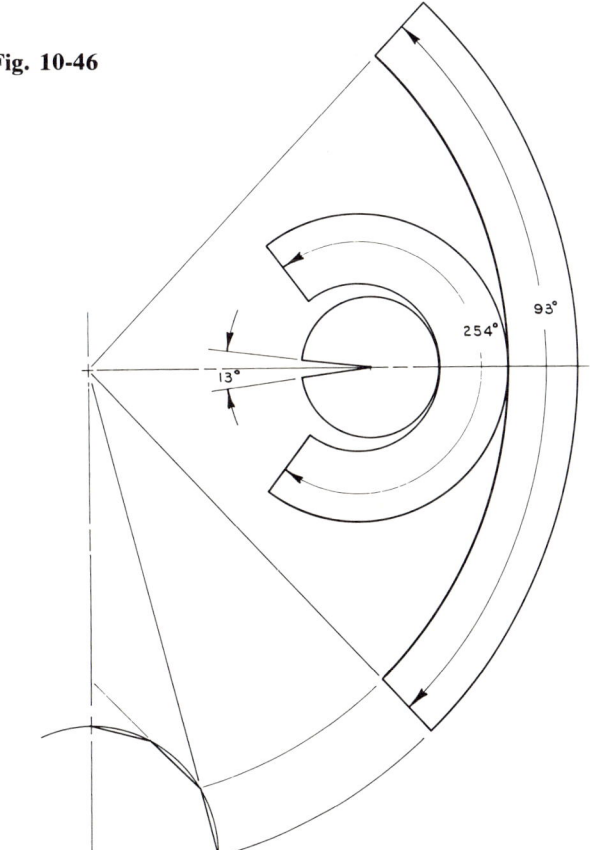

Fig. 10-46

178 SHEET METAL DRAFTING

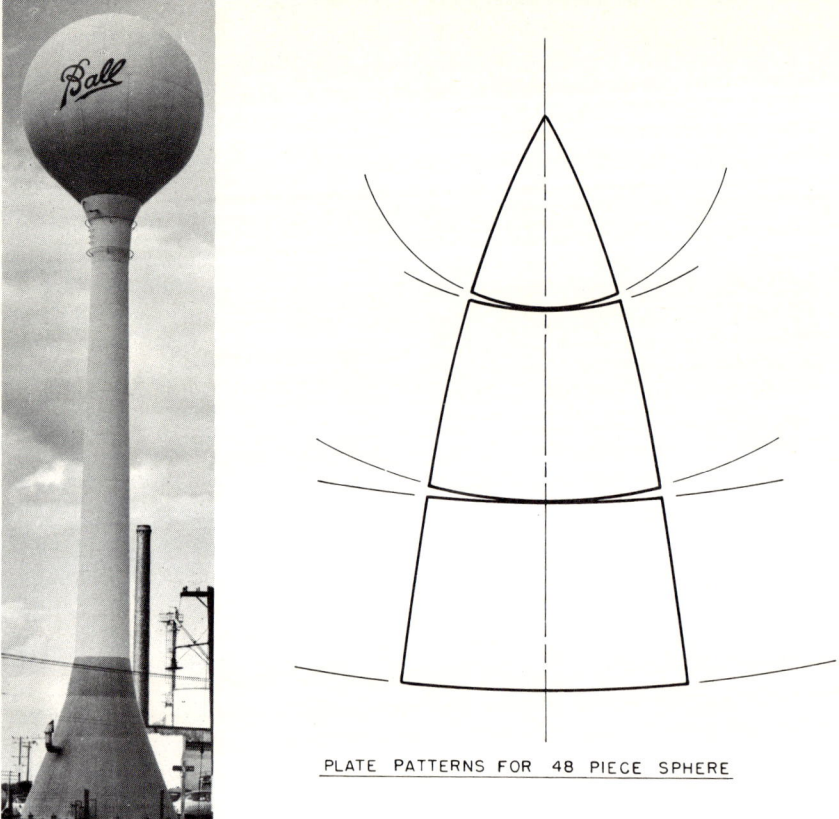

PLATE PATTERNS FOR 48 PIECE SPHERE

Fig. 10-47 Plate patterns for water sphere. (*Ball Bros., Hillsboro, Ill.*)

As the *gore* method is a cylindrical approximation of a sphere, and the *zone* method a conical approximation, the two methods may be combined, as shown in Figure 10-47, to provide patterns for heavy sheet or plate that are to be given a double curvature by hammering, forging, or stretching. The plates are then fabricated by welding or riveting to form a sphere, as shown in the figure.

THE PYRAMID

10-11 The pyramid compared to a cone

Cones and pyramids are very closely related geometric shapes. A pyramid may be considered as "a cone of a limited number of sides," or a cone can be thought of as "a pyramid of infinite sides." Indeed, large conical shapes in heavier metal are often formed by braking as for a many-sided pyramid (see Chapter 7, problem 9). Thus pyramids and cones have similar characteristics, and when developing patterns for them, similar requirements. They also have *differences*

RADIAL-LINE DEVELOPMENT

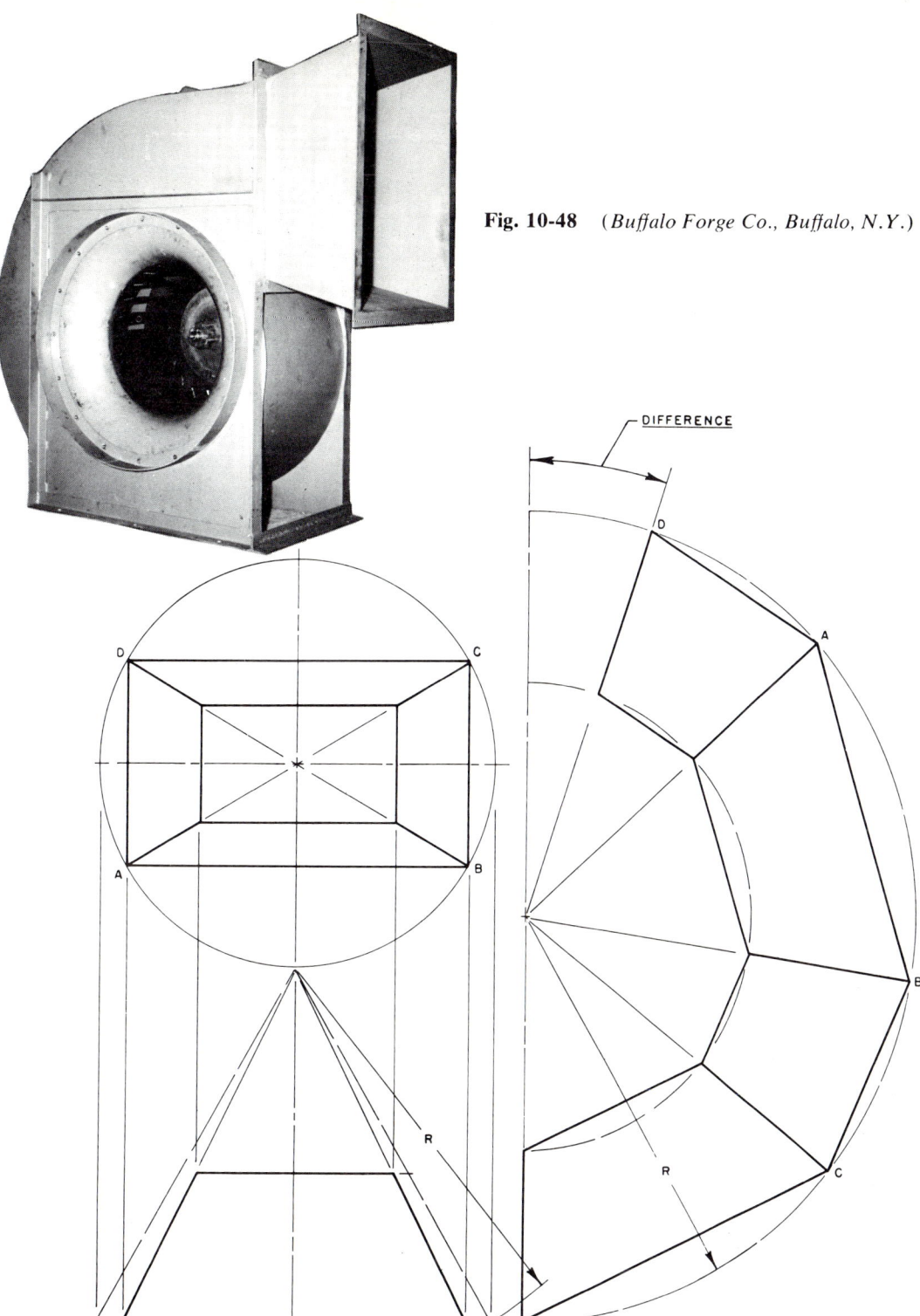

Fig. 10-48 (*Buffalo Forge Co., Buffalo, N.Y.*)

which must be recognized in order to avoid mistakes in development. The base of a cone is a circle (or an ellipse); the base of a pyramid is usually a regular polygon. Development for both is similar, in that a front and top view completely describe the shape. In addition, the front view furnishes the true length of elements or edges which are parallel to the front plane of projection. The corners of a pyramid must therefore be parallel to, or rotated until parallel to, the front projection plane V in order to establish true length. The top view shows the true shape of the base, the true circumference of a cone, and the true perimeter or girth of a pyramid.

Thus, one primary difference, as regards development, is the difference that exists between the *circumference* and the *girth* of a base. The student must be aware of the resulting difference in pattern sweep for a pyramid as compared to the pattern sweep of a cone which envelops the pyramid. This difference is illustrated in Figure 10-48. Increasing the number of sides of the pyramid would tend to increase the sweep angle on the pattern, when compared to the cone envelope pattern. For example, in Figure 10-49 the pattern for the cone of Figure 10-48 is compared with the pattern for an equivalent right decagonal pyramid. Notice that the

Fig. 10-49

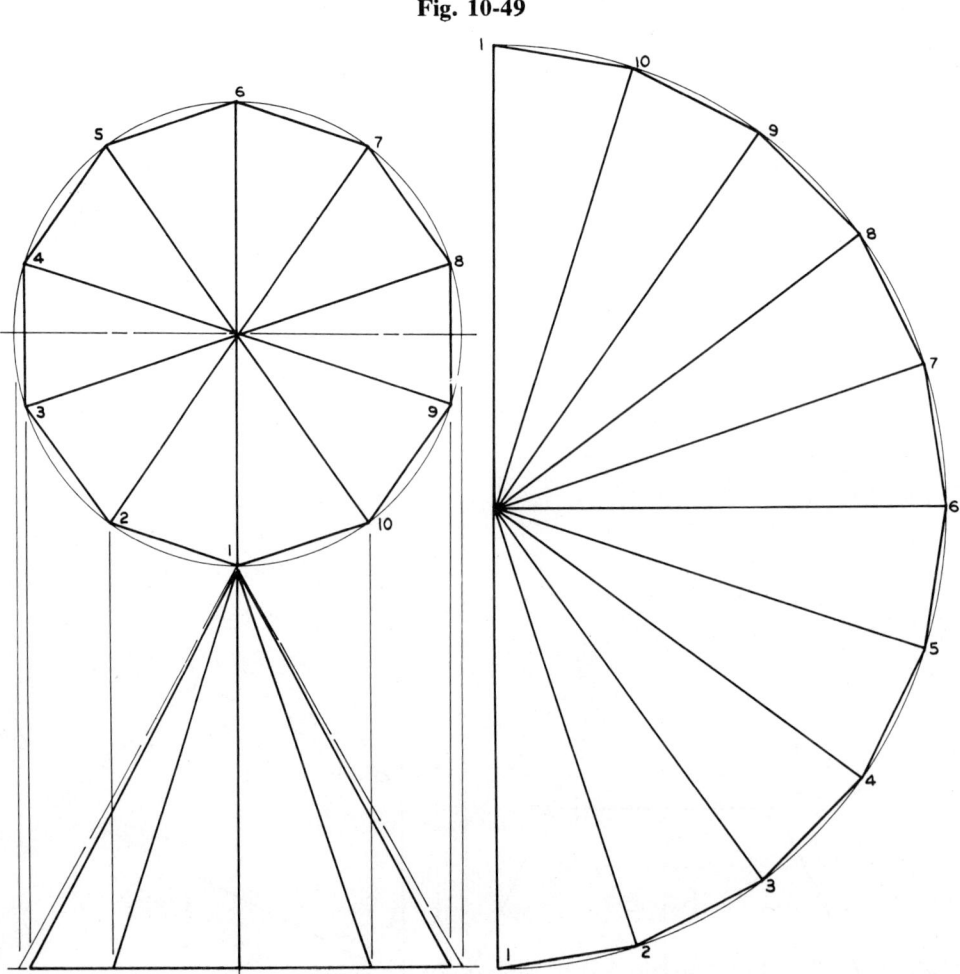

pyramid, due to the increase in number of sides, now encompasses approximately the same angular sweep as the cone envelope.

In Figure 10-48, the blower fitting is drawn as a true pyramid. Thus the small and large openings are both rectangles of similar proportions. In some installations, however, the blower fitting may serve as a *tapered transition* in which the intake and exhaust openings are still rectangles but of different proportions. Figure 10-50 shows top and front views drawn to meet the new requirements. Notice that the sides are no longer a part of a single pyramid, but are represented by apexes 1, 2, 3, and 4 at two different elevations. The identical sides *D-2-C* and *A-4-B* represent part of one pyramid. The sides *D-3-A* and *C-1-B* represent part of a second pyramid. In Figure 10-50, the required pattern is contrasted with an equivalent cone pattern in order to illustrate errors that might occur were the designer unaware of these differences. The pattern for a tapered transition therefore would best be developed by means of auxiliary views (Chapter 7), or triangulation (Chapter 11).

10-12 Truncated right regular pyramid

The hopper, shown in Figure 10-51, represents a right hexagonal pyramid truncated by an inclined plane. The problem is similar to the cone problem of Figure 10-16 except

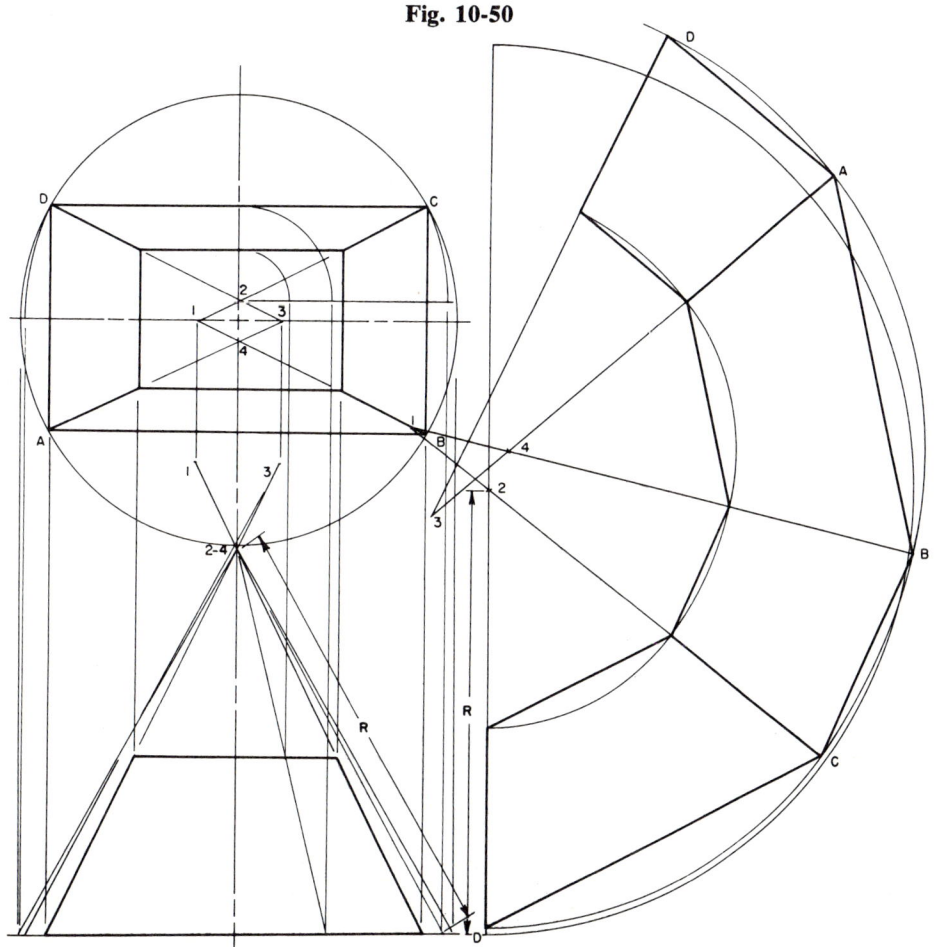

Fig. 10-50

Fig. 10-51

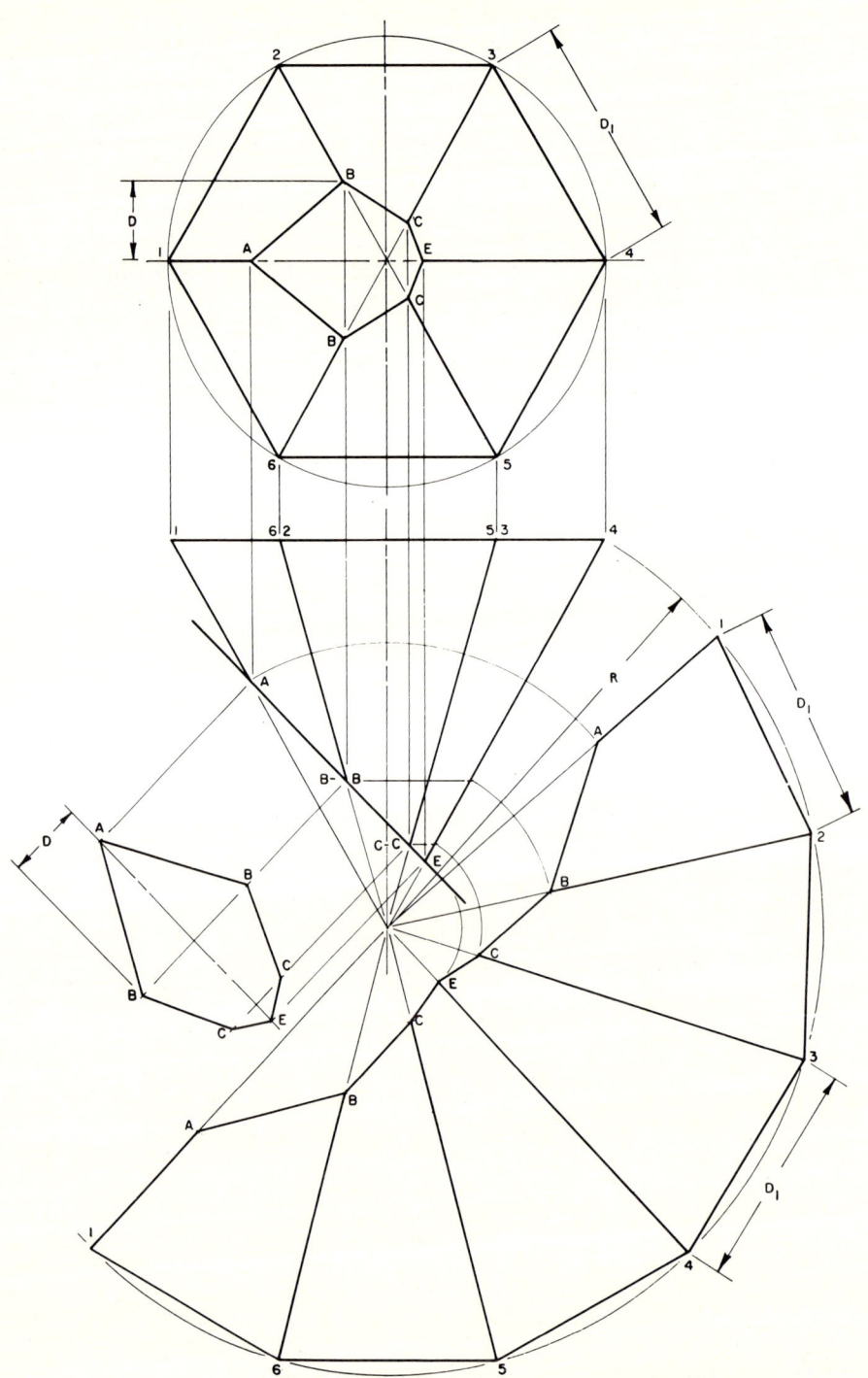

RADIAL-LINE DEVELOPMENT

Fig. 10-52 Development of an oblique pyramid.

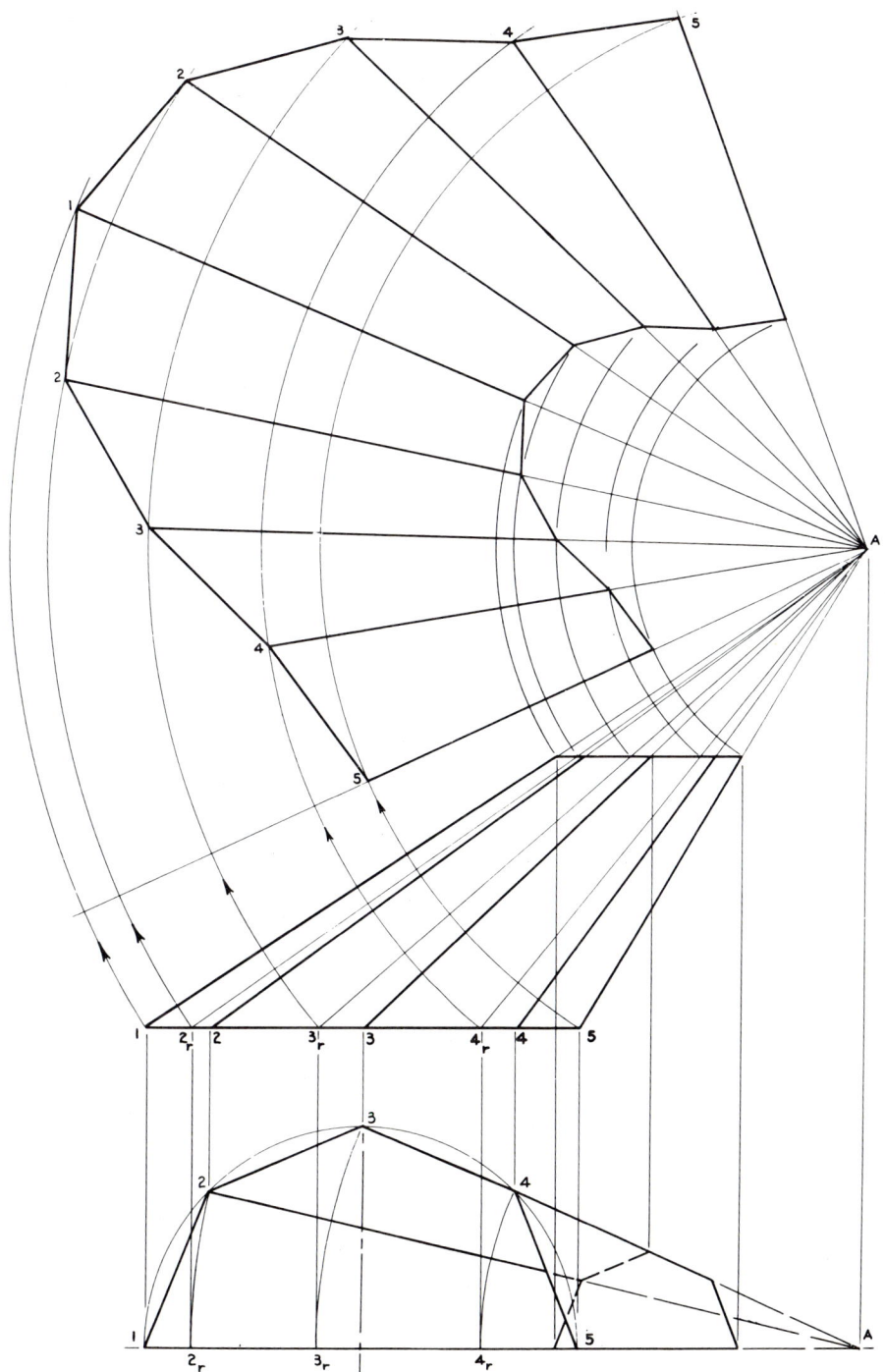

for the differences between cones and pyramids previously mentioned. To determine the true size and shape of the plane opening and to develop a pattern for the hopper: (1) Draw the top and front views of the hexagonal pyramid and the front view of the plane as an edge. The front view thereby locates the points at which pyramid edges 1, 2, 3, 4, 5, and 6 pierce the plane (the piercing points are labeled A, B, C, and E). (2) Project the piercing points to their corresponding edges in the top view. (3) By projection and reference as shown by the sample dimension D, locate the points in the auxiliary view. (4) Connect the points in the auxiliary view to complete the true size and shape of the plane opening. (5) Since the radius R is equal to the true slant height of one corner of the pyramid (corner 1 or 4 in the front view), the hopper is developed in a manner similar to a right circular cone. (6) The cord D_1, from the top view, is used to strike off the six base dimensions, thereby locating the corners on the pattern and establishing the total sweep of the stretchout. Notice that the truncated corners 1-A and 4-E are the only corners true length in the front view. The truncated corners 2-B, 3-C, 5-C, and 6-B must be rotated in order to determine true length for transfer to the pattern.

10-13 Oblique pyramid with regular base

The *octagonal offset reducer* shown in Figure 10-52 is developed in a manner similar to the conical offset reducer of Figure 10-19. (1) A front and partial bottom view are drawn of the required oblique pyramid. (2) Number the corners of the octagon (1 to 5 and back to 1). (3) Five different true lengths from apex to base are required for the eight corners of the pyramid. Corners 1 and 5 are true length in the front view; corners 2, 3, and 4 are rotated until they appear true length in the front view at points $2r$, $3r$, and $4r$. (4) Using the apex of the front view as a center, sweep the true lengths into the development area, as shown by the arrows in Figure 10-52. (5) To establish the points on the pattern, set the compass equal to the distance between corners (as taken from the partial bottom view). Starting with corner 5 (the shortest one), strike the arc to cross the sweep arc from 4_r. This locates point 4 on the pattern. Using 4 as a center, strike the arc to cross the sweep from $3r$. Repeat the procedure until all eight corners have been located. (6) Draw the corners or brake lines on the pattern and terminate them for the small opening when they intersect their respective sweep arcs.

10-14 Low-altitude pyramid

Low-altitude pyramids serve as designs for sheet metal covers, ventilating hoods, hoppers, etc. The rectangular cover in Figure 10-53 is an example of the development of this sort of fitting.

To develop a pattern: Draw the top and front views of the required cover. (Place the seam AC on a flat surface where it will be as short as possible.) The lines *not* true length in the top or front views are the pyramid corners AB, AD, AE, AF, and the seam AC. Rotate AC about a vertical axis until it is true length in the front view AC_r. Rotate a typical edge AB until it is true length AB_r. Select a center point above the top view and use AC_r and AB_r as radii. Draw the construction arcs shown. Project the true length of the base line FE to the construction arcs and transfer, by compass, the true lengths FB, DE, BC, and CD. Connect points to complete the pattern. Notice that the side FE is tangent to arc R' and that the opposite side BD is found by striking BC and DC to intersect the arc of R'; in this manner, the amount of cutout CC is established on the pattern.

RADIAL-LINE DEVELOPMENT

10-15 Pyramidal intersections

Pyramids, when used in combinations to form sheet metal fittings, intersect in straight lines in all views. Since pyramids contain a number of oblique surfaces, it would be impossible to secure a single view in which all the surfaces of one pyramid would appear edgewise. Hence, the best method for establishing the points where the edges of one pyramid pierce the surfaces of another is the *cutting-plane method*. As an example, study the plane and line situation illustrated in Figure 10-54. The point at which line 5-4 pierces the plane 11-12-8-9 is desired. A cutting plane passed vertically through the line 5-4 in the top view will cut a "trace" on the plane 11-12-8-9. The trace (identified as *xy*), when projected to the front view of 11-12-8-9, crosses 5-4 at point *C*; therefore, *C* is the point at which line 5-4 pierces the 11-12-8-9 plane. To apply the cutting-plane method to a sheet metal problem, study the ventilating hood of Figure 10-55 in which a

Fig. 10-53 Development of rectangular cover.

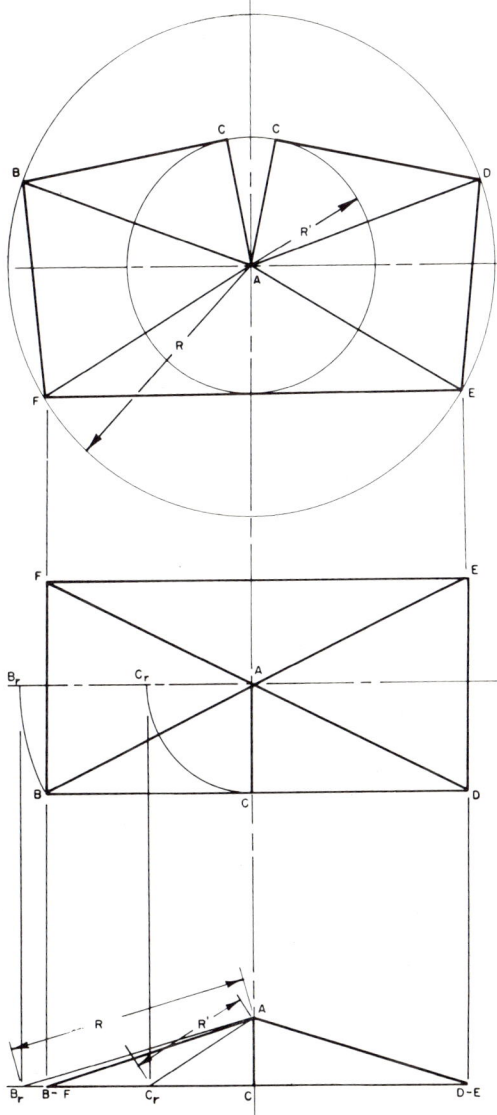

Fig. 10-54 Finding a piercing point by means of a cutting plane.

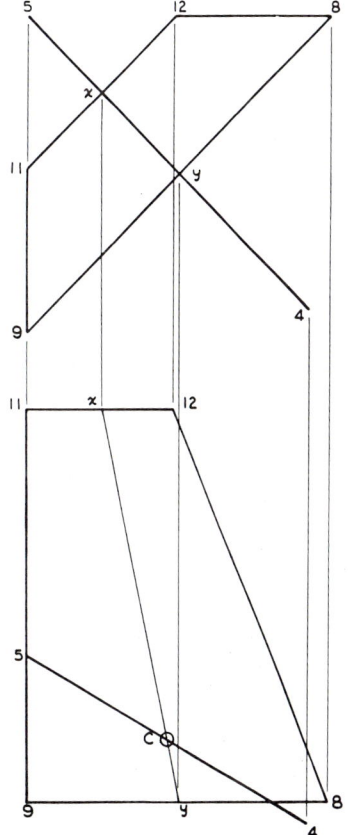

186   SHEET METAL DRAFTING

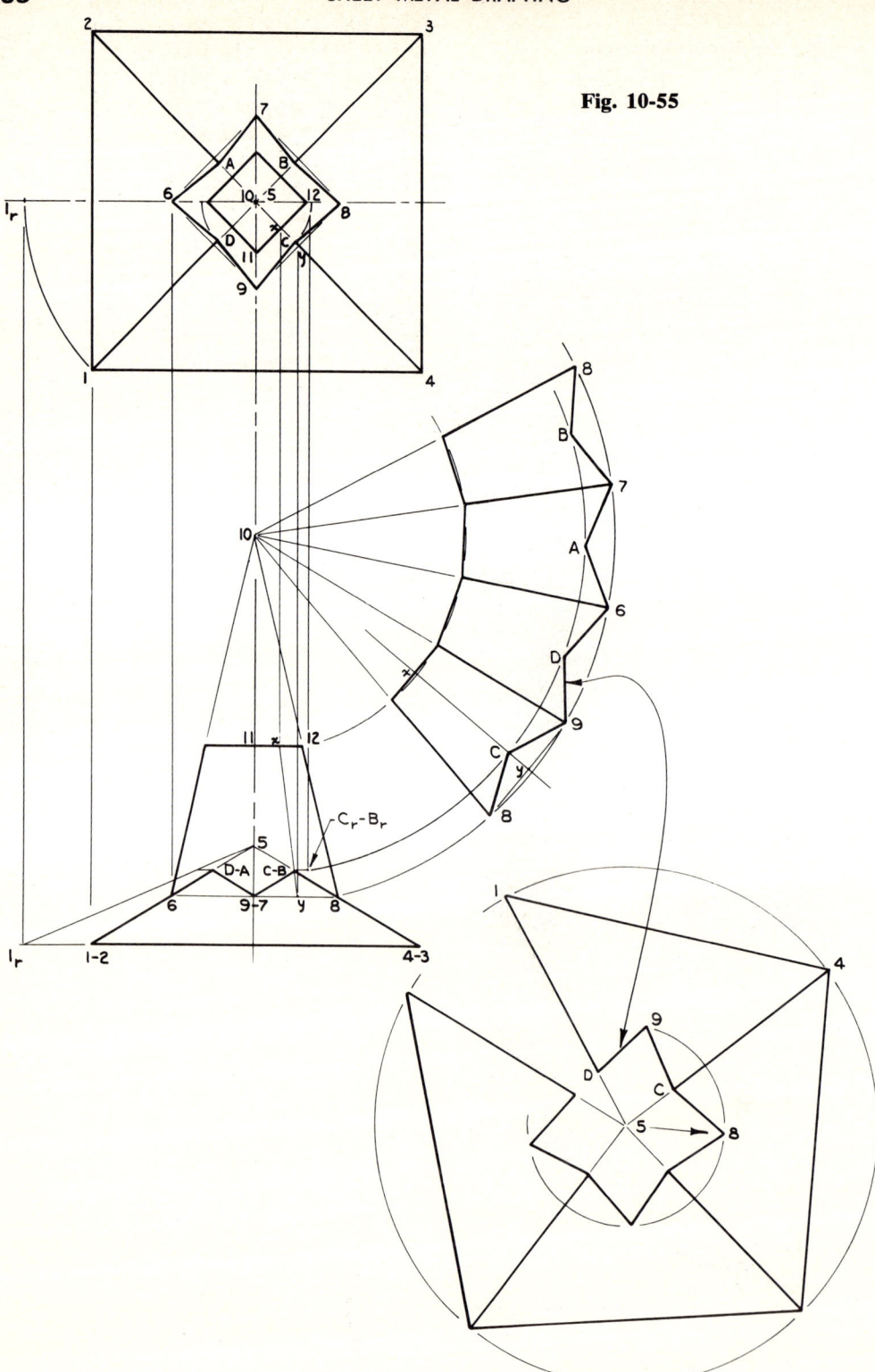

Fig. 10-55

RADIAL-LINE DEVELOPMENT

high square pyramid intersects a low pyramid. To make the layout and development: (1) Draw the top and front views of the low pyramid. Since all corners of this pyramid appear foreshortened, it is necessary to rotate a typical corner until it appears true length in the front view. (2) Rotate the corner 1 until it is true length as shown at 1_r. (3) The pattern for this pyramid now may be developed (as for any right square pyramid), minus only the notching for the intersection. (4) Draw the top and front views of the high square pyramid. (5) Typical corners 6 and 8 of this pyramid are true length in the front view; therefore, the pattern for this pyramid frustum also may be drawn minus only the notching for the intersection. (6) Apply the method described for Figure 10-54 to locate point C (which is also point B in the front view and equivalent to points A and D). (7) The true slant height of C must be located by rotating C until positioned as shown by $C_r B_r$. (8) The points for notching the pattern at A, B, C, and D may now be located as shown. (9) The true slant height of the notching for the low pyramid pattern is already established in the front view by the typical points 6 or 8; therefore, the notching on this pattern may now be located at points 6, 7, 8, and 9 as shown.

Figure 10-56 illustrates the development of a square hopper which intersects a cylin-

Fig. 10-56 Pyramid and cylinder intersection.

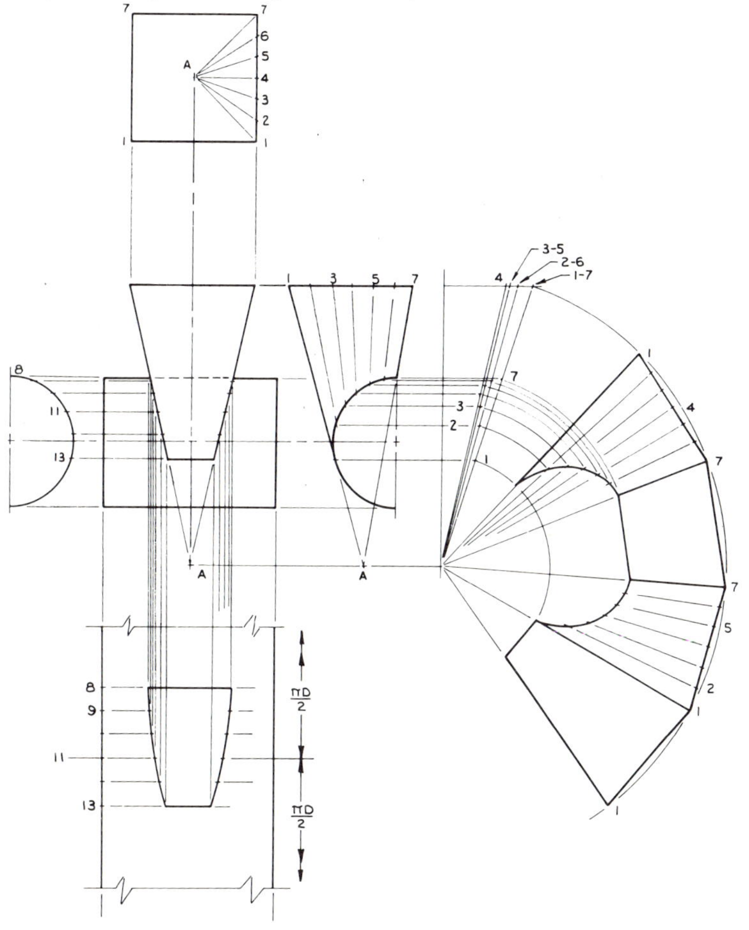

der. Use a partial left-side view (of the cylinder only) to uniformly space elements on the cylinder. The elements then pierce the pyramid in the front view and the hole in the cylinder is developed using parallel-line methods as described in Chapter 9.

To develop a pattern for the body of the hopper: (1) Divide the two surfaces of the pyramid which cut a curve on the cylinder into a convenient number of elements (7 in Figure 10-56). (2) The right side view locates the points at which each element pierces the cylinder. (3) From the top view, using the measurements A-1, A-2, A-3, A-4, etc., construct a true-length diagram (see section 8-3). (4) From the true-length diagram and from the top view of pyramid base, develop the pattern for the hopper body as shown. Note that the true-length diagram is used in this case, in preference to rotation because it separates the points involved in the radial-line development from the other set of spacings involved in the parallel-line development of cylinder and hole.

Figure 10-57 illustrates the intersection of two pyramids to form a *square tee reducer*.

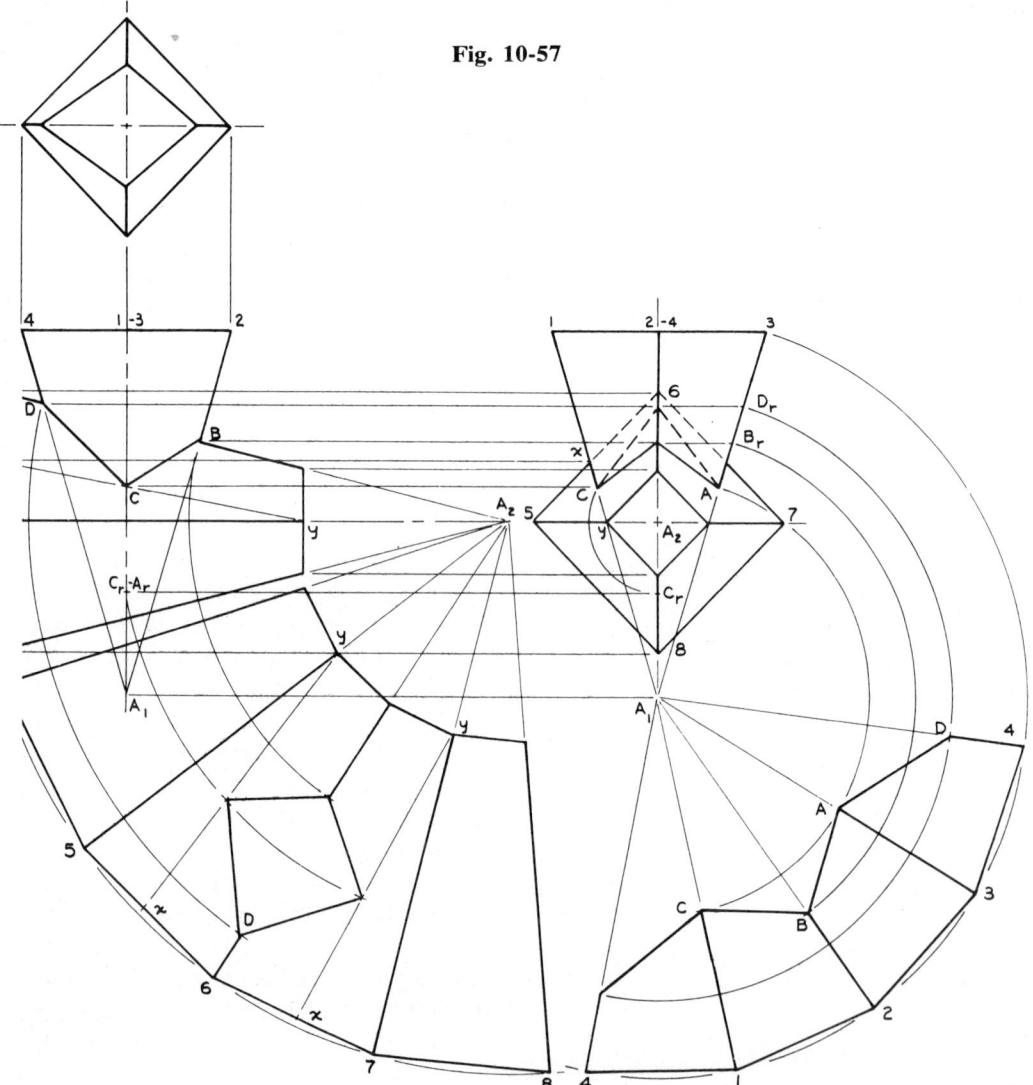

Fig. 10-57

RADIAL-LINE DEVELOPMENT

To develop patterns for each piece: (1) Draw the front and right-side views of both pyramids and a partial top view (of vertical square pyramid only). (2) All points on the line of intersection are easily located by projection except points C and A. (3) Point C (the same as A) may be located by finding where the number 1 element of the vertical pyramid pierces the other surface. This is done using a cutting plane in a manner simi-

Fig. 10-58 Pyramid and cone intersection.

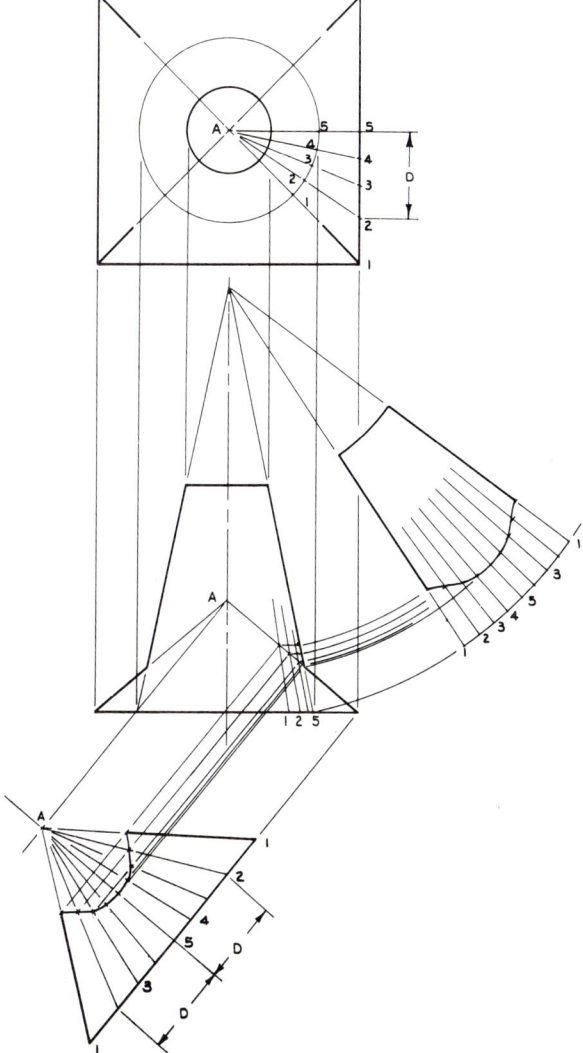

lar to that used in Figure 10-54. The trace xy produced on surface 5-6 by the cutting plane crosses at C. Notice that the corners A, B, C, and D must be rotated until true length, then the pattern is developed in the same way as previous right square pyramids.

10-16 Pyramid and cone intersections

Except where a surface lies in the plane of the cone axis, the line of intersection between pyramids and cones will be a series of curves. The curves must be accurately plotted before the patterns for each piece can be developed. Again, the best methods for locating a series of points on the curve depend upon the particular problem involved. Probably the piercing-point or the cutting-plane methods are best suited to problems of this type. As a sample problem, examine the reducer and hood intersection of Figure 10-58. To make the drawing: (1) Draw skeleton top and front views of the pyramidal hood and cone reducer from specifications. (2) Divide one-eighth of the cone base into five equally spaced elements 1, 2, 3, 4, and 5 and project them to the front view. (3) Draw the elements in the front view, thus locating five points on the pyramid and cone surfaces. (4) Develop a cone pattern using the equally spaced elements and their respective piercing points rotated to a true-length position, as shown in Figure 10-58. The one-fourth cone pattern thus developed represents the intersection with one side of the hood. The pattern may be repeated for the remainder of the cone. (5) Develop one panel of the pyramidal hood as an auxiliary view. Notice that lines radiating from the apex of the pyramid in the top view would pass through the piercing points previously located in the front view. Hence, if the spacings along the base line in the top view are transferred or duplicated in the auxiliary view, the points may be projected from the front view to their respective lines in the auxiliary view as shown in Figure 10-58.

Many variations in pyramid and cone intersections occur in industrial situations; therefore, a second example is given in Figure 10-59. In this example, the axis of the pyramid is perpendicular to the axis of the cone. The problem, as presented, represents a square hopper feeding into a conical reducer. Several methods for solving the problem may be selected but those shown in Figure 10-59 represent the application of the piercing-point, cutting-plane, rotation, and other basic methods which can be applied to similar but different problems.

The pyramid and cone axes are drawn as specified and the front and right-side views of both bodies are skeletonized: (1) In the right-side view, divide the upper half of the cone base into a suitable number of points for cone elements (1 to 7 and back to 1 in Figure 10-59). (2) In order to complete the elements in the right-side view connect the points with apex A. (3) Project the points to the base of the cone in the front view and draw the elements there so as to pierce the two pyramid sides BEA_1 and CDA_1. (4) Project the apparent piercing points back to the elements in the right-side view, thus establishing a visible curve for surface CDA_1 and a hidden curve for surface BEA_1. Notice that a special element (such as x in the figure) is sometimes needed to locate the limits of the curve in the development of the cone opening. (5) Since the sides CBA_1 and DEA_1 appear as the edge view of planes in the right-side view, are parallel to, and do not intersect the cone axis, the line 8-9 on these surfaces would be a part of a hyperbolic curve. The curve in this case lies near the extremities of the hyperbola and for all practical purposes is a straight line. This

Fig. 10-59

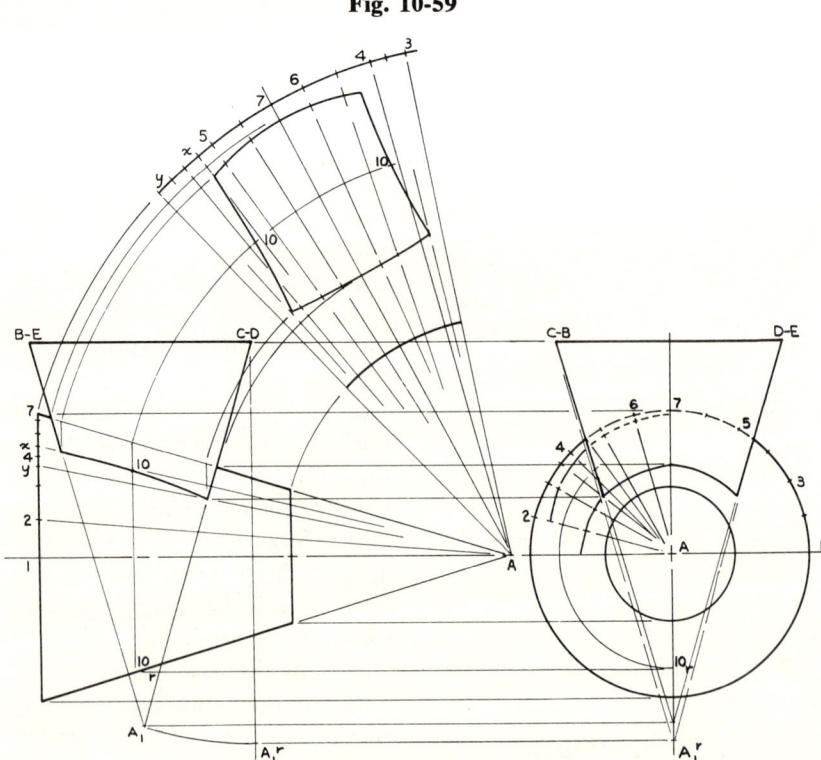

situation exists for most problems of this type. In order to check the amount of curvature involved, however, the method of Figure 10-32 was applied at point 10 in Figure 10-59. (6) To develop the cone opening, develop a portion of the right circular cone sufficient to encompass the extreme cone elements involved in the curve. Rotate the piercing points as shown and sweep them into the pattern as for previous cone problems. (7) To develop the pattern for the hopper, rotate the apex A_1 to the position A_1^r; thus the corners A_1^r, C, B, D, and E are true length in the side view. Sweep the pattern about the apex A_1^r as for any right square pyramid. (8) The curves on the pyramid sides CDA_1 and BEA_1 are developed by rotation and transfer as shown in detail in Figure 10-60. Notice that the slant height is transferred from the front view to the pattern; then depth distances (as shown by sample dimension D) are transferred to the pattern from the side view in order to complete the location of points for the curve.

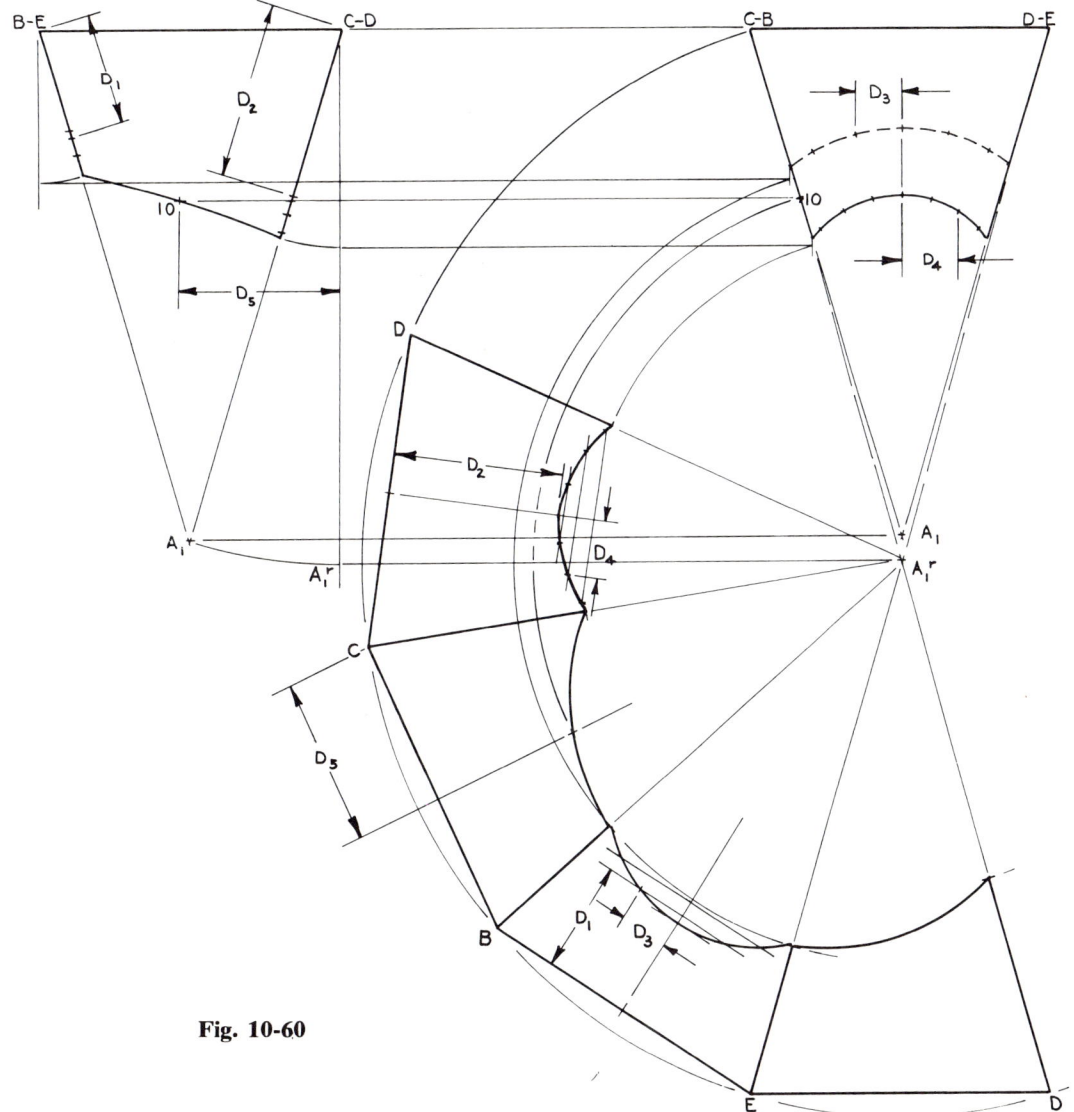

Fig. 10-60

SHEET METAL DRAFTING

EXERCISES

1. Develop a pattern for a flaring circular pan to meet the following measurements: diameter of bottom 12 in [305 mm]; diameter of top 14 in [356 mm]; vertical side height 3 in [76 mm].
2. Develop patterns for the funnel as specified in Figure 10-61.
3. Develop patterns for the .95 liter flaring measure of Figure 10-62.
4. Develop a pattern for the cone portion of a pitched round cover to meet the following measurements: diameter of base 24 in [610 mm]; altitude of cone 2½ in [64 mm].
5. An electrical fixture store needs a pattern for a conical glass fiber lampshade of the following measurements: diameter of shade 20 in [508 mm]; slope angle between base and sides 15 degrees. Provide a pattern for the shade.
6. A Christmas tree as a table decoration is made of chicken wire formed as a cone, then covered with evergreen tips. Provide a pattern for cutting the chicken wire. The tree dimensions are as follows: diameter of base 10 in [254 mm] (cut from plywood); height of tree from base to tip 15 in [381 mm]. (The tree may be fitted with a center shaft and placed in a flowerpot.)
7. Develop patterns for the two cones involved in the fabrication of the rain hood of Figure 10-63.
8. Develop a pattern for the stack collar in Figure 10-64.
9. Develop a pattern for the stack collar in Figure 10-65.

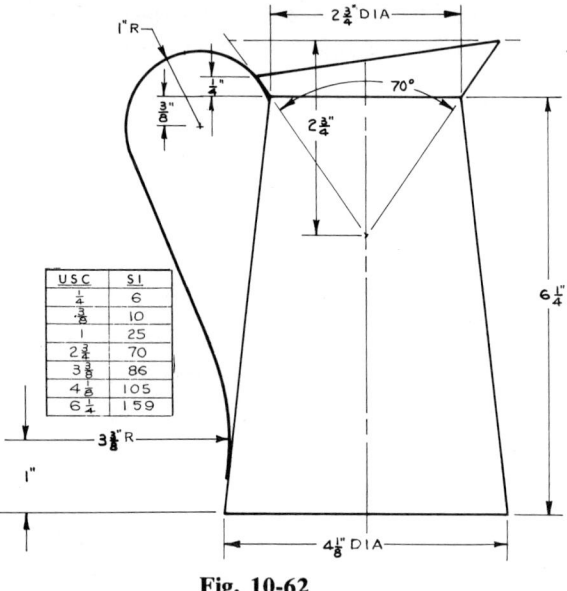

Fig. 10-62

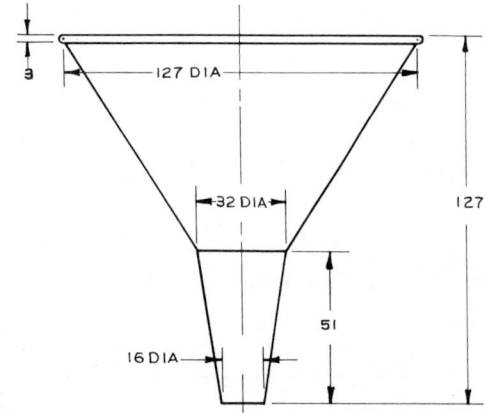

Fig. 10-61
DIMENSIONED IN MILLIMETERS

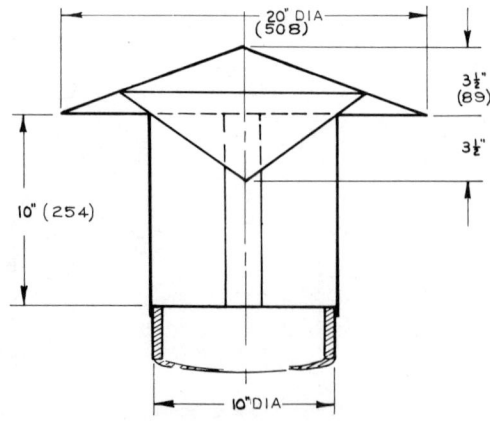

Fig. 10-63   Rain hood.
DIMENSIONED IN INCHES AND MILLIMETERS

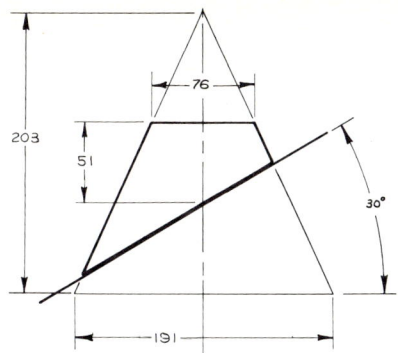

DIMENSIONED IN MILLIMETERS

Fig. 10-64 Stack collar.

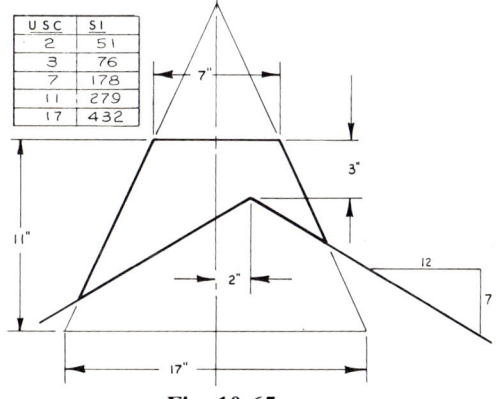

USC	SI
2	51
3	76
7	178
11	279
17	432

Fig. 10-65

Fig. 10-66

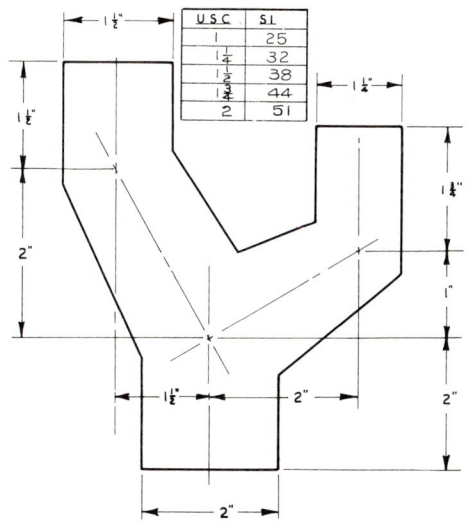

USC	SI
1	25
1¼	32
1½	38
1¾	44
2	51

10. Prepare engineering data in the form of computations suitable for the development of pieces for a cyclone separator similar to that shown in Figures 10-11 and 10-12 but computed for the following sizes: $A = 24$ in [610 mm]; $B = 48$ in [1 220 mm]; $C = 8$ in [203 mm]; $D = 48$ in [1 220 mm]; $E = 24$ in [610 mm]; $F = 4$ in [102 mm]; $G = 76$ in [1 930 mm].

11. Develop the pattern for a conical offset reducer to meet the following specifications: diameter of large pipe 3 in [76 mm]; diameter of small pipe 2 in [51 mm]; amount of offset 3 in [76 mm] run of fitting 3 in [76 mm].

12. Develop patterns for the following conical reducer elbow: large diameter 3 in [76 mm]; small diameter 1½ in [38 mm]; elbow to be four-piece 90-degree bend. Centerline radius, 4½ in [114 mm].

13. Develop patterns for the following conical reducer elbow: large diameter 3 in [76 mm]; small diameter 1 in [25 mm]; elbow to be three-piece 75-degree bend. Center-line radius 5 in [127 mm].

14. Develop patterns for the breeching of Figure 10-66.

15. Develop a hyperbolic curve by cutting a cone with a plane parallel to the cone axis. The diameter of the cone base is 3½ in [89 mm]. The curve of the hyperbola shall have a span of 3 in [76 mm] and a rise of 3½ in [89 mm].

16. Develop patterns for the two intersecting cones shown in Figure 10-67.

Fig. 10-67

DIMENSIONED IN MILLIMETERS

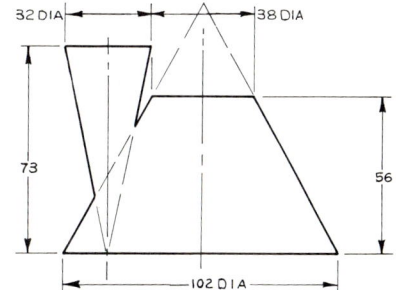

SHEET METAL DRAFTING

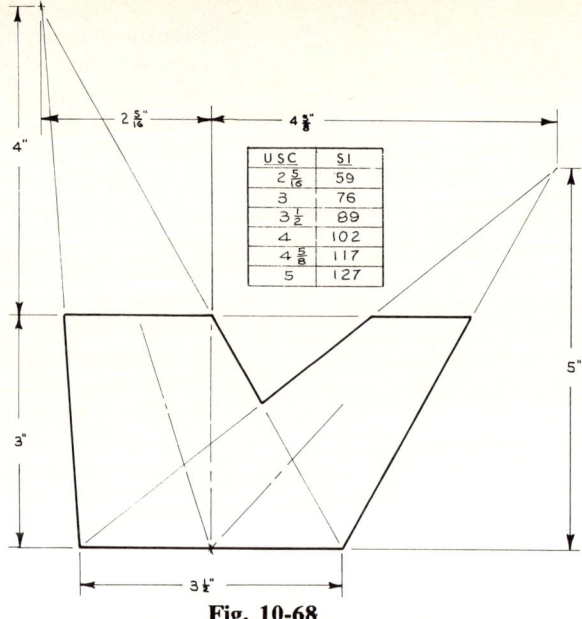

USC	SI
2 5/16	59
3	76
3 1/2	89
4	102
4 5/8	117
5	127

Fig. 10-68

Fig. 10-69

DIMENSIONED IN MILLIMETERS

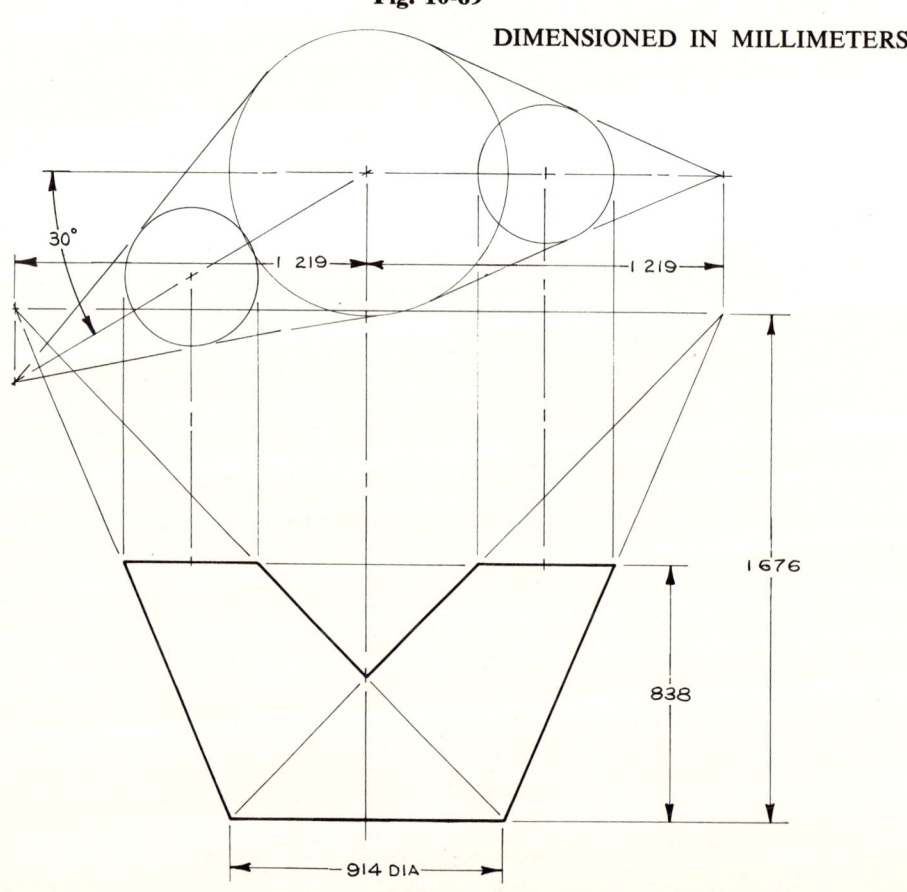

17. Develop patterns for the two intersecting cones shown in Figure 10-68.
18. Develop patterns for the cones shown in Figure 10-69.
19. Develop patterns for the two intersecting cones shown in Figure 10-70.
20. Develop patterns for the two intersecting cones shown in Figure 10-71.
21. Using the polycylindric or gore method, develop a pattern for a twelve-gore sphere of 4-in [102-mm] diameter.
22. Using the polyconic or zone method, develop a pattern for an eight-zone sphere of 3½-in [89-mm] diameter.
23. Using the gore method, develop a pattern for an eighteen-gore parabolic motor nacelle to be placed over a fan motor in a supersonic wind tunnel. Specifications as shown in Figure 10-72.
24. Develop the pattern for a hopper to meet the following specifications: square hopper with inlet opening of 6'-0" [1 829 mm]; outlet opening of 2'-0" [610 mm]; height of hopper 8'-0" [2 438 mm].
25. Develop a pattern for a hexagonal pyramid to be used as a part of a finial on a roof peak. The diameter of the hexagonal base (across flats) is 3 in [76 mm]; the altitude of the pyramid is 5 in [127 mm].
26. Develop a pattern for that portion of the octagonal pyramid which is above the roof planes in Fig. 10-73.
27. Develop patterns for the four parts of the sand hoist redesigned as shown in Figure 10-74. Use the conversion factor 25.4 mm = 1 in; change all dimensions to SI.

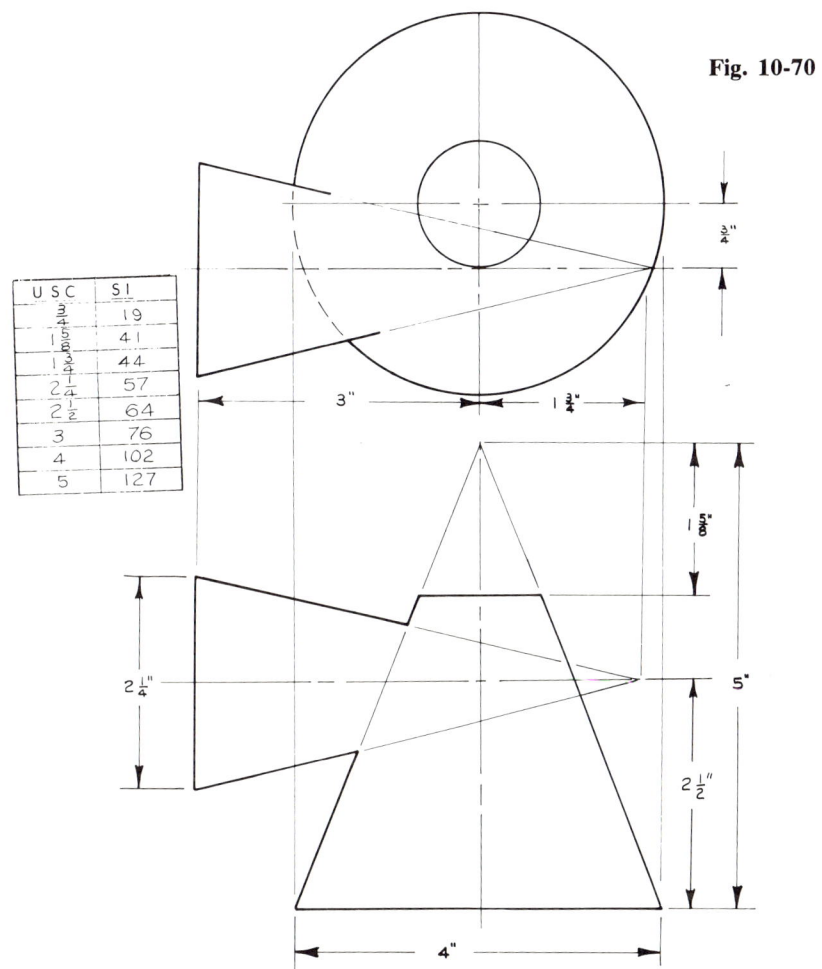

Fig. 10-70

SHEET METAL DRAFTING

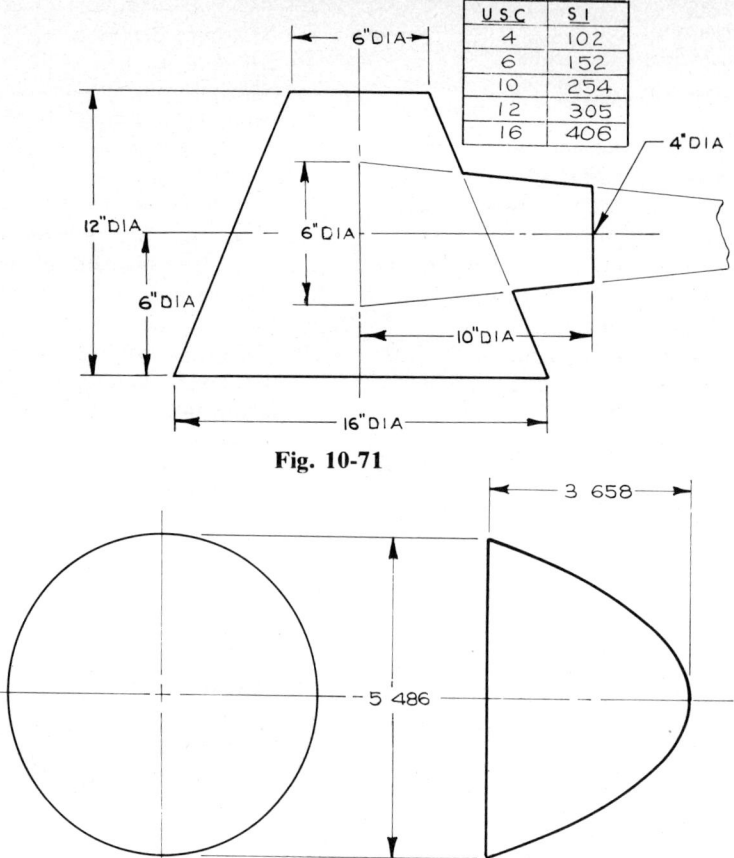

USC	SI
4	102
6	152
10	254
12	305
16	406

Fig. 10-71

DIMENSIONED IN MILLIMETERS

Fig. 10-72

Fig. 10-73

DIMENSIONED IN INCHES AND MILLIMETERS

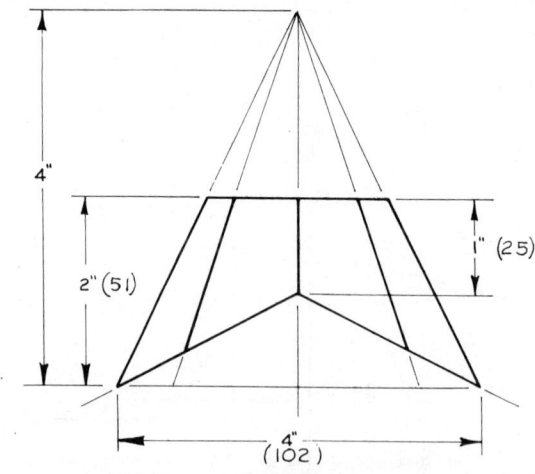

RADIAL-LINE DEVELOPMENT

Fig. 10-74 Sand hoist. (*Ohio Art Co., Bryan, Ohio.*)

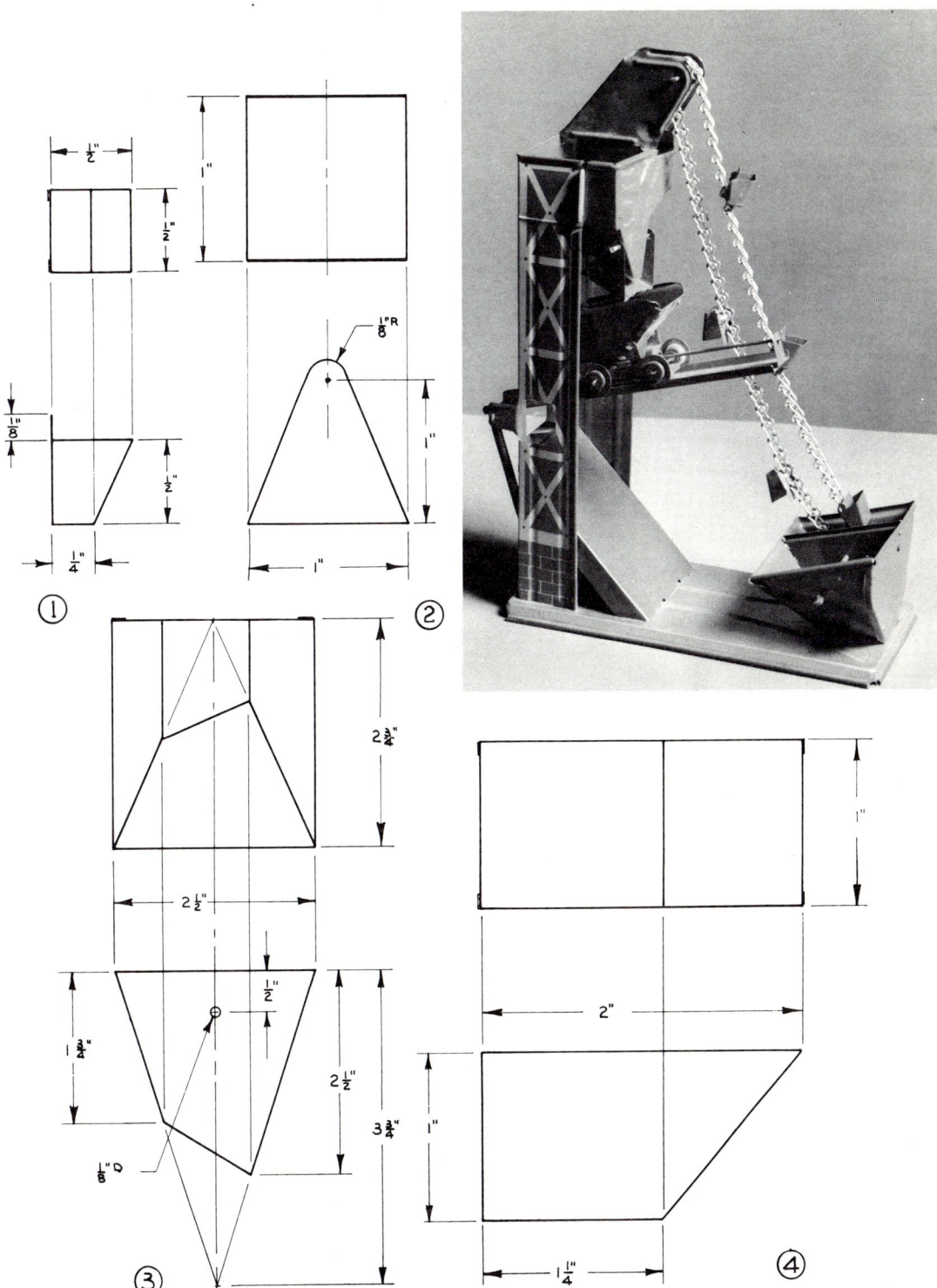

28. Develop a pattern for the hexagonal offset indicated in Figure 10-75.
29. Develop a pattern for a square pitched cover to suit the following measurements: size of base 17 inches; amount of rise 2½ inches.
30. Develop a pattern for a rectangular pitched cover to suit the following measurements: base 5 by 12 inches; amount of rise 1½ inches.
31. Develop a pattern for the hopper and for the cylinder opening in Figure 10-76.
32. Develop a pattern for the hopper and for the duct opening in Figure 10-77. Convert measurements to SI.

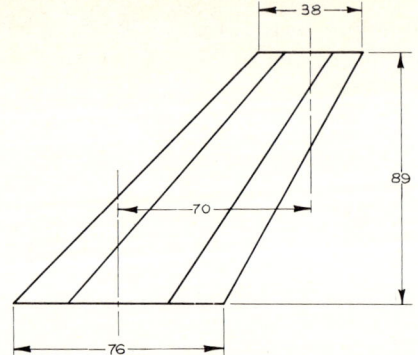

DIMENSIONED IN MILLIMETERS

Fig. 10-75

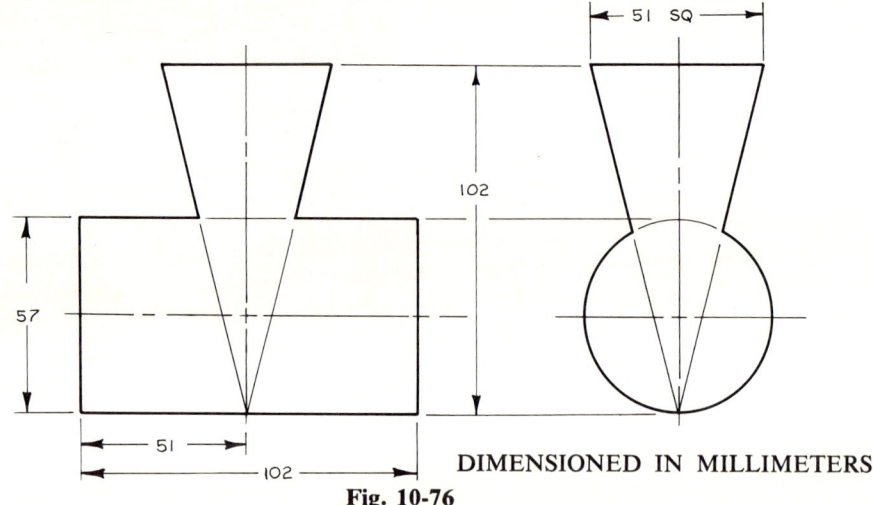

DIMENSIONED IN MILLIMETERS

Fig. 10-76

Fig. 10-77

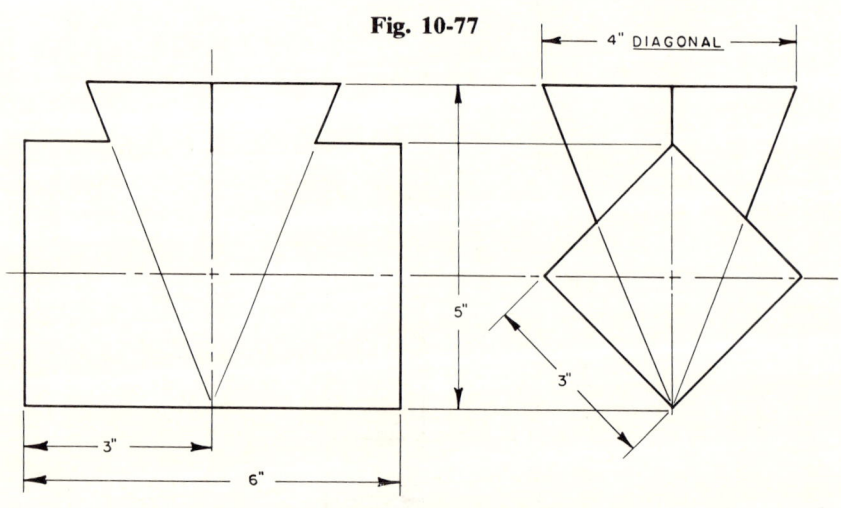

RADIAL-LINE DEVELOPMENT

33. Develop a pattern for the square reducer and for the hood opening as shown in Figure 10-78.

Fig. 10-78

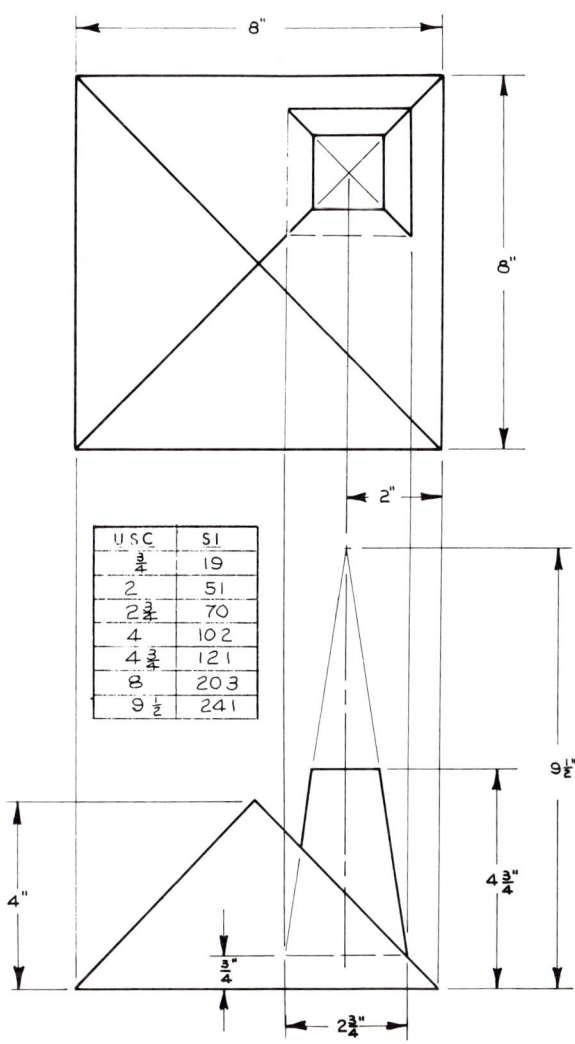

34. Develop a half-pattern for the square duct breeching shown in Figure 10-79. Place the seam on line 1-2.

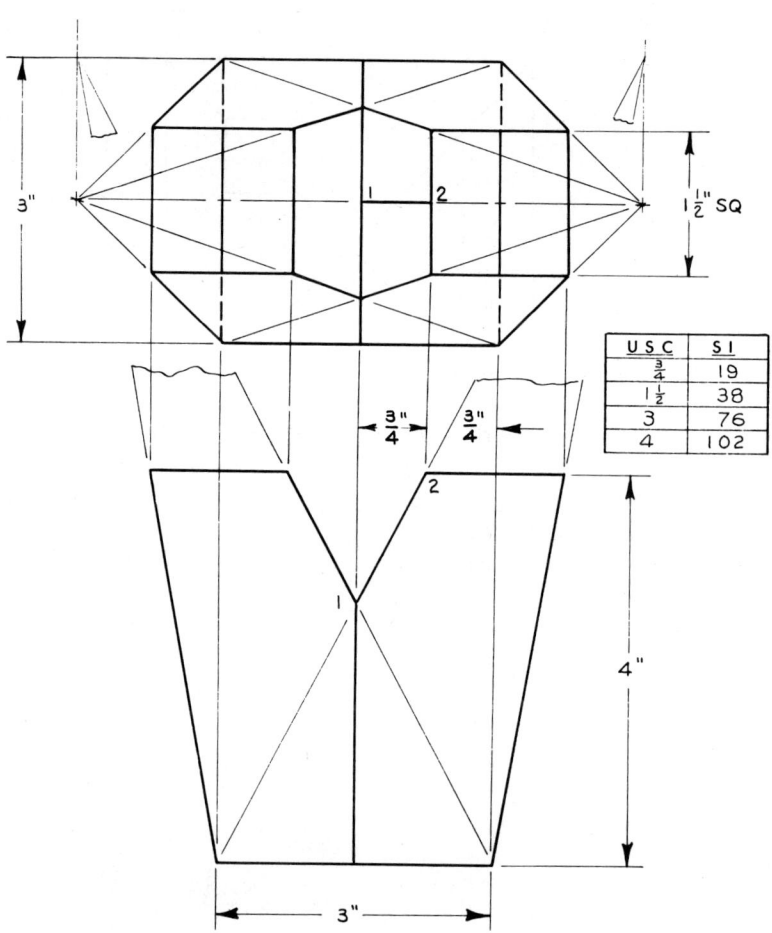

Fig. 10-79

chapter 11

TRIANGULATION

At the beginning of the two preceding chapters, the types of surfaces to be developed by the methods discussed in those particular chapters were defined. Therefore, at the beginning of this chapter on *triangulation,* we will, using the knowledge gained from the preceding chapters, further define and classify surfaces and forms.

Most of the sheet metal surfaces thus far studied (prisms, pyramids, cylinders, cones, etc.) are of a type which are generated by moving a straight line (the straight lines being elements or corners of the object thus generated). Since sheet metal is a plane surface to begin with, it is logical to assume that the sheet metal draftsman is concerned primarily with these surfaces which are called *ruled surfaces*. They are classified into three general types as follows: (1) planes; (2) single-curved surfaces; (3) warped surfaces. Notice in Figure 11-1 that plane and single-curved surfaces are those formed by intersecting or by parallel elements. Note also that the parallel forms discussed in Chapter 9 and the radial forms of Chapter 10 are all of the plane or single-curved variety (with the exception of the sphere and paraboloid which are double-curved surfaces). The problems presented as parallel-line or radial-line developments, being plane and single-curved surfaces, could have been developed by triangulation; however, the problems presented in this chapter, which involve *warped surfaces, must* be developed by this method.

Fig. 11-1

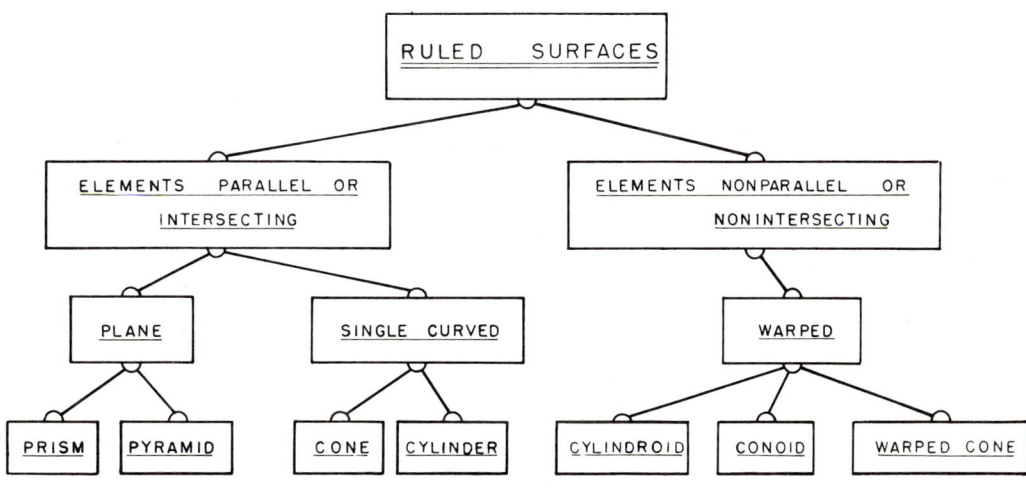

201

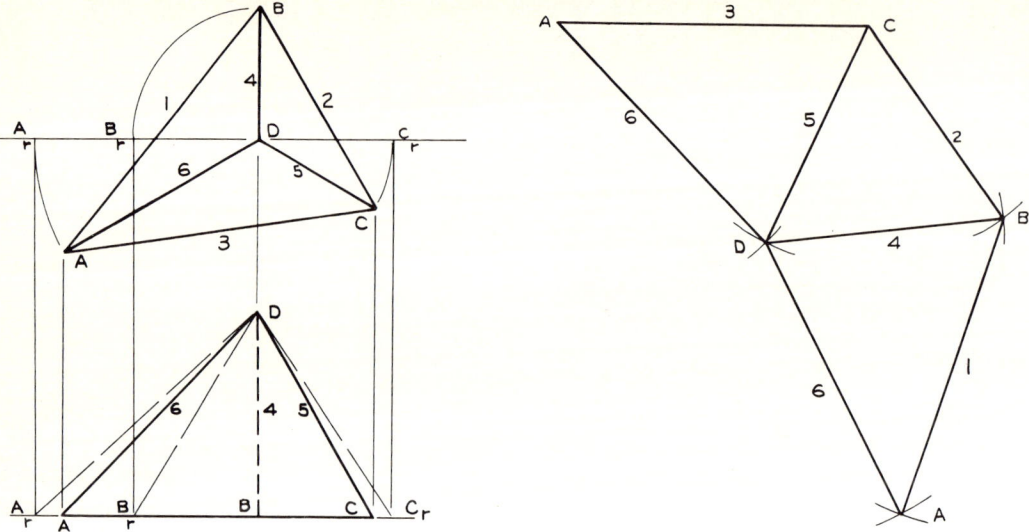

Fig. 11-2

11-1 Basic principles

Strictly speaking, triangulation is the act of making a triangle from three given lines. More liberally defined, however, triangulation usually refers to the operation of dividing a surface into triangles, thereby making it possible to transfer it to a new position, or to duplicate its shape a number of times. (See section 5-11.) As an example of the basic principles of triangulation (triangle construction), study the development of the pattern for an aluminum monument cap, shown in Figure 11-2. The top and front views are given. Notice that lines 1, 2, and 3, being horizontal, are therefore true length in the top view. The base points A, B, and C are rotated so that lines 4, 5, and 6 become true length in the front view. To *triangulate* the pattern, using the true length of line 6, establish points D and A. Using line 1 as radius with A as center, and line 4 as a radius with D as center, strike intersecting arcs to locate point B on the pattern. Using line 2 as radius with B as center, and line 5 as radius with D as center, strike intersecting arcs to locate point C. In a similar manner, use lines 3 and 6 to complete the pattern.

The surfaces thus developed in this example are plane surfaces. The same basic principles, if a sufficient number of triangles are assumed, must be applied to warped surfaces. Actually, a warped surface divides into triangles of which one or more sides are curved lines. Thus, the more curvature involved, the more triangles required in order to keep the approximation to a minimum.

As an illustration, develop a pattern for the warped surface $ABCD$ of Figure 11-3:

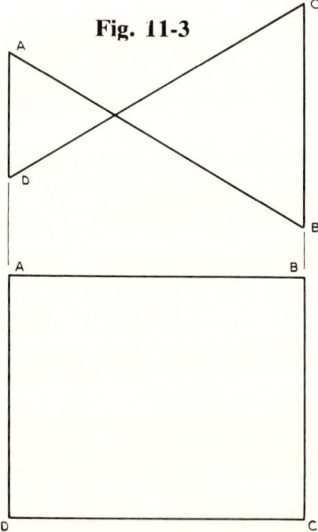

Fig. 11-3

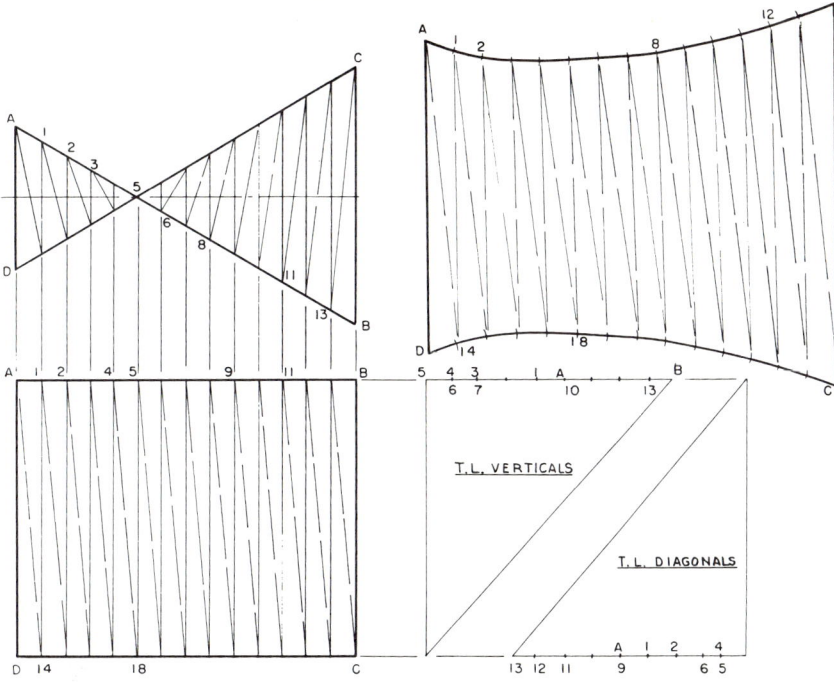

Fig. 11-4

(1) Divide the edges AB and DC into the same number of equal spaces, as shown in Figure 11-4 (one set of points at 5-18 represents the only true-length line in the front view). (2) Construct diagonals between points so as to divide the warped surface into a series of triangles. The diagonals may be either right or left, that is, from A to 14 or from D to 1. Since the pattern becomes symmetrical, the true length of A-14 and D-1 are equal and may be used to maintain symmetry when triangulating the pattern (see pattern of Figure 11-4). (3) By means of the distances taken from the top view and the height projected from the front view, establish a true-length diagram for the lines between points and also a true-length diagram for the diagonals (section 8-3). (4) Using the true distance AD, D-14, and A-14, construct the first triangle of the pattern. (5) Using the distances A-1 and 14-1, construct the second triangle of the pattern (now A-14 should equal D-1). (6) Using the true lengths, as taken from the diagrams, continue to add triangular segments until all twenty-eight sections have been triangulated. (7) Sketch a freehand curve through the numbered points to complete the pattern.

All future problems, developable by triangulation, apply the basic principles thus far explained and demonstrated. There is very little that is new or difficult about triangulation. The main difficulty exists in the area of design, the area having to do with the creation of the shapes needed for job situations. Once the lines and surfaces of a particular fitting have been determined and triangles assumed upon the surface, the actual pattern construction becomes routine —a routine which must be alertly applied, however, lest wrong measurements be used. To demonstrate the design and development of fittings suitable for triangulation, twists, transitions, and breechings will be considered.

TWISTED PIPES

11-2 Rectangular-to-rectangular 90-degree twist

The problem shown in Figure 11-5 involves pipe openings of similar shape but of different measurements arranged perpendicularly. Since the edges of the upper opening are parallel to the edges of the lower opening, the pipe openings may be connected by using four quadrilateral planes as shown. The slant height in the front view represents the true length of a typical seam xy; therefore, begin the pattern as an auxiliary view of X-Y-8-4 (X-4 and Y-8 are true distances, taken from the top view). By means of a diagram, determine the true length of an assumed diagonal 5-4 or 8-1 (they are identical). Triangulate the surface 4-8-5-1. Determine the true length of the diagonal 6-1 or 5-2. Triangulate surface 5-1-6-2. By triangulation, duplicate surface 4-8-5-1 as surface 6-2-3-7, and duplicate X-Y-8-4 as surface 3-7-X-Y; in this way you will complete the pattern.

Fig. 11-5

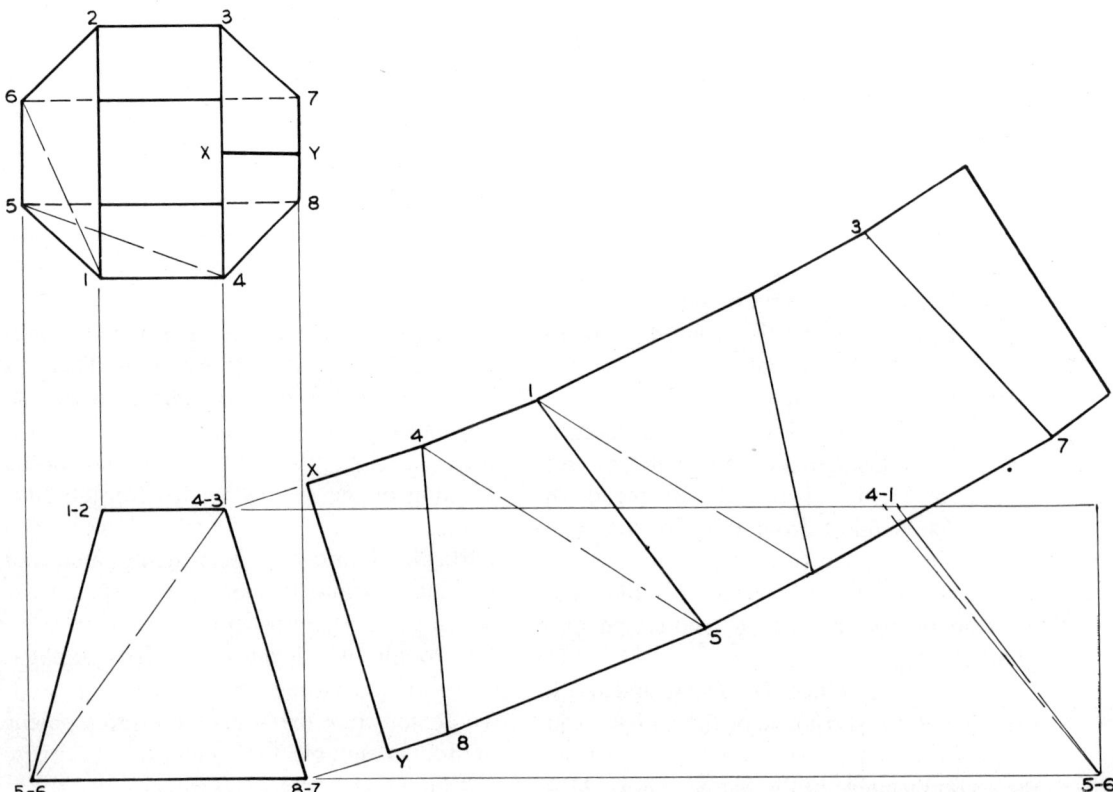

11-3 Square-to-rectangular 45-degree twist

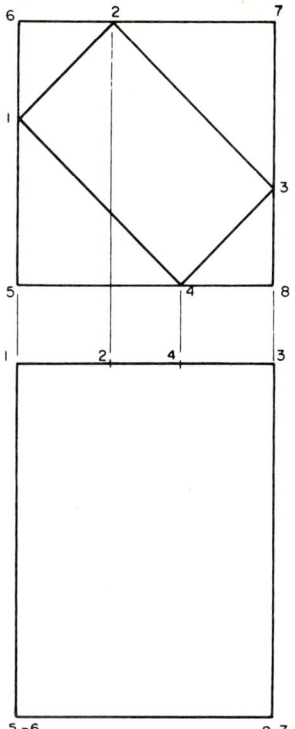

To meet the specifications, first draw a top and a front view of the required square and rectangular openings properly placed as regards position, size, and height (see Figure 11-6). A system of numbers or letters should always be used to identify corners, thus making the actual design of the fitting easier. The rectangular opening is numbered 1, 2, 3, 4 and the square opening 5, 6, 7, 8 in Figure 11-6. At this time in designing a fitting, carefully analyze the openings and the surfaces required to connect them. One might reason as follows: Each opening has four corners and four edges; but the edges of one opening are not parallel to the edges of the other, as they were in the previous problem. Therefore, a corner of one opening, if connected with two corners of the other opening, will

Fig. 11-6 Square-to-rectangular twist transition.

Fig. 11-7

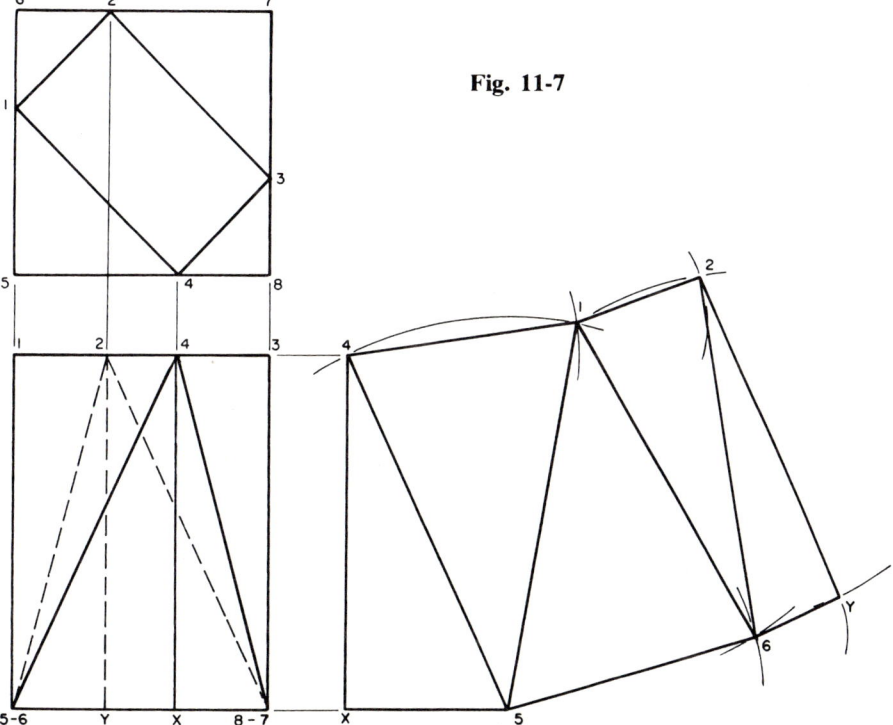

form a triangular plane surface between them. Each edge of each opening must therefore become one side of a triangle. The corner opposite, on the other opening, will be the third point of the triangle. Thus the fitting becomes a series of alternately upright and inverted triangles connecting the openings, as shown by the top and front views of Figure 11-7. The triangular surfaces are numbered 1-4-5, 5-8-4, 4-3-8, 8-7-3, 2-3-7, 7-6-2, 2-1-6, and 6-5-1.

To develop the triangles into a pattern, the true length of each line is needed. Since this particular fitting is symmetrical, all lines either appear true length or have *an equivalent line* which is true length in one of the given views. For example, lines 5-1 and 7-3 are the same length as lines 5-4 and 7-2 which appear true length in the front view; lines 8-3 and 6-1 are the same length as lines 8-4 and 6-2 which also appear true length in the front view. All opening edges, being horizontal, are true length in the top view. The pattern developed in Figure 11-7 is a half-pattern, using the lines 2-*Y* and 4-*X* as seams. This places the seams on a flat surface, and also conserves metal when cutting the two halves.

11-4 Square-to-rectangular 45-degree twist offset

The problem shown in Figure 11-8 is similar to the previous one in that the edges of the openings are not parallel. Thus this fitting also consists of eight triangular surfaces alternately upright and inverted. Notice that, in the true-length diagrams of Figure 11-8, lines having similar top views are grouped as one because they are equal in length. In developing the pattern for a fitting wherein the top and front views, relative to number of surfaces visualized, are as difficult to read as this particular drawing, it is wise to tabulate the surfaces and check them off as they are developed. Tabulation may be

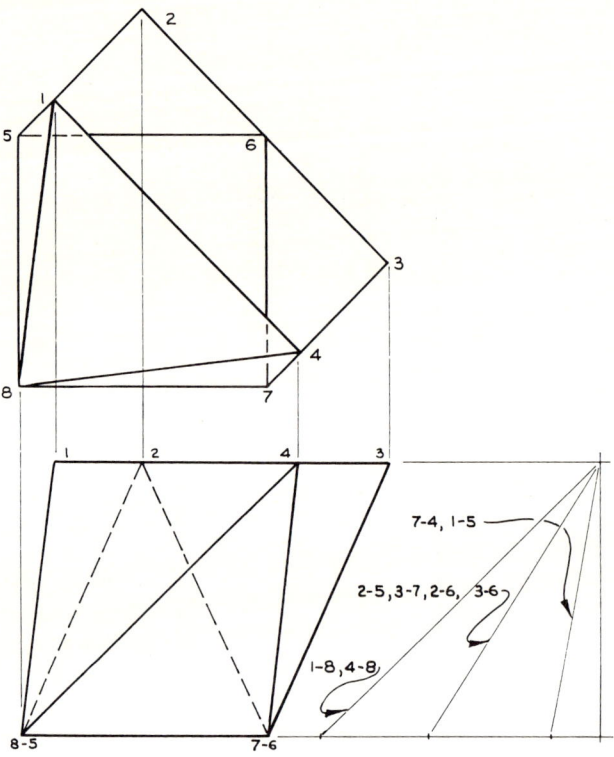

Fig. 11-8 Offset twist transition.

Fig. 11-9 Tabulating surfaces.

4	1	2	3
7	8	5	6
8	5	6	7

UPRIGHT TRIANGLES

INVERTED TRIANGLES

1	2	3	4
4	1	2	3
8	5	6	7

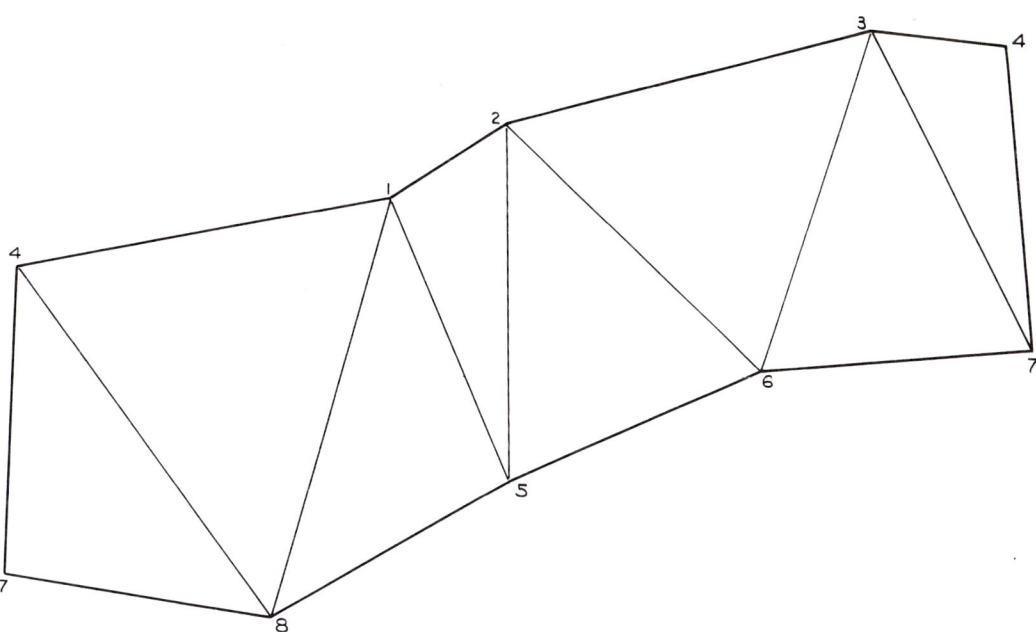

Fig. 11-10

done in a systematic manner, as shown in Figure 11-9. Start with any two visible adjoining surfaces, such as upright surface 7-8-4 and inverted surface 4-1-8. Since both openings are numbered clockwise in the top and front views of Figure 11-8, all members may now be supplied horizontally in the tabulation by following a clockwise order for the upper or lower opening. The completed development is shown in Figure 11-10. Notice that the clockwise numbering of Figure 11-8 is triangulated in a reverse direction on the pattern. The result is an "inside up" view, which is preferred in the shop because of the limitations imposed by the bending machinery.

TRANSITIONS

11-5 Square-to-hexagonal transition

This fitting connects an opening of four edges with an opening of six edges. Skeletonize the views and number the corners, as shown in Figure 11-11. To analyze the circumstances for connecting surfaces, notice that edge 7-10 of the square and edge 1-6 of the hexagon lie in the same plane. Edge 8-9 lies in the same plane as 3-4. Then each of these coincident situations forms a trapezoidal plane surface connecting the openings on the sides. To connect the back (or the front) of each opening, note that three edges are involved. Two of these are on the hexagon and one is on the square, hence three triangular surfaces can be used to make the connection. The completed top, front, and half-pattern are shown in Figure 11-12. Note that the seam is shown placed on a triangular surface, but it could also be located just as well on the trapezoidal surface, or on a corner.

11-6 Cylindroid

The triangulation problems considered thus far have involved plane surfaces only. Actually, triangulation is used most frequently to approximate the surfaces of warped ruled shapes. The *cylindroid* is a warped shape which may be used to connect cylindrical pipes. A cylindroid is defined as a shape created by constructing all elements parallel to a plane (called the plane director) and in contact with two different curves at each end (usually circles or ellipses) called the curved directrices. The top and the front views of a cylindroid, used as an offset to connect two pipes, is illustrated in Figure 11-13. Notice that, in the front view only, the elements appear parallel. They are therefore all parallel to any plane *which appears as an*

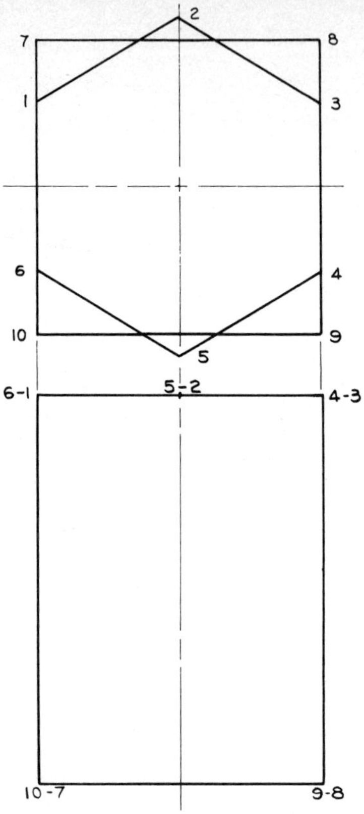

Fig. 11-12

Fig. 11-11

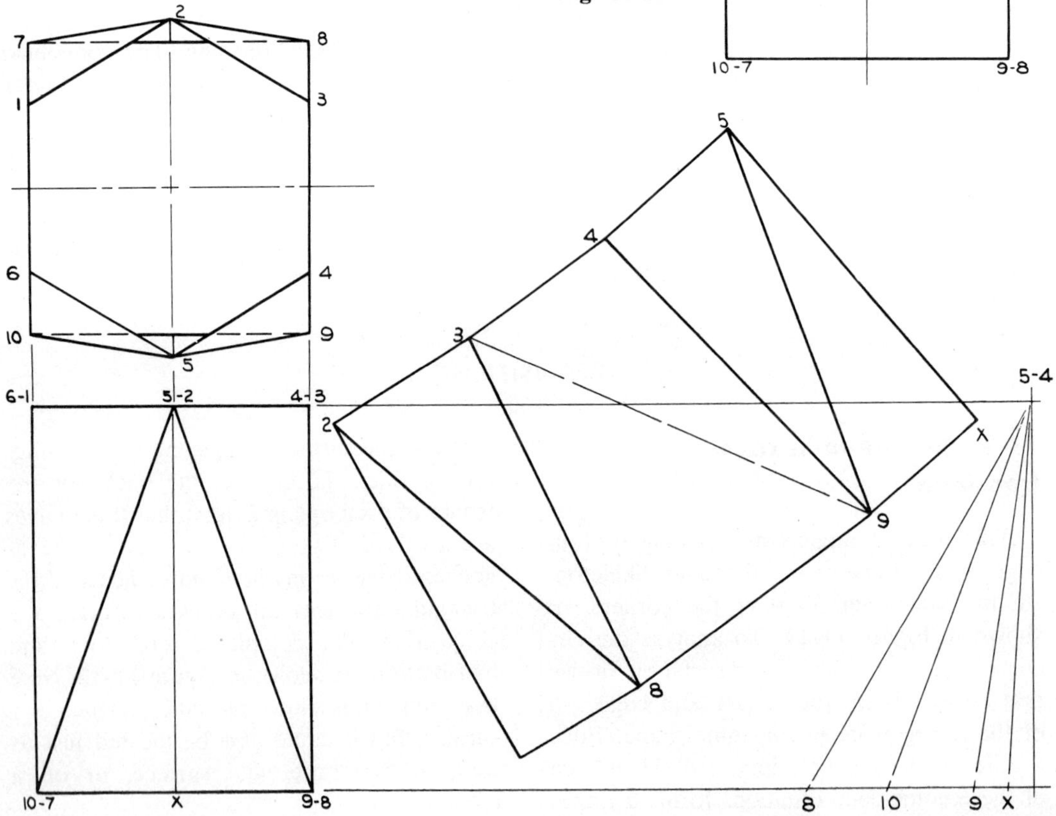

TRIANGULATION

edge in that view and is parallel to any one element. *Since the elements do not appear parallel in the top view they are not parallel to each other.* The offset is therefore a warped ruled surface. The curved directrices are represented by the circles of the top view. Directrix 1-17 is a circle and directrix 2-18 is an ellipse.

To construct the top and front views of the fitting: (1) Draw the center lines for the pipes according to specifications. (2) Let the larger pipe be cut as shown and divide the circle into a uniform number of points (odd numbers 1 to 17 and back to 1). (3) In the front view, draw the elements parallel to the previously established center line. The contour elements of the offset intersect the contour elements of the small pipe at X and Y. (4) Draw the miter line XY. (5) Continue the elements to the top view of the small pipe and number the points (even numbers 2 to 18 and back to 2). As shown in Figure 11-14, three items of construction must be completed before the pattern can be triangulated: (1) The surface must be divided into triangles by drawing diagonals between elements. Triangulation is easier to follow if the diagonals are drawn so as to create a zigzag which follows the numbered points. (2) The true girth of the elliptic curve 2-18-2 must be plotted as an auxiliary view. (3) The true lengths of all elements and diagonals must be determined by means of a true-length diagram, as shown. Notice that each even-numbered point has two true lengths involved, one for an element and one for a diagonal.

Figure 11-14 shows one-half the pattern triangulated as follows: (1) Begin with element 1-2 which is true length in the front

Fig. 11-13

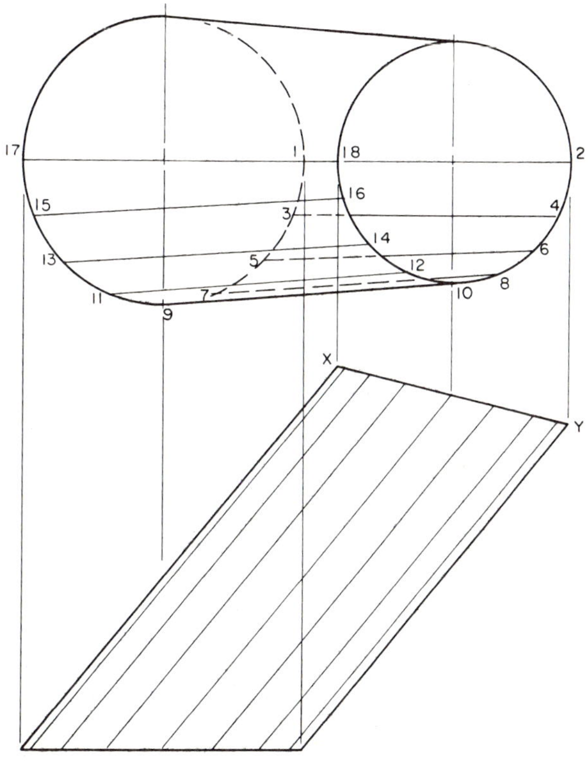

SHEET METAL DRAFTING

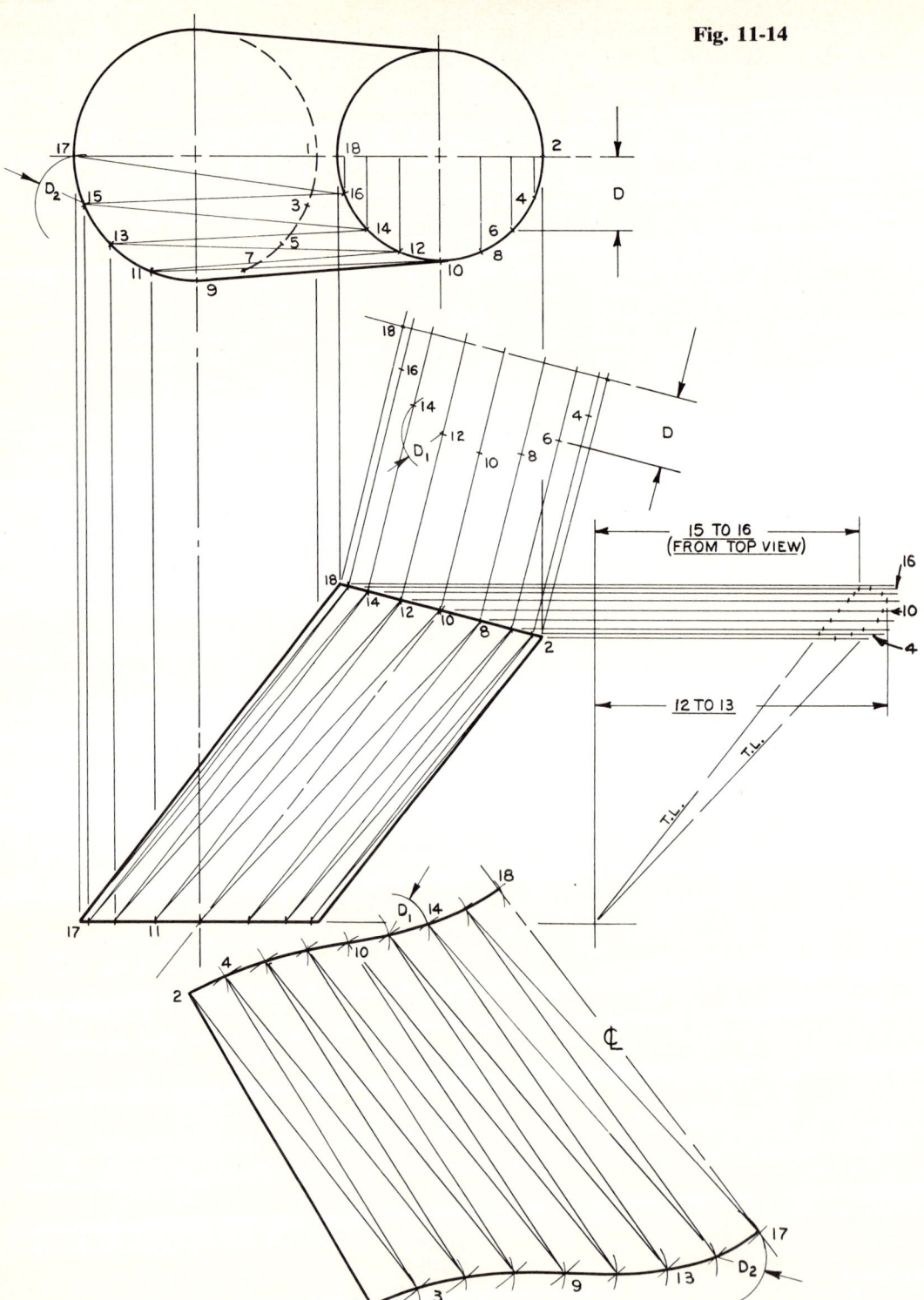

Fig. 11-14

view. (2) Set a compass for the true distance between odd numbered points, as shown in the top view, and from point 1 on the pattern strike an arc for the distance 1-3. From point 2, cross that arc with another compass setting representing the true distance 2-3, thereby locating point 3 on the pattern. From point 2 strike an arc using the true distance 2-4 (as taken from the auxiliary view). Cross that arc with a setting equalling the true distance 3-4, thus locating point 4 on the pattern. (3) Continue the procedure until one-half the pattern has been triangulated.

11-7 Warped cone

Figure 11-15 shows a conical form which is, because of the change in opening shapes,

Fig. 11-15 Round-to-elliptical transition.

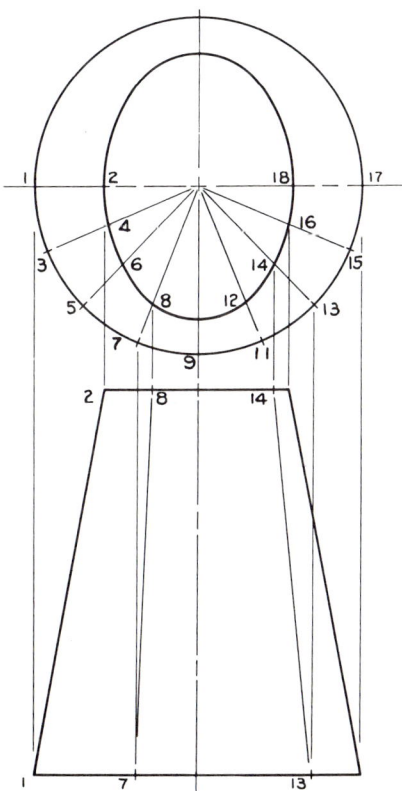

a warped surface. Such a shape is called a *warped cone*. The sheet metal fitting represented in Figure 11-15 is a *transition* from round to elliptical pipe. Observe that the elements 1-2, 3-4, 5-6, 7-8, 13-14, etc., all intersect the central axis of the cone, *but,* each element intersects at a different point. The cone therefore does not have a single apex and must be developed by triangulation. As shown in Figure 11-16, the numbering procedure, true-length diagrams, and pattern development are similar to those of previous problems. Notice that two true-length diagrams are constructed. One is for elements 1-2, 3-4, 5-6, etc., and the other is for the diagonals 2-3, 4-5, 6-7, etc. Such an arrangement is desirable in order to select them when they have all been constructed or to refer back to when checking. The warped cone elements are spaced uniformly on the base circle but the spacing on the elliptical opening *is not uniform.* Be sure to use the true spacings for each curve, as taken from the top view, when developing the pattern.

The more common arrangement of lines on the surface of the fitting is shown in Figure 11-17. In this example, both circular and elliptical openings are spaced evenly so as to provide the same number of spaces at each opening. This is a convenient arrangement for development, as two bow instruments may be set separately for each spacing and the settings used repeatedly. Identical fittings are shown in Figures 11-16 and 11-17. The pattern results are the same by either method, the only difference being that the triangles are arranged differently on the warped conical surface.

11-8 Square-to-round transition

The square-to-round transition, as shown in Figures 8-4 and 11-18, is a very common fitting which is developed by triangulation. The top view of the openings is drawn from

SHEET METAL DRAFTING

Fig. 11-16

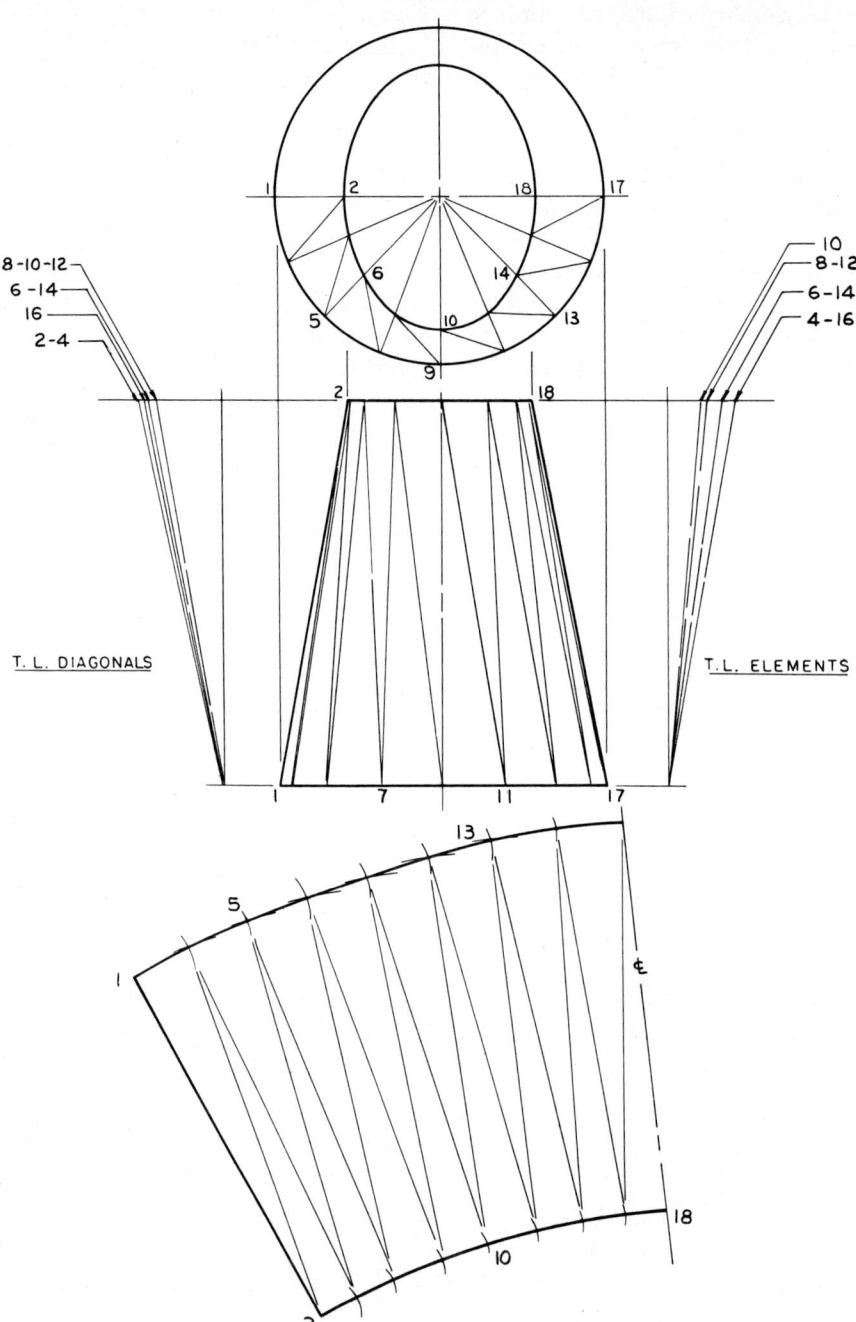

TRIANGULATION

Fig. 11-17

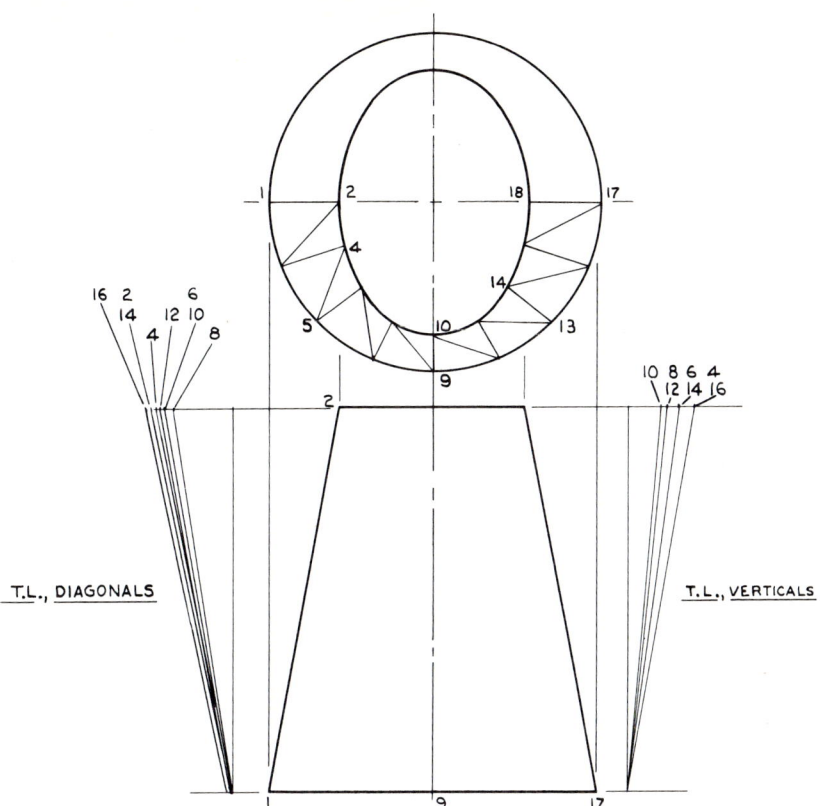

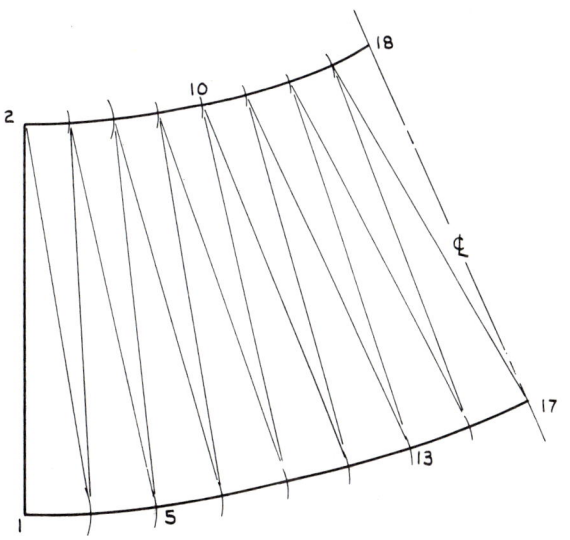

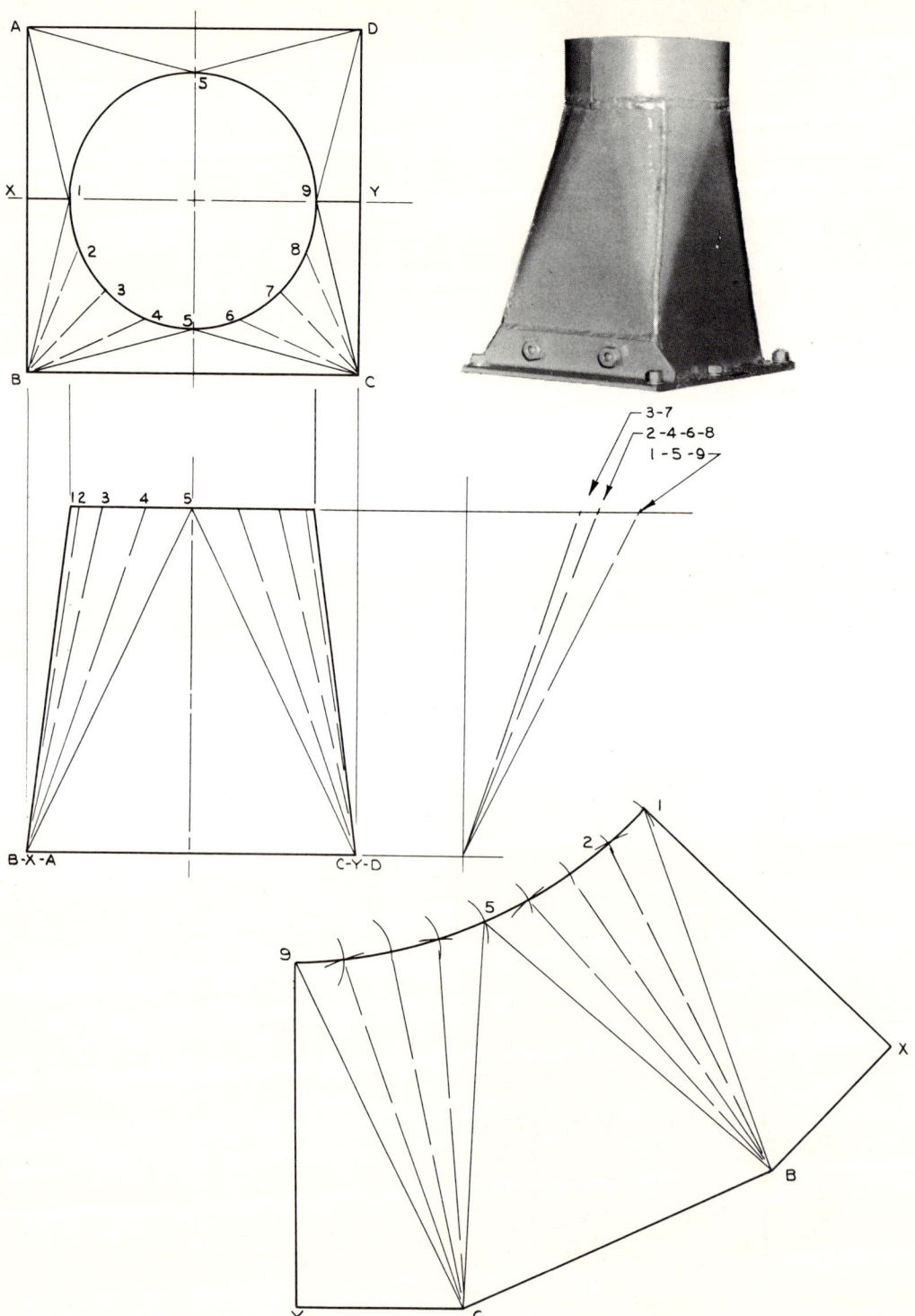

Fig. 11-18 Square-to-round transition. (*Prater Pulverizer Co., Chicago, Ill.*)

specifications and they are shown as lines in the front view. Divide the top view of the circular opening into a convenient number of points for elements (1 to 9 and back to 1 in Figure 11-18). Since the square opening *ABCD* presents four straight edges, four triangular plane surfaces are represented in Figure 11-18 by the points *A-B*-1, *B-C*-5, *C-D*-9, and *A-D*-5. Construct a true-length diagram (see Figure 8-4). The shortest seam, and one that is on a flat surface, is a vertical seam through the mid-point of one of the triangular surfaces. Begin the pattern with the line 1-*X* (which is true length in the front view). Add the base line *XB* perpendicular to 1-*X*. *XB* is recognized in the top view as being perpendicular to 1-*X* because of a descriptive geometry theorem which states that *lines which are perpendicular in space appear perpendicular in a view in which at least one of them is true length*. They appear perpendicular in the top view, and *XB* is true length there also. Triangulate the segments 1-2-*B*, 2-3-*B*, 3-4-*B*, and 4-5-*B* (one-fourth of an oblique, ellipti-

cal cone). Triangulate *B*-5-*C*. Duplicate the previously developed elliptical cone segment 1-5-*B* as 5-9-*C*. Duplicate the previously developed right triangle 1-*B-X* as 9-*C-Y*. Half the pattern is thus complete, as shown in Figure 11-18.

11-9 Square-to-round transition with offset

The square-to-round transition shown in Figure 11-19 is drawn and developed in a manner similar to that shown in the previous problem. The only difference being that the center lines of the two pipes are offset; hence measurement *A* is required for the fitting from specifications.

11-10 Oblong-to-round transitional offset

Figure 11-20 shows the top view of the required openings for the fitting, the numbered points, and the front view of the lines on the surface. Observe that a numbering system is again used which will facilitate the triangulation of a zigzag path when developing the pattern. The numbering system shown also allows the draftsman to divide and keep track of the various true lengths, using two true-length diagrams, rather than to crowd them into a single one. The true lengths to the left of the front view are for lines numbered increasingly from even to odd, while the true lengths to the right of the front view are for lines numbered increasingly from odd to even. For example, line 6-7 is diagrammed to the left; line 5-6 to the right.

Once the top and front views of the fitting have been established, as in Figure 11-20, the procedure for triangulation is similar for all problems. The major area of study should be directed toward fitting design and the breakdown of surfaces into developable triangles.

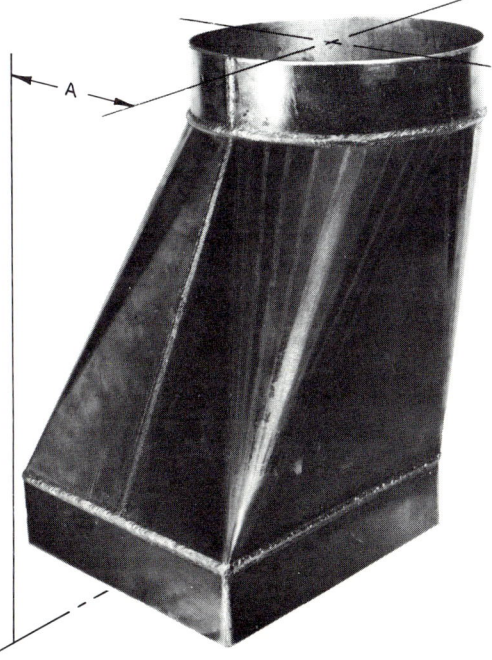

Fig. 11-19 (*Naylor Pipe Co., Chicago, Ill.*)

Fig. 11-20 Oblong-to-round offset.

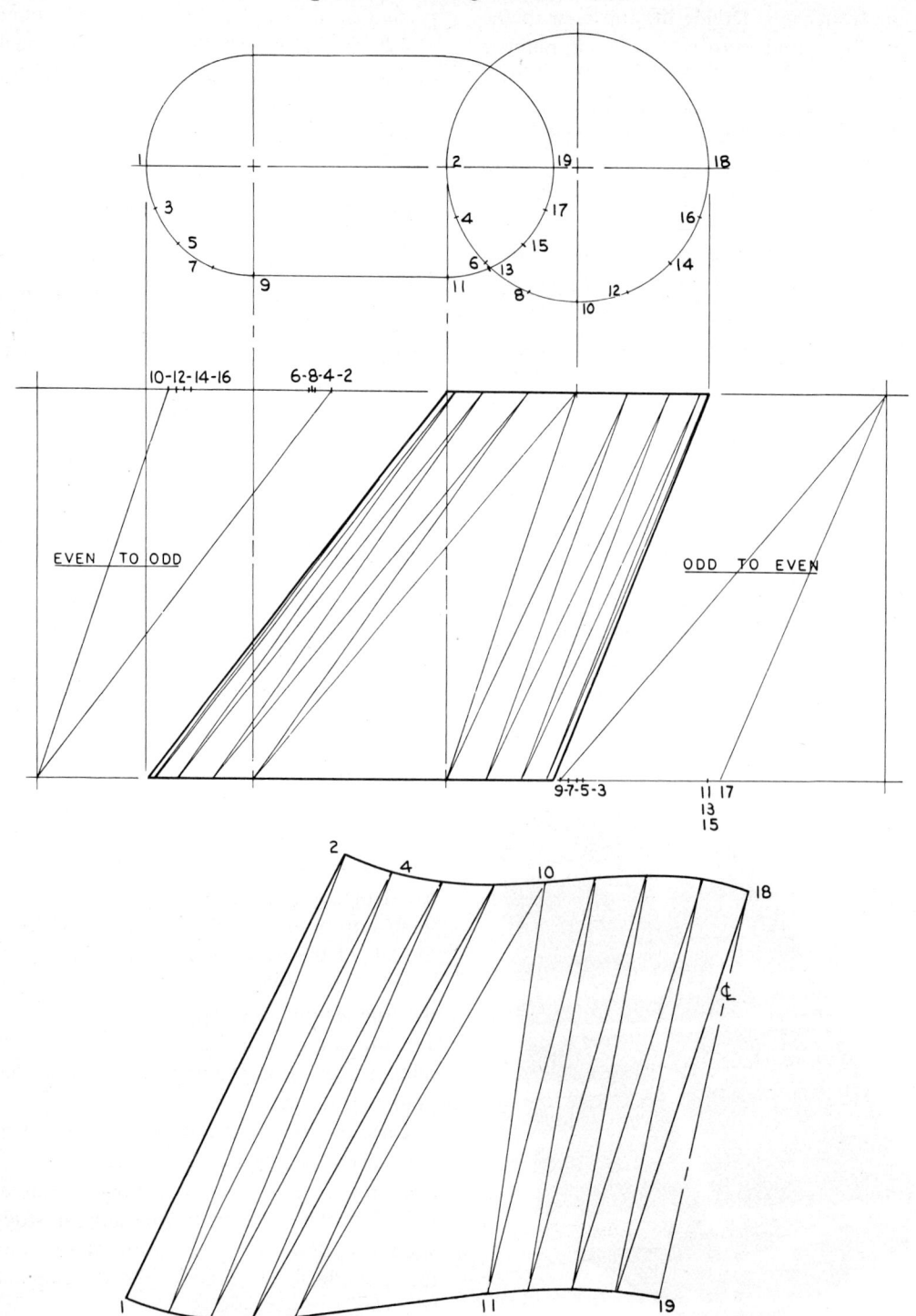

TRIANGULATION

Fig. 11-21

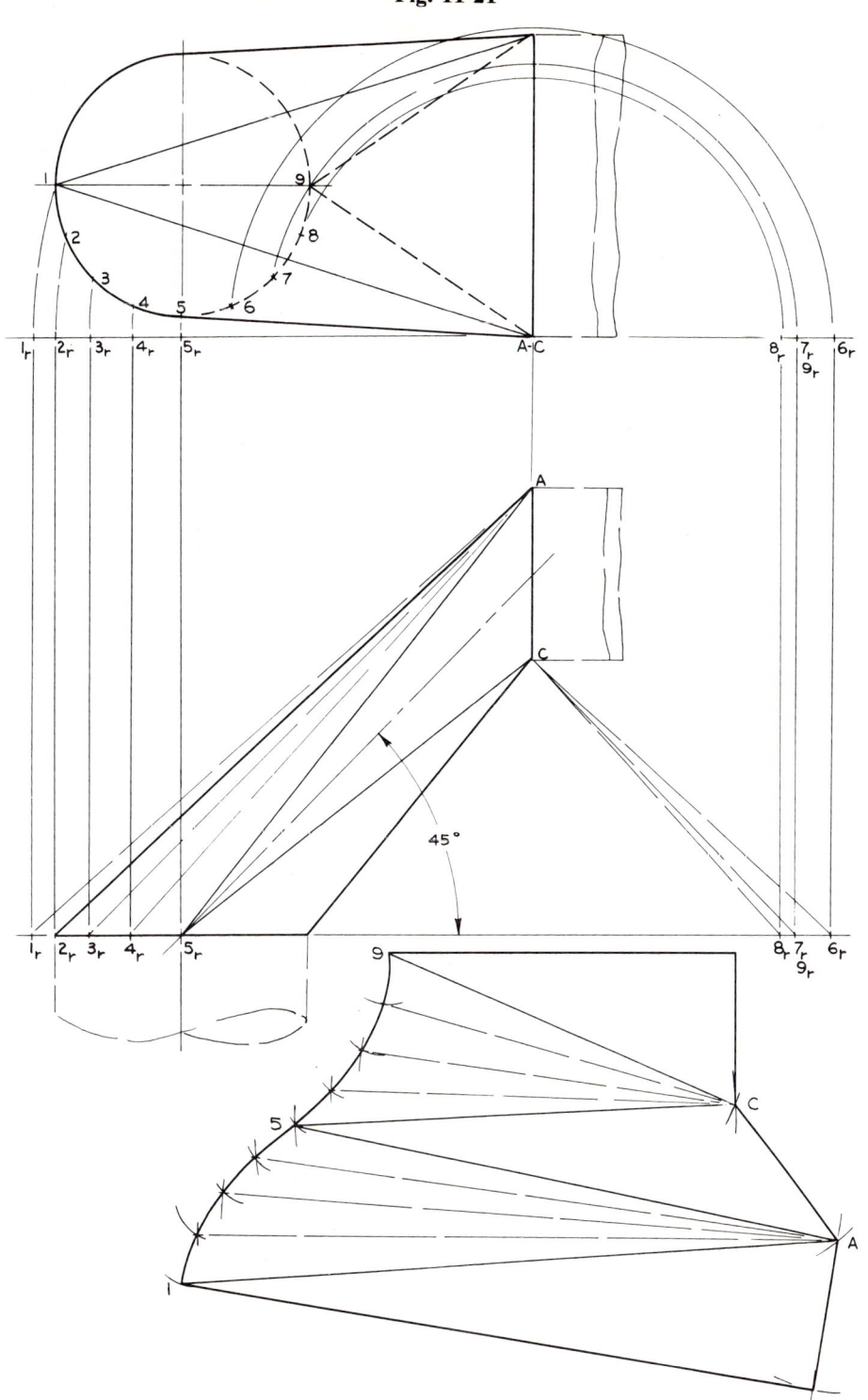

11-11 90-degree rectangular-to-round transition

In furnace work, this fitting may be called a *boot* or a *take-off*. As such it is designed for rather tight quarters and is used as a register take-off or as a vertical stack boot. The fitting shown in Figure 11-21, however, is designed for an open installation. Thus, the throat of the elbow may be made large enough to avoid excessive restriction to air flow in the fitting. Notice in Figure 11-21 that the center line of the fitting is at 45 degrees with both connecting pipes, which distributes the amount of restriction and equalizes each end.

To develop a pattern, draw the top and front views and identify the points necessary to create triangles on the surface. In this problem it is easy to secure the true length of each of the elements of the elliptic conical segments by rotation. Observe that points 1, 2, 3, 4, and 5 are rotated about a vertical axis through *A* in the top view until parallel with the front plane of projection. Thus the lines *A*-1, *A*-2, *A*-3, etc., become true length in the front view. The points 5, 6, 7, 8, and 9 are rotated clockwise about a vertical axis through *C* in the top view and are stopped in the same frontal plane as *C*. Consequently, the lines 5-*C*, 6-*C*, 7-*C*, etc., become true length in the front view. Since the triangulation procedure is the same as for previous problems, only a partial pattern is illustrated in Figure 11-21. A semipattern would be required and two pieces formed in opposite hand to construct the fitting, with seams at points 1 and 9.

11-12 Transitional elbows of several pieces

Transitional elbows involving square or rectangular openings transforming into cylindrical or elliptical openings may be designed as an elbow of several pieces, as pictured in Figure 11-22. Fittings of this type are complex and costly in time and labor. They can be avoided by using constant shape elbows with transition made separately. On occasion, however, such fittings are required, as illustrated by the five-piece rectangular-to-round elbow in Figure 11-22 (made from 8-gauge [4-millimeter] steel). A study of transitional elbows is worthwhile because they lead into a series of problems in which the ends of each part of the fitting are not in parallel planes. Since the initial steps in the drawings of a transitional elbow are similar to those in the design of a conical elbow, the student should review section 10-7 and Figure 10-22.

After review of the reference, consider as a typical example the design and development of a three-piece 90-degree square-to-round transition. Each piece of the elbow is to be treated as a separate problem consisting of top, front, and necessary auxiliary views. Before each piece can be drawn, however, *a master drawing of the transition as a straight fitting must be made.*

Begin the master drawing from specifications, as shown in Figure 11-23: (1) Draw a base line and a vertical center line. (2) On the base line, center and measure the edge view of the square opening (3′-0″ [914

Fig. 11-22 (*Naylor Pipe Co., Chicago, Ill.*)

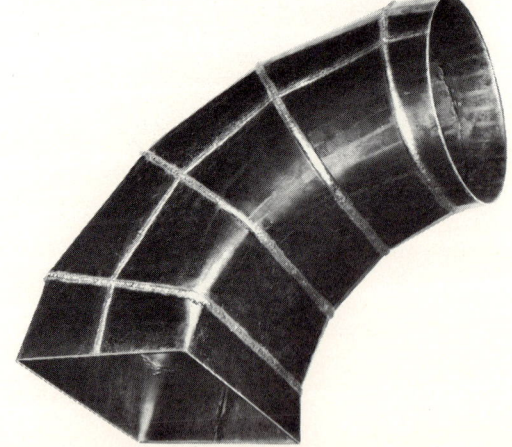

Fig. 11-23
DIMENSIONED IN MILLIMETERS

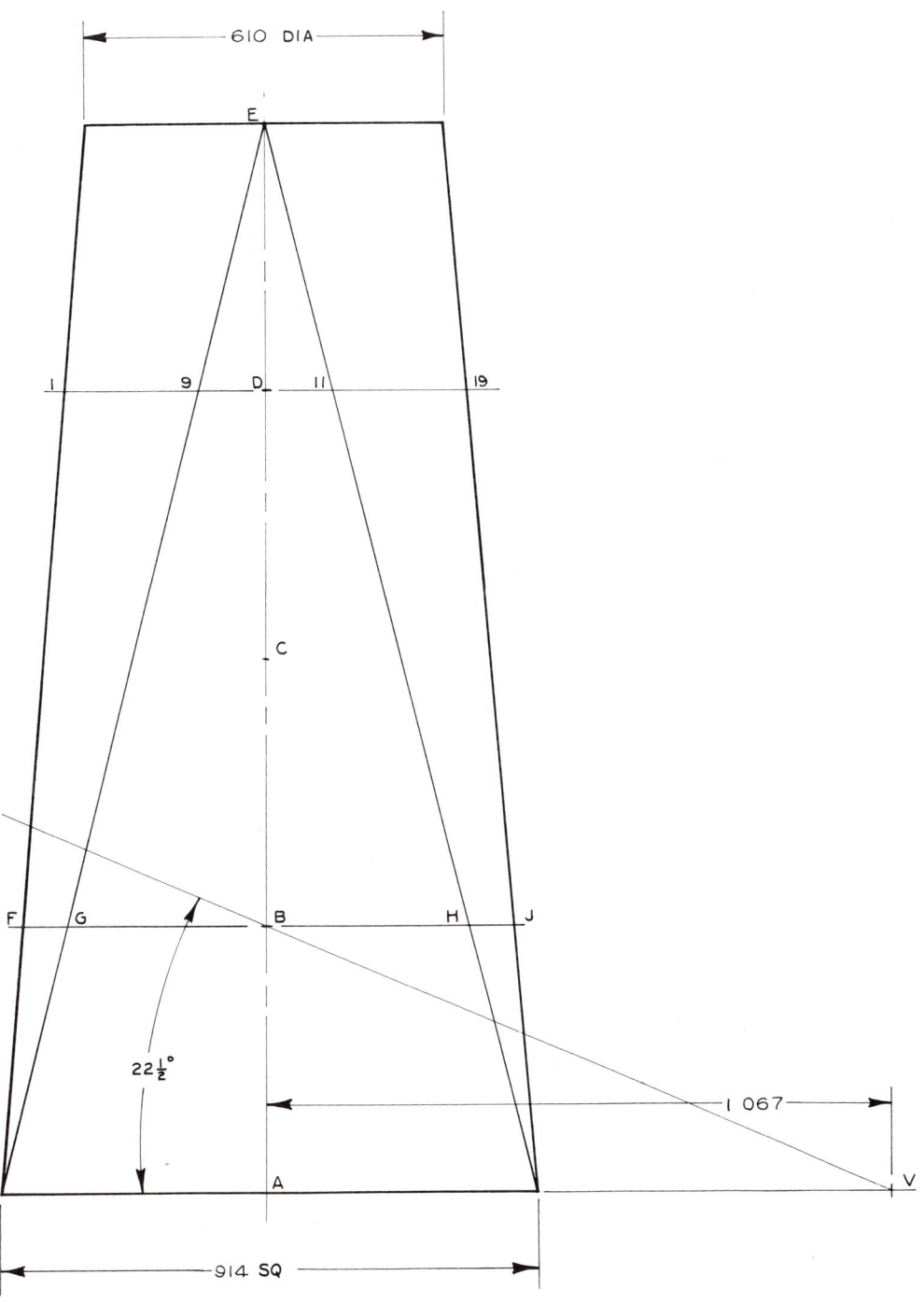

mm]). (3) Measure the specified center-line radius (3'-6" [1 067 mm]) to locate the center of the elbow V. (4) Calculate the miter angle (same as for a cylindrical elbow). Thus, in Figure 11-23, the miter angle equals 22½ degrees. The angle drawn through V determines the center-line space AB. (5) Calculate the number of spaces needed for the entire fitting (spaces = $2N - 2$). (6) Since four spaces are needed in the example, AB is duplicated as BC, CD, and DE. (7) Through E, center and measure the edge view of the round opening (2'-0" [610 mm] diameter). (8) Complete the view of the straight transitional as shown. (9) Two joints of the fitting pass through points D and B; therefore, *right sections* taken through these points result in *square openings with rounded corners.* Such openings, being symmetrical, will still match up when rotated 180 degrees. Hence, the section established at D may be used as the base of a transitional piece whose center line is DE. The right section established at B may be used as the base of a transitional piece whose center line is AB. Likewise, the right sections established at D and B may be used to create a transitional whose center line is DB (the center piece of the fitting).

In making the drawings of the individual pieces, care must be taken relative to three items being transferred from the master drawing. First, the bases must be the proper center-line distance apart. Second, the bases must be at the proper angle to one another or to the center line (E and D are at 22½ degrees to each other, D and B are at 45 degrees to each other). Third, the bases used *must be those established as right sections of the master transition.*

To draw the necessary views of the end AB, as shown in Figure 11-24: (1) Construct base line, center line, miter angle, and edge view of the square opening as before. (2) On the miter line through B, locate points F, G, H, J, as taken from the right section on the master drawing. (3) By means of auxiliary projection, construct a half view of base B in true size and shape, as shown. This is easily done because the right section is square with rounded corners, and the radius for the corners is projected from the front view. (4) Divide the rounded corners of the auxiliary view into a suitable number of equal spaces and project the points back to the front view. (5) Complete the front view. (6) Construct a half top view of both bases (base A is a square; the curves of base B are plotted by projection from the front view and by reference transfer from the auxiliary view). Complete the views of the fitting and identify the points. Notice that the front face of the pipe is divided into two triangles by the line 1-H; this is done to avoid a warped surface; the front face then becomes two triangular plane surfaces. (7) Establish true-length diagrams for the corner elements and line 1-H as shown. (8) Triangulate a half-pattern for piece AB as follows: Starting with the true length of MN, as taken from the front view, construct two lines perpendicular to MN on the pattern. Measure the true lengths of N-4 and MJ, as shown in the top view, and transfer them to the perpendiculars on the pattern. Triangulate the remaining segments of the pattern as for previous problems. Observe that true distances between points on the curves *must be obtained from the auxiliary view.* The completed semipattern is shown in Figure 11-24.

To draw the necessary views of center piece BD, place the right section 1-D-19 horizontally, as shown in Figure 11-25. Thus the center line is 22½ degrees off vertical and the right section through B is at 45 degrees to the horizontal section. This represents the best position for constructing the partial top and front views. The points are numbered and arranged as for previous problems so that the pattern can be zigzagged in numerical order. Also, the true-length diagrams are divided so that lines increasing as

TRIANGULATION

Fig. 11-24

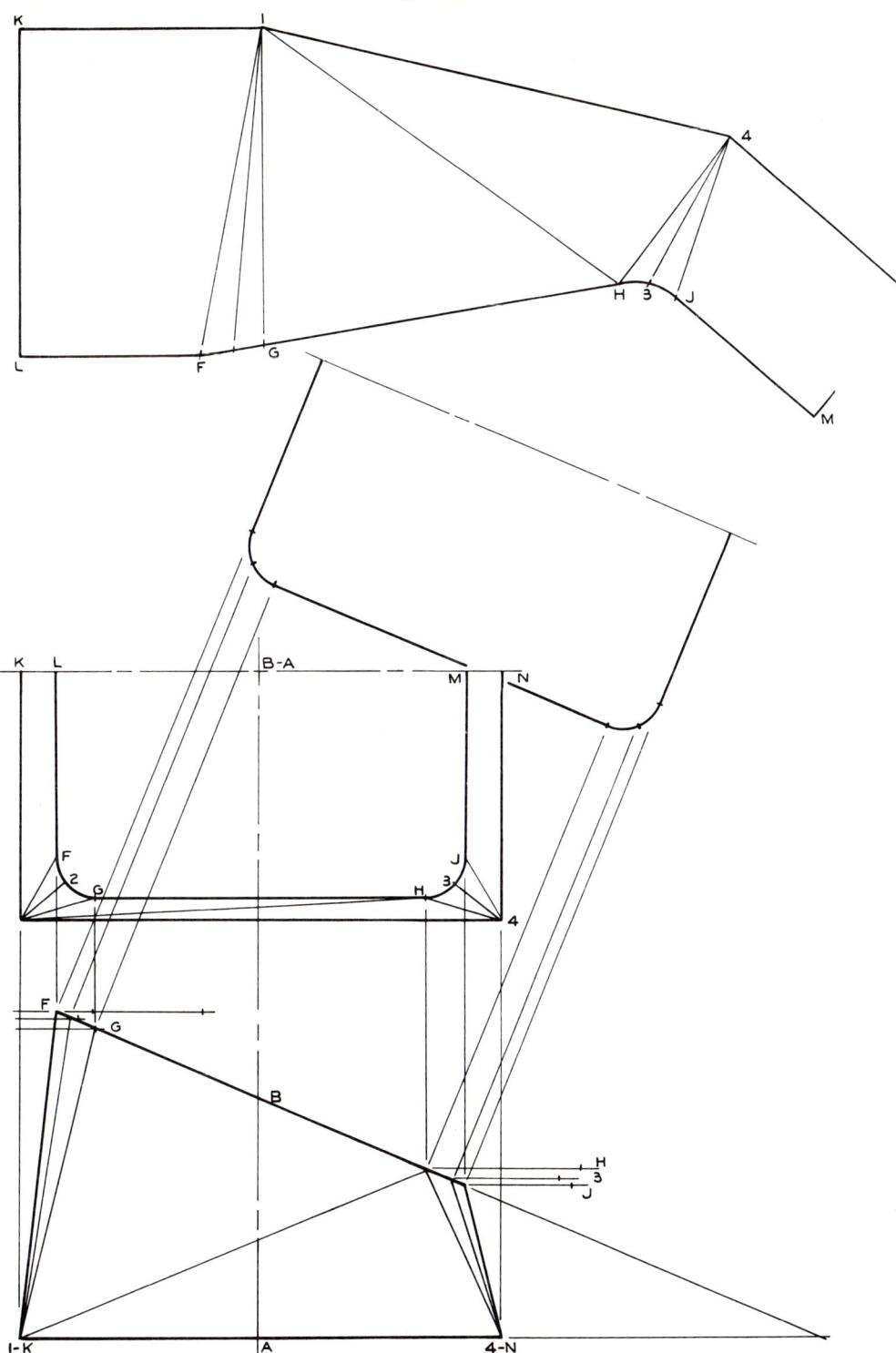

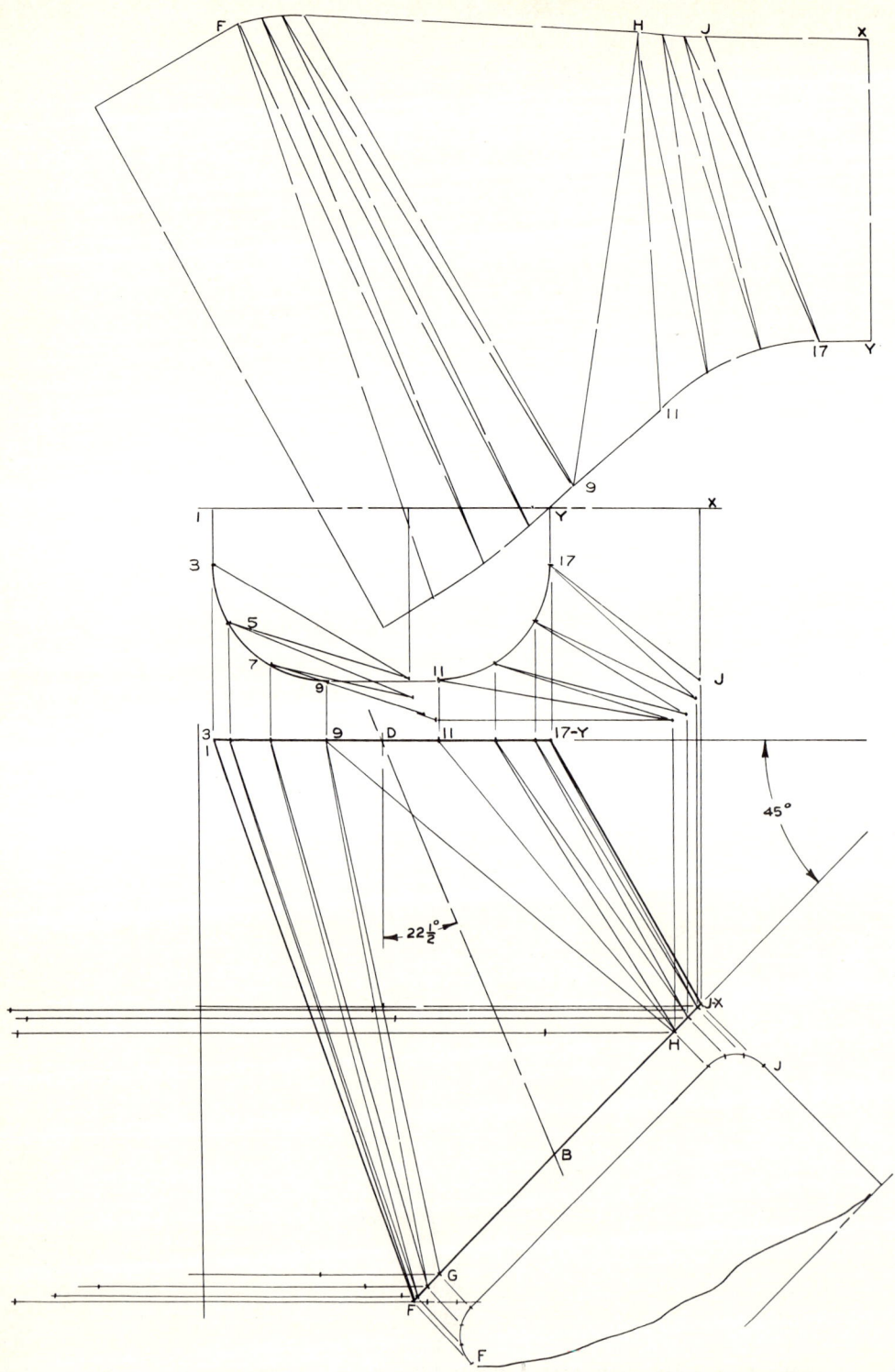

TRIANGULATION

even to odd numbered are to the left while lines increasing as odd to even numbered are to the right. The pattern is started with the shortest element *XY*. Again note carefully that the true distances between points on the curves of the horizontal section are obtained in the top view while the *true distances between points on the curves of the inclined section are obtained from the auxiliary view.*

Figure 11-26, in a similar manner, shows the treatment for development of the *DE* piece of the elbow. The three pieces assembled to form the complete transitional elbow are shown in Figure 11-27. Notice that the transformation is uniform on all surfaces of

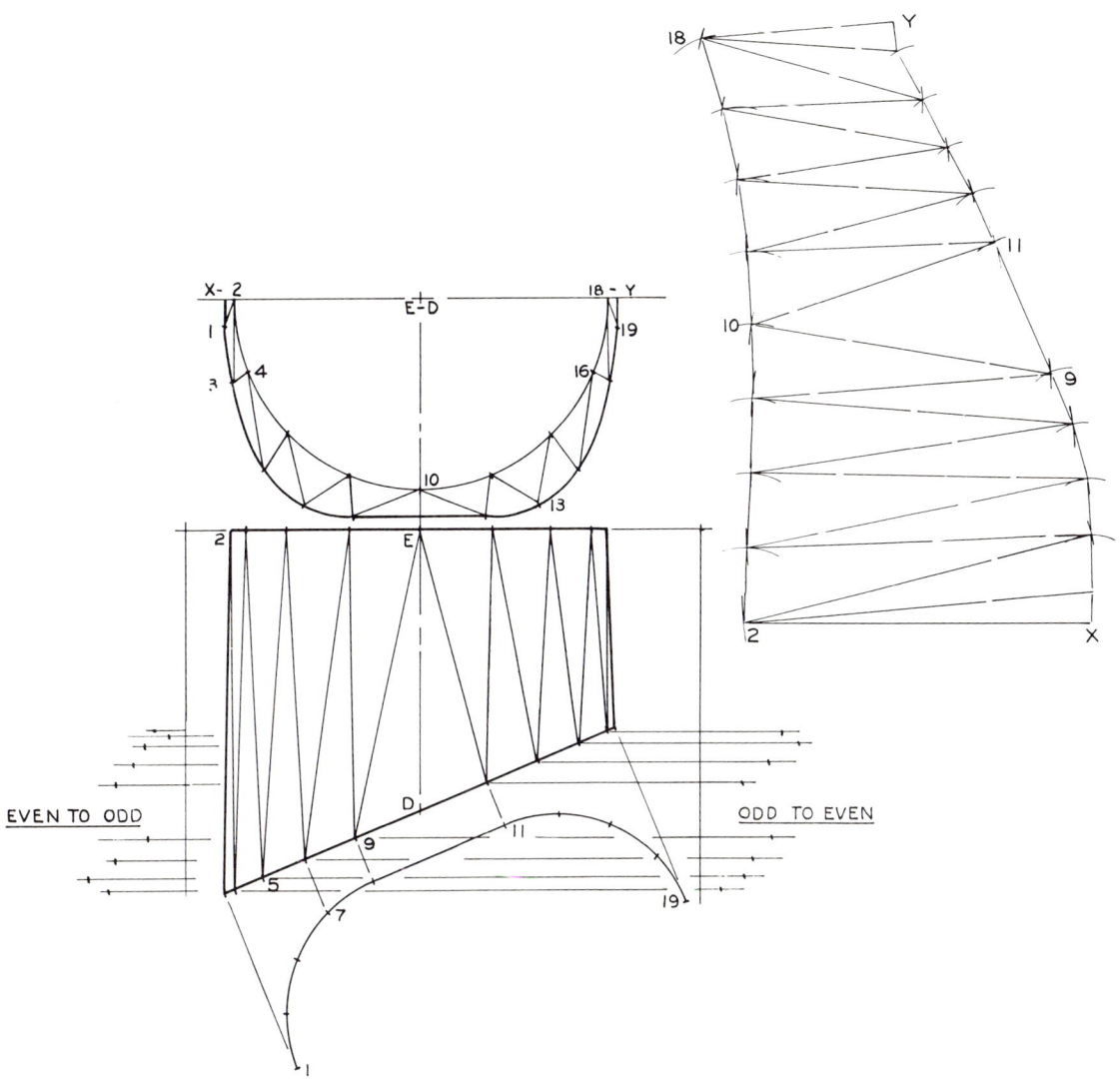

Fig. 11-26

DIMENSIONED IN MILLIMETERS

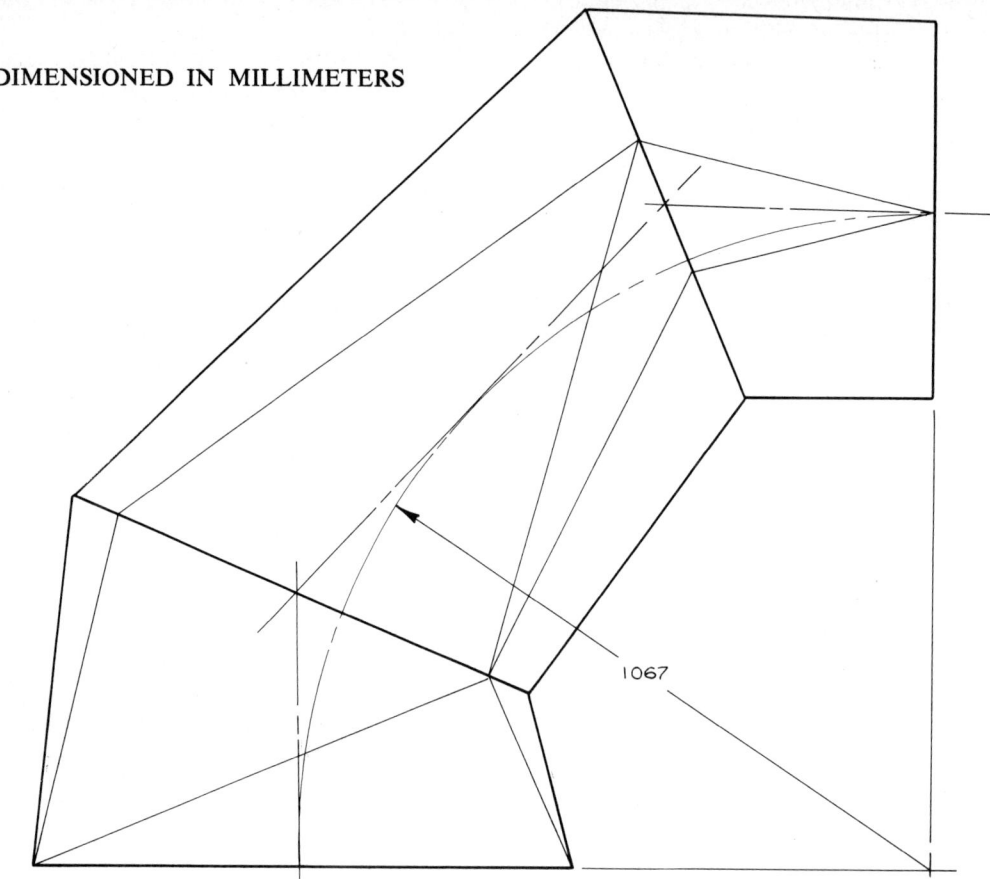

Fig. 11-27 Square-to-round 90-degree transition elbow.

the fitting and that the center-line radius (3'-6" [1 067 mm]) is maintained throughout.

11-13 Rectangular-to-round transitional elbow

If the transitional elbow involves an opening in which the dimensions are different in the side view, the additional view is required as a part of the master drawing. For example, consider a rectangular-to-round five-piece 90-degree elbow. The necessary views on the master drawing are constructed as shown in Figure 11-28. The side view shows that the radius for the rounded corners of a right section is the same in both side and front views (measurement B) *as long as the top opening is a true circle.* The side view also furnishes the measurements C, D, and E used in completing right sections when drawing the individual pieces (see Figure 11-29). Section AA of Figure 11-28 is a typical right section of the fitting and is used as the miter joint between pieces 3 and 4. For the necessary views and development of piece 3, see Figure 11-29. Note that the miter angle for a five-piece 90-degree elbow is 11.25 degrees; hence the angle between the two miter lines on piece 3 is 22½ degrees. Four additional pieces for the complete fitting are developed in the same manner as piece 3 in the example.

TRIANGULATION

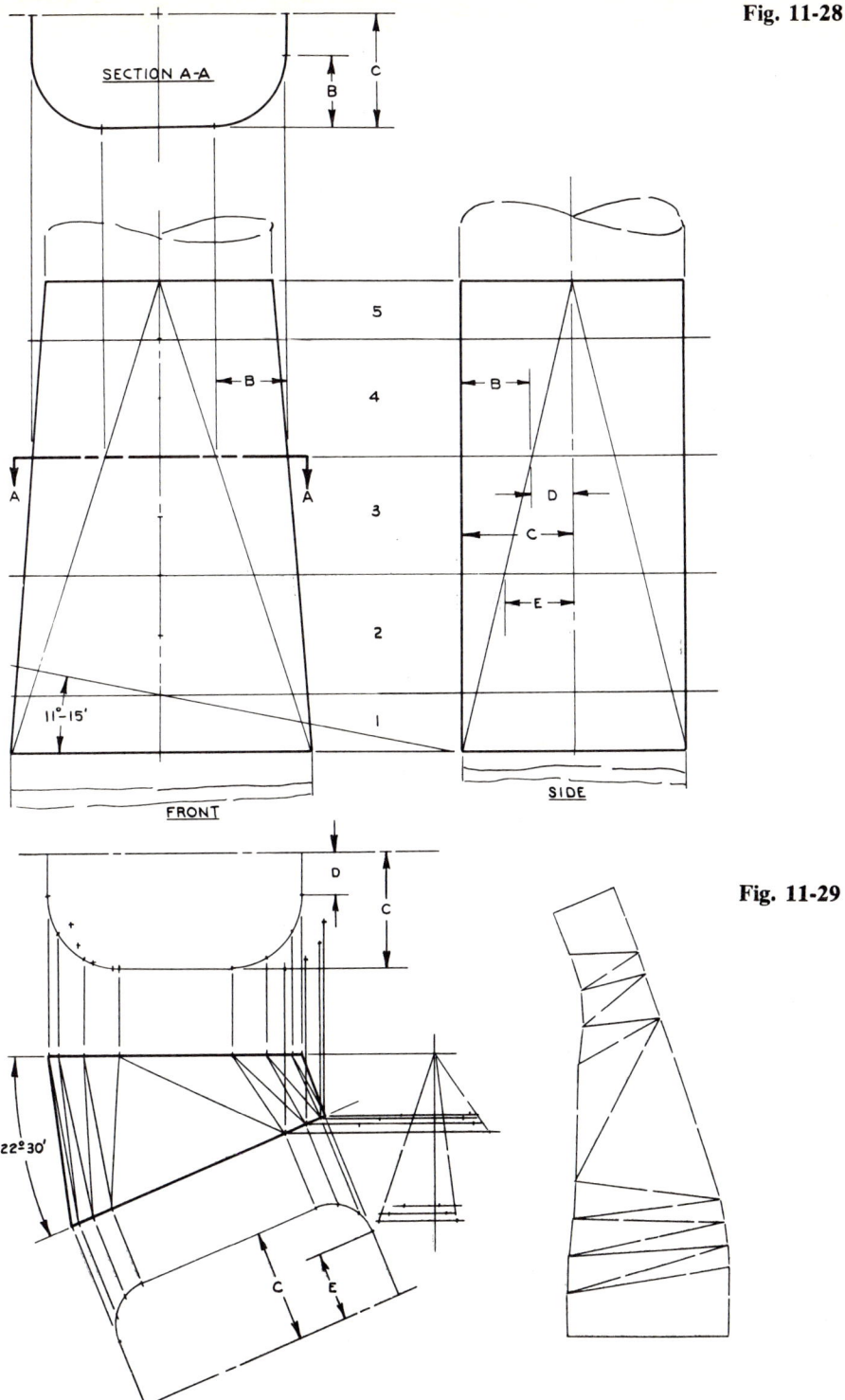

Fig. 11-28

Fig. 11-29

11-14 Round-to-elliptical elbow

The procedures established for the previous transitional elbows may also be applied when developing a pattern for a round-to-elliptical fitting. For example, in the master drawing of Figure 11-30, the front and side views of a round-to-elliptical transition are cut in right sections so as to provide miter lines for a three-piece 75-degree elbow. Figure 11-31 shows the views necessary for the development of the center piece of the elbow. The miter planes are established at 37½ degrees to one another (18.75 degrees to center line). Notice that the miter sections are ellipses and that the major and

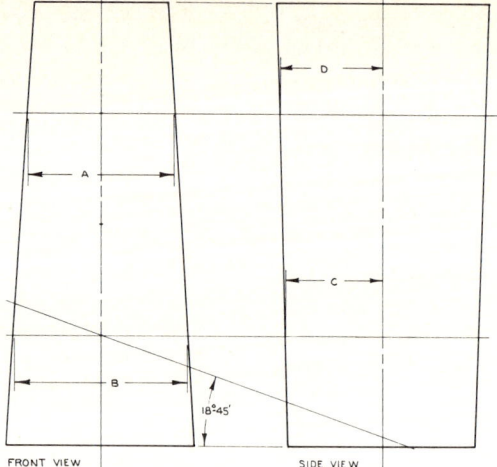

Fig. 11-30

Fig. 11-31

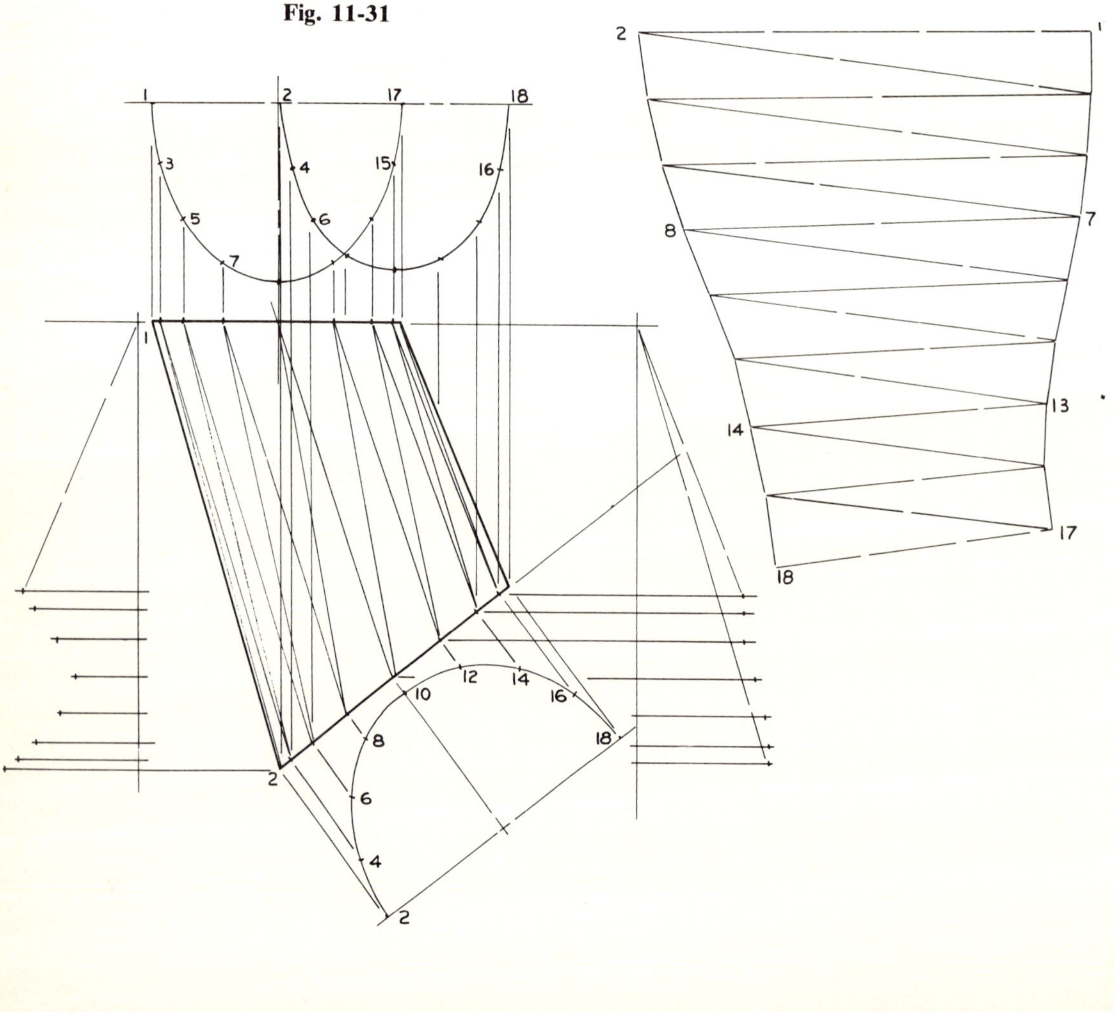

TRIANGULATION

minor diameters are obtained from the master drawing as indicated by dimensions A, B, C, and D. From the major and minor dimensions, the elliptical curves are constructed by trammel (see section 5-14). Divide each curve into a suitable number of equal parts, as shown. The usual numbering system and true-length diagrams furnish the information needed in order to triangulate a pattern for the center piece.

11-15 Transitional elbows in ductwork

Transitional elbows are sometimes needed in heating and air-conditioning ductwork, as shown in Figure 11-32. The fitting shown is part of a rectangular plenum (6'-0" by 7'-0" [1 829 by 2 134 mm]) which is transformed into a horizontal duct (2'-6" by 5'-0" [762 by 1 524 mm]). The top and front views of the transitional elbow are shown in Figure 11-33. While the fitting could be designed with a continuous transformation on the cheeks, such a design would require a continuously warped surface. The company engineers involved in the example of Figure 11-32 wisely designed the fitting so that the transition on the cheeks takes place by means of triangular plane surfaces A, B, C, and D. Such a design also takes into consideration standard sheet sizes since some seaming is necessary. With surfaces A and D true size and shape in the front view, all of the cheek transformation is made in two steps by B and C. On the heel and throat strips, the transition is continuous, as shown by the developments in Figure 11-34. The development of heel and throat strips is a parallel-line problem (Chapter 9). Parts A and D of the cheek pattern are directly transferable from the front view (with seam allowances added). The few true lengths needed for the triangulation of B and C are easily determined by rotation (see Chapter 8).

Fig. 11-32 90-degree square-to-rectangular transition off plenum. (*York and Shipley Co., York, Pa.*)

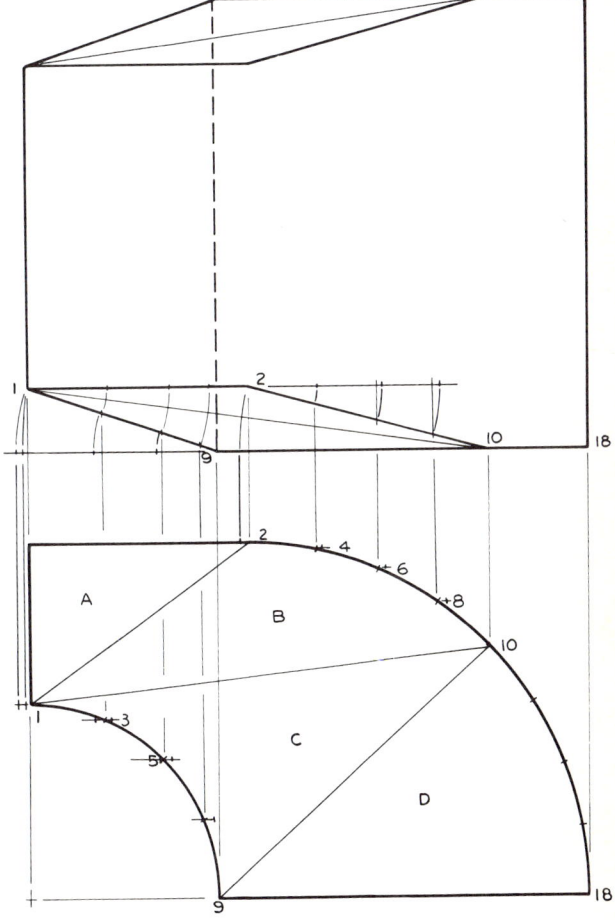

Fig. 11-33

228 SHEET METAL DRAFTING

Fig. 11-34

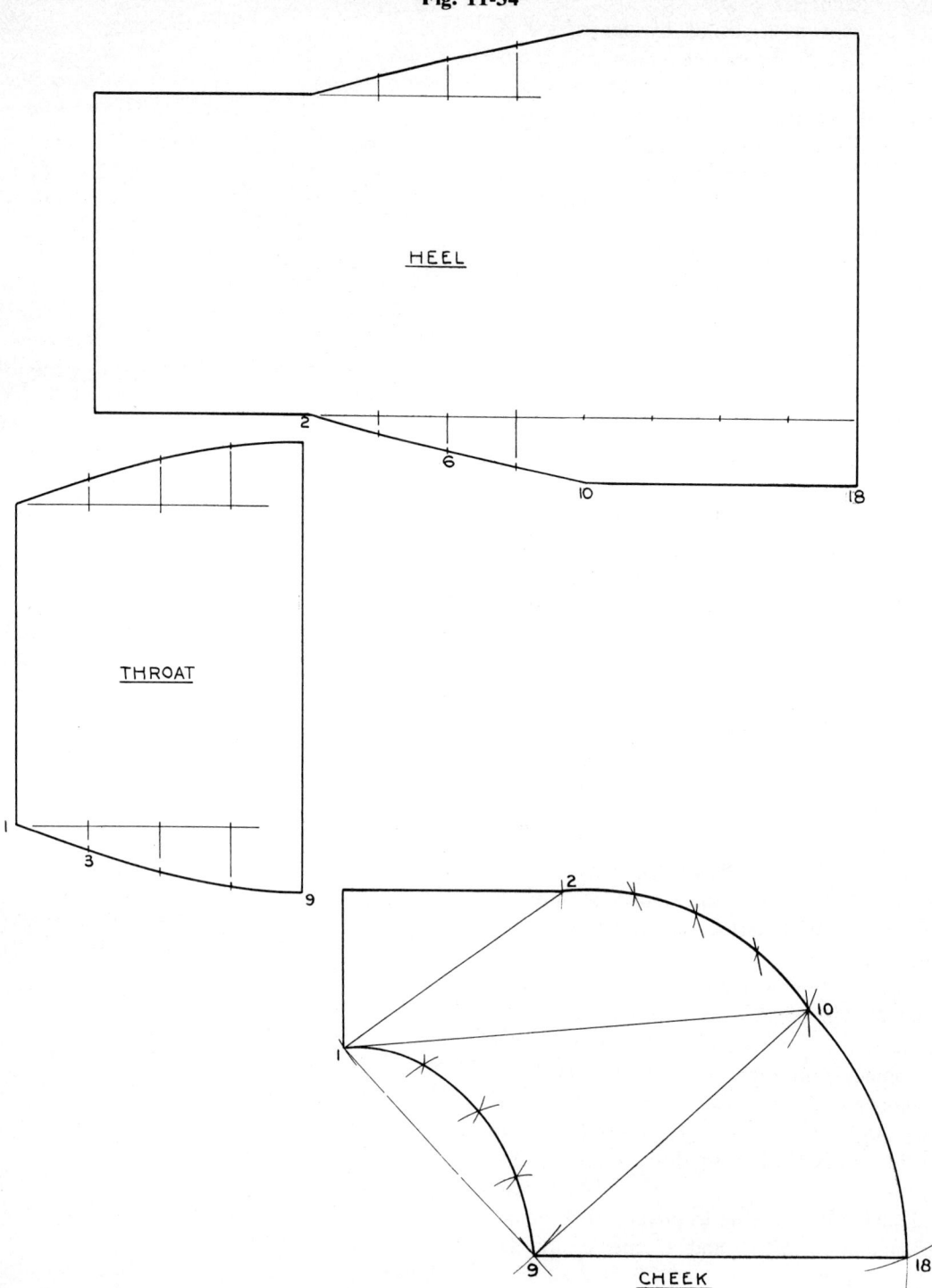

TRIANGULATION

BRANCH FITTINGS

11-16 Symmetrical two-pronged oblique cones

Branch fittings are sometimes needed in round and square pipe or in various combinations of them. Such fittings, while needed only occasionally in practice, do present challenging problems to the student. The development of patterns for Ys, or breechings as they are sometimes called, is a continuation of the triangulation procedures previously applied. However, the line of intersection between the various intersecting branches must be determined before development can be made. Branch fittings are composed principally of oblique cones, warped cones, transitions, or combinations of the above.

To begin with an elementary example, the oblique cone frustum for a tapered offset, described and developed in section 10-6, is used in the design of a branch fitting. In Figure 11-35, the front and partial bottom views drawn from specifications indicate that the fitting is symmetrical with both branch pipes at the same angle and size. Therefore the line of intersection of the two oblique cones is a vertical miter line on center. Since both branches are the same, only one need be drawn and developed, as shown in Figure 11-36. To make the drawing and development: (1) From specifications draw the

Fig. 11-35

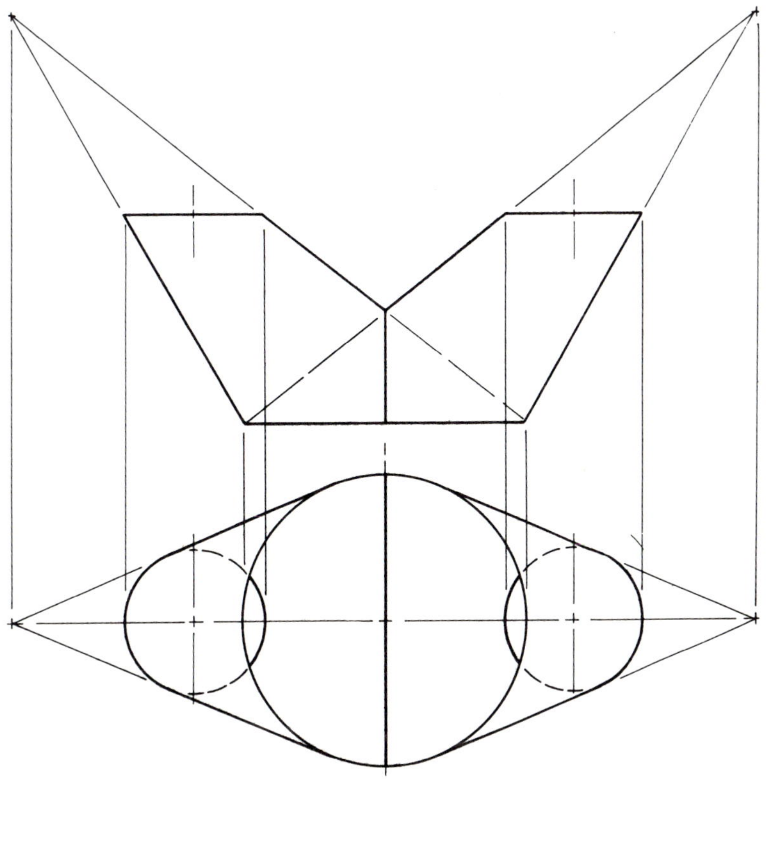

SHEET METAL DRAFTING

Fig. 11-36

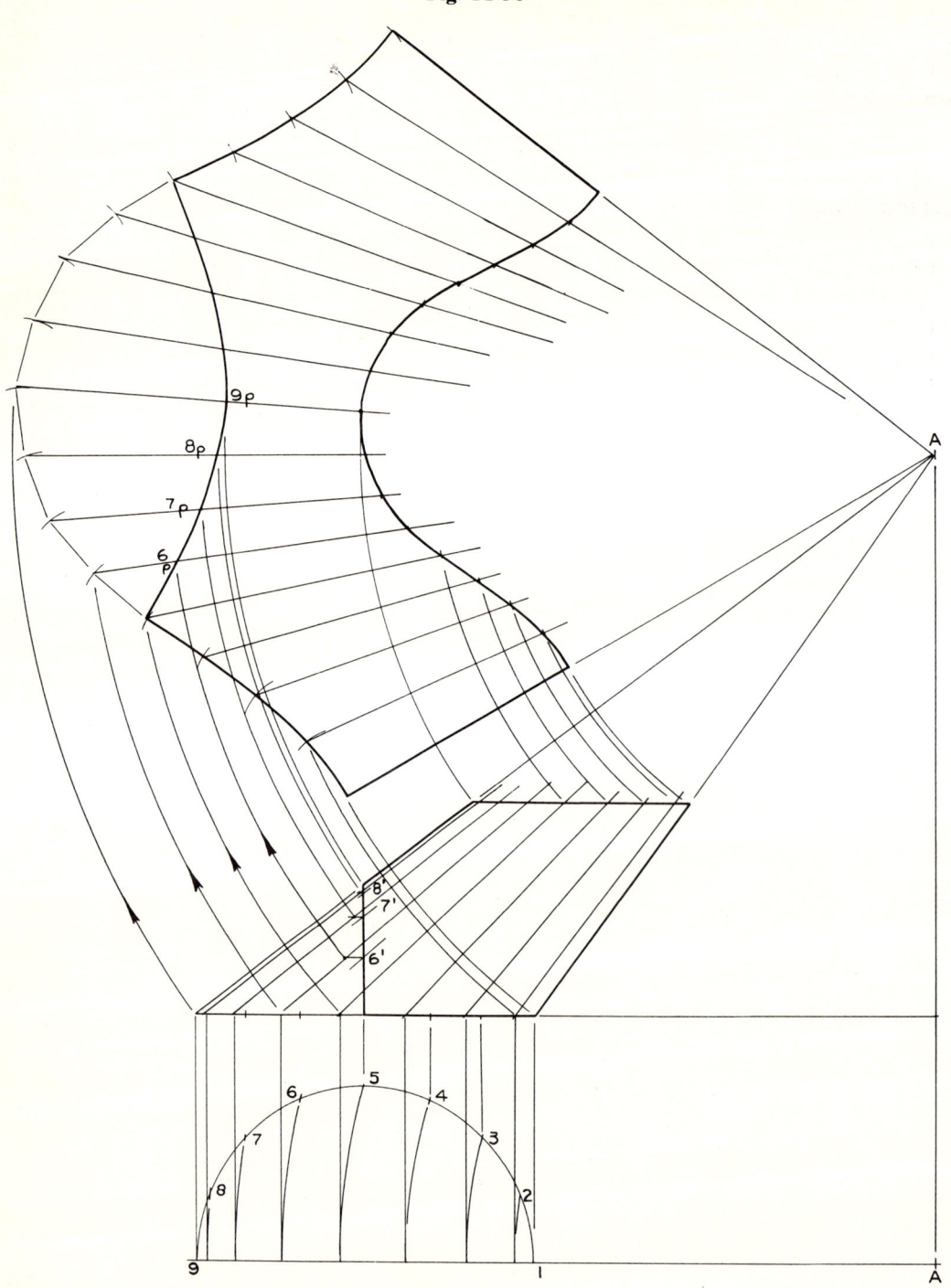

front view and vertical miter line of one prong of the fitting. Extend the cone to its apex, as shown in Figure 11-36. (2) Draw a half-bottom view of cone base and apex only. (3) Divide the base into a suitable number of equal spaces and number the elements thus established (1 through 9 and back to 1). (4) Project the elements to the base of the front view and, as if drawing the elements in the front view from base to apex, *make only short lines across the miter line.* This locates the points on the miter line where each element terminates at the joint 6'-7'-8'. (5) Using the apex as center, rotate the elements *in the bottom view* until each is parallel to the front plane of projection. (6) Project the *rotated* elements to the front view. (7) Draw the *rotated* elements in the front view, where each *is now true length.* (8) Using the apex as center, sweep the true lengths into the pattern development area, as shown by the arrows of Figure 11-36. (9) To triangulate the pattern before notching for the miter joint, set the compass equal to the true distance between elements (as taken from the bottom view) and starting at any place on the arc from element 1, strike the compass setting to cross arc 2; using this point as center, strike the setting to cross arc 3, etc. Repeat this until points for all elements on the pattern have been located. (10) Draw the elements on the pattern. (11)

The piercing points previously located (6'-7'-8') are now rotated in the front view by following a *straight* path *parallel* to the base until they join their respective true-length element (see Figure 11-36). They are then swept into the pattern to locate points 6_p, 7_p, 8_p, as shown. (12) Draw the curve for the pattern. (13) The points for the top curve of the pattern are swept onto the development in a similar manner. Notice that the development of this problem is essentially the same as for the tapered offset of Figure 10-20, the difference being that the piercing points of the elements with the miter plane must first be determined before they can be rotated to the true length position in order to complete the pattern.

11-17 Two-pronged symmetrical branch

Pattern development of branch fittings, once the necessary surface triangles have been determined, is similar to previous triangulation procedures. The student should therefore study the design possibilities relative to a given situation and the factors which will affect the final surface appearance of the fitting. Items such as cross-sectional areas, shapes, and surface classifications are typical factors affecting final appearance.

Relative to pipe areas, it has been established that it is desirable to maintain velocity pressure in the branch pipes equivalent to that in the main or trunk line. Thus the volume delivered by each branch is uniform and a minimum loss of energy occurs. To maintain equivalent velocity pressure, the total cross-sectional area of the branches must be approximately the same as that of the main pipe. For example, in Figure 11-37 the main has a diameter of 4 inches [102 mm] and a cross-sectional area of 12.56 square inches [8 103 square millimeters] ($A = 0.7854D^2$). To determine a suitable size for the branch pipes, because there are two

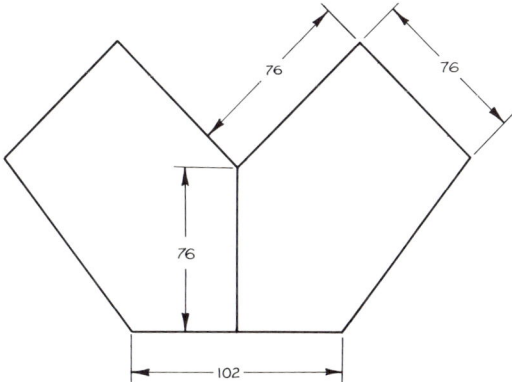

Fig. 11-37

DIMENSIONED IN MILLIMETERS

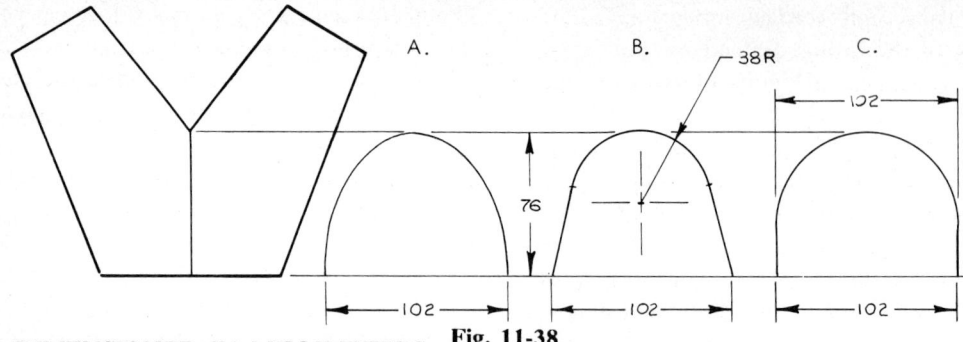

DIMENSIONED IN MILLIMETERS Fig. 11-38

branches, use one-half the main pipe area (6.28 sq in [4 052 sq mm]). Using the formula $D = \sqrt{A \times 1.27}$, the diameter of a branch pipe is computed to be 2.83 in [72 mm]. The next nominal size larger (3 in [76 mm]) is therefore indicated in Figure 11-37.

In addition to the opening sizes as factors in the design of a fitting, one must also consider the branch angles and the resulting size and shape of the fitting at the miter joint. An included angle of 10 degrees or less is necessary for high-velocity piping in order to minimize friction and turbulence losses. While the size at the miter joint may be set arbitrarily (the height may vary, thus regulating the area) a good rule of thumb for height is to make it about equal to the branch diameter. The area of the joint section should be approximately midway between main and branch area values. For example, in Figure 11-37 the measurements for a semielliptical area are 2 in [51 mm] semiminor diameter, and 3 in [76 mm] semimajor diameter. The area of a full ellipse is computed using the formula A = semimajor diameter $\times$ semiminor diameter $\times$ π; therefore, $3 \times 2 \times 3.14 = 18.84$ sq in [$76 \times 51 \times 3.14 = 12\ 170$ sq mm]. The joint section of the fitting, being a half ellipse, is therefore $18.84/2 = 9.42$ sq in [$12\ 170/2 = 6\ 085$ sq mm].

The shape of the joint section may also be established arbitrarily and the surfaces developed to produce the desired shape. It is preferable, however, to determine a shape which resembles the true cross section of the branches. Figure 11-38 illustrates possible joint sections to comply with the measurements of Figure 11-37. The elliptical section A is the preferred joint section because it produces a warped surface with roundness which gradually transforms half the main flow into each branch. Example B is a joint section easily established using the branch diameter as a semicircular top with flat tapered sides. The front view of a branch using a joint junction shaped like B appears as shown in Figure 11-39. Example C is to be avoided, except for low velocities, because the shape results in flats parallel to the large diameter. The flats produce a sudden expansion chamber within the fitting. A sudden expansion area or contour which produces nonuniform flow will increase the

Fig. 11-39

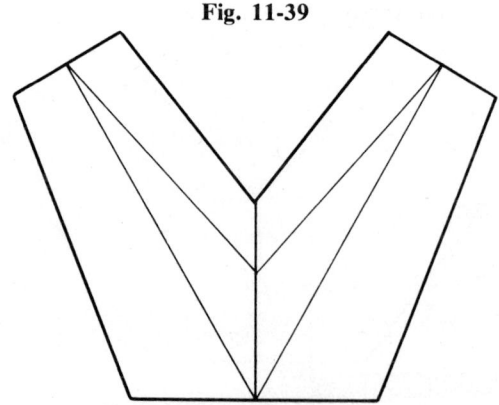

losses in the circuit. Figure 11-40 illustrates a branch design (breeches) that is satisfactory for low velocity when the branch angles are large.

The triangulation and development of the designs thus far mentioned are illustrated by the pattern for one branch of a typical fitting, as shown in Figure 11-41. The pattern is developed as follows: (1) Draw a front view of one branch to meet the specifications of Figure 11-37. (2) Construct half views of the openings (partial auxiliary and bottom views respectively). To save space, it is good practice to construct them attached to the front view, as shown. (3) Construct an elliptic curve to represent a partial side view of the miter joint. This is drawn using the trammel method of Figure 5-25. (4) Divide the half view of the branch opening into eight equal parts. (5) Divide the half view of the semicircular base into four equal spaces. (6) Divide the side view of the elliptic curve into four equal parts. Note that the total number of spaces on the curves of the miter and base is equal to the total number on the branch opening. (7) Project all points to the front view. (8) Number the points, as shown, so that the pattern can be started with line 1-2 and triangulated in numerical order. (9) Since no top view is required, the true length of each line is constructed from information in the front and partial views. To construct a true-length diagram involving only lines increasingly numbered even to odd, set off vertically the depth dimensions for points 3, 5, 7, and 9 (as shown for point 5 by the sample dimension A). Measure the distance between points *in the front view* and transfer to the base line of the diagram (as shown for line 4-5 by the sample dimension B). Erect a vertical in the diagram for each even numbered point and transfer the depth dimension (as shown for point 4 by the sample dimension C). Thus the true length of the even-to-odd line 4-5 is found on the diagram (as shown by the measurement TL). The remaining even-to-odd lines are set off in the diagram in the same manner. (10) Construct a diagram for the lines increasingly numbered odd to even. (11) Begin with line 1-2 (which is true length in the front view) and triangulate a half-pattern. End with line 17-18, which is also true length in the front view. Remember to take the distances between curve points from the partial views where the distances are true length.

11-18 Two-pronged unequal branches

When each branch may vary as to angle and/or size, a pattern must be developed for each piece. This results in a considerable number of lines and twice as many true-length constructions. For the student, the construction can be clarified by first making a master drawing of the fitting, then making separate drawings of each branch for development of the respective patterns. The master drawing is designed in a manner similar to previous problems. A typical situation

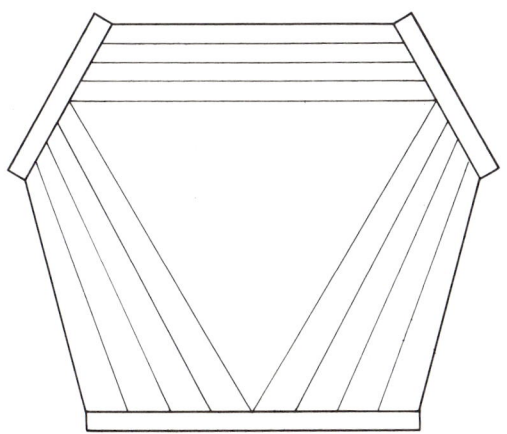

Fig. 11-40

SHEET METAL DRAFTING

Fig. 11-41

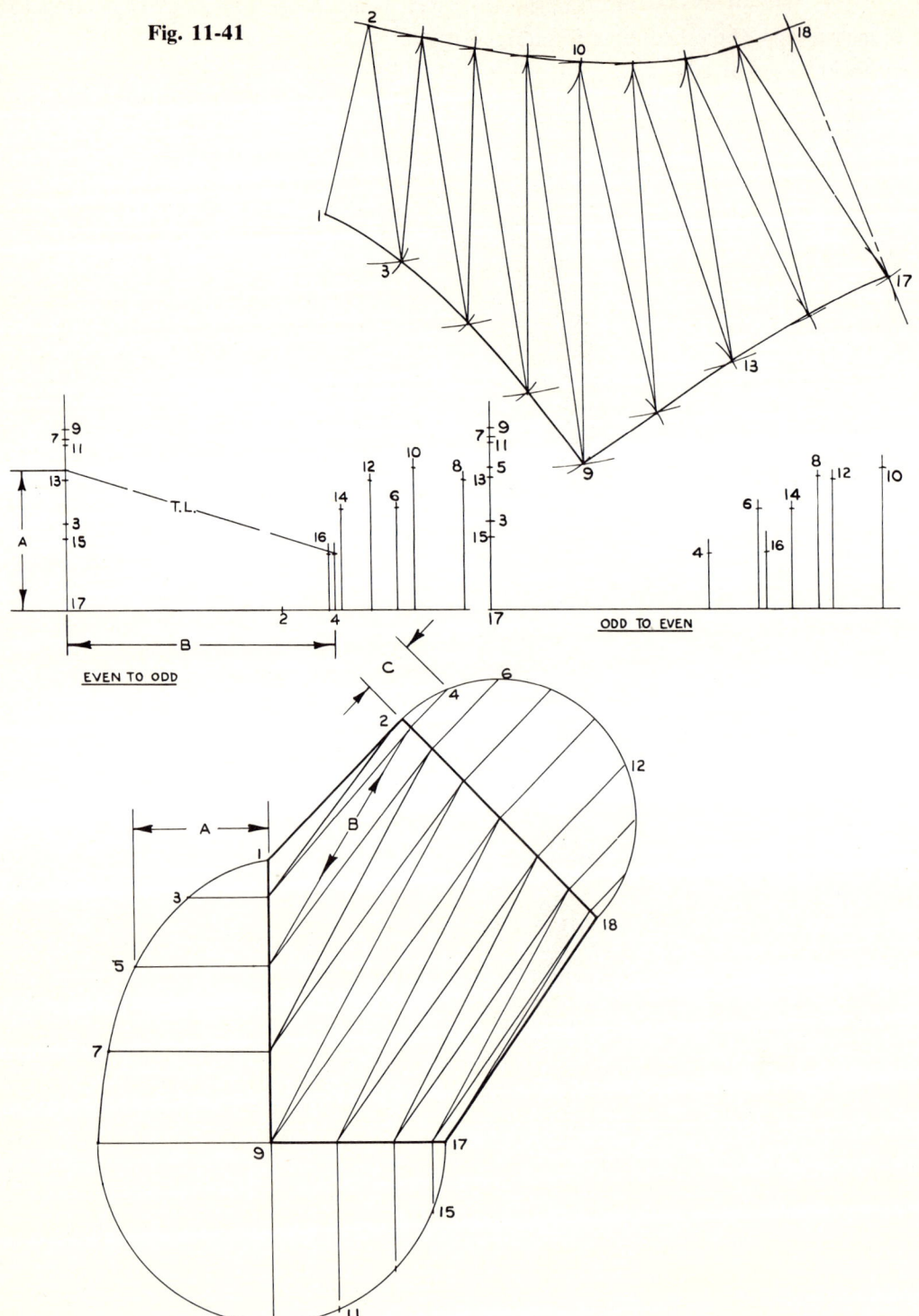

TRIANGULATION 235

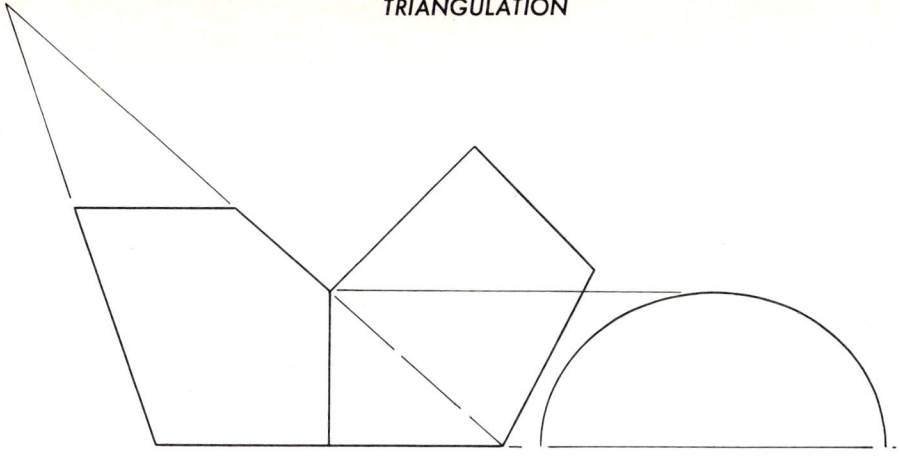

Fig. 11-42

is shown in Figure 11-42. It will be noted that the design is a combination of the oblique cone of Figure 11-35 and a prong similar to that of Figure 11-37. While the one prong is conical, the shape of the joint at the junction of the prongs can be established arbitrarily, provided both patterns are triangulated (rather than develop the conical one by radial methods). The ellipse, as constructed previously, again is the most practical shape at the joint junction. Also notice in Figure 11-42 that the height of the joint is determined by the intersection of the oblique cone with the vertical plane of the joint. The usual procedure is applied to develop patterns for each branch.

As a related commentary, it is interesting to notice that the true shape of the joint section, as determined by cutting the oblique cone with the vertical plane of the joint, is required in order to match pipes if the cone is developed by radial methods. In fact, the true shape may be the preferred shape, even when triangulating both patterns. The true section cut from the cone by the vertical joint plane is constructed using a front and top view to place elements on the cone. The piercing points projected and transferred to a side view represent the true section. In all probability, the resulting curve is a hyperbola (see sections 5-15 and 10-8d).

11-19 Square mains, round branches

Square or rectangular trunk lines may be connected to round branches by means of offset transitions, thus forming a Y branch as shown in Figure 11-43. To develop the fitting, top, front, and partial auxiliary views

Fig. 11-43

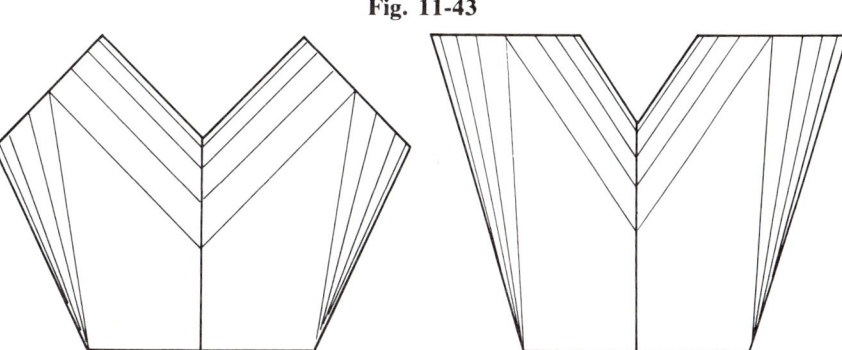

are constructed for one branch, as shown in Figure 11-44, just as for any square or rectangular-to-round offset transition. A true-length diagram is drawn for the elements radiating from corner X. Since the elements beginning at corner Z pierce the joint plane, a second diagram for those elements is desirable. Figure 11-45 illustrates a half-pattern developed as for one complete branch before notching for the joint. To locate the points for the notchout, transfer the true-lengths Z to piercing points, to their respective lines on the pattern. The true length of each is determined in the Z diagram of Figure 11-44 by projection, as shown. The logic for such a technique may be stated as follows: If the diagram, in effect, represents rotation of the line about a vertical axis, then any point on the line rotates in a plane and is represented by a horizontal path. (See section 8-3.)

11-20 Round main, square branches

Figure 11-46 illustrates typical branch fittings in which a round trunk is connected to duct branches. Observe that the joint section may be arbitrarily established as for previous problems, and that the joint shape controls the final appearance of the fitting surface. Again, good judgment must be exercised in establishing the joint shape. To answer the question of whether the shape

Fig. 11-44

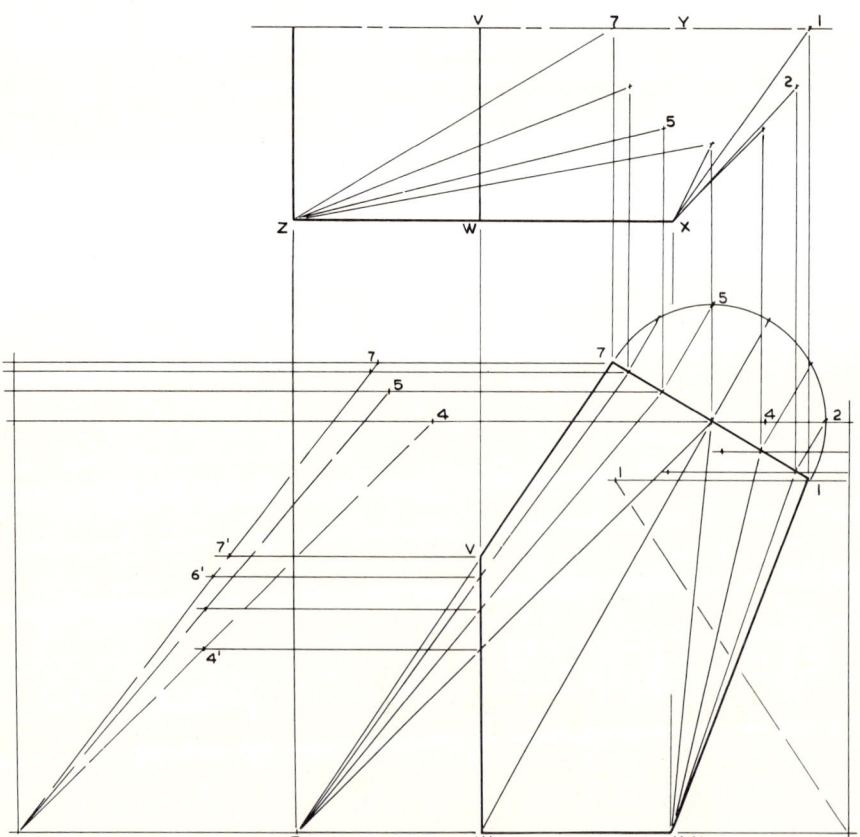

TRIANGULATION

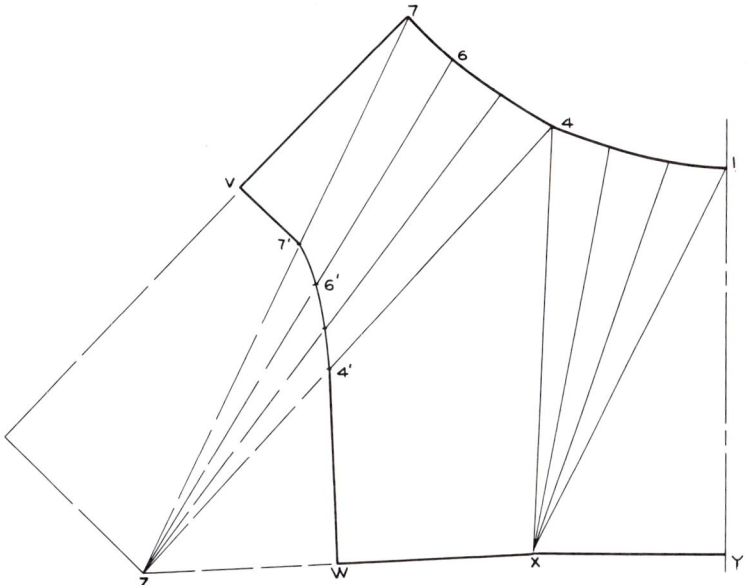

Fig. 11-45

Fig. 11-46

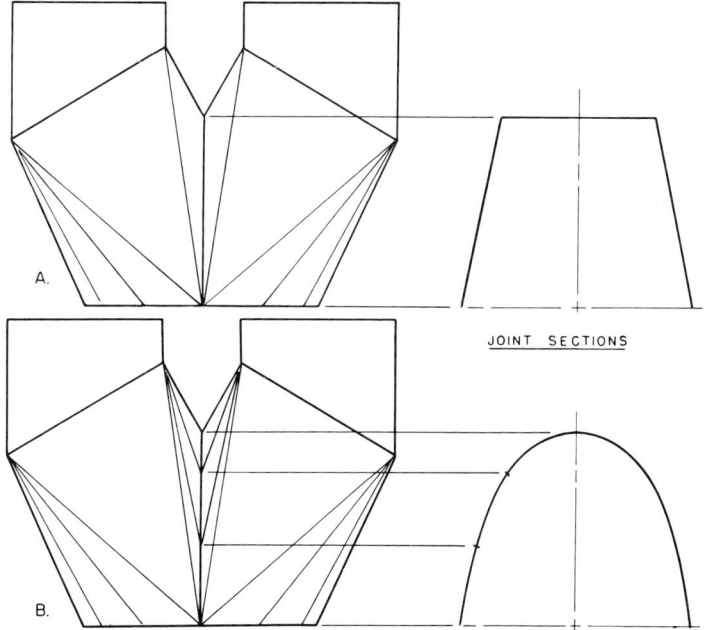

JOINT SECTIONS

should be predominantly flat sided as *A* of Figure 11-46, or should have circular or elliptical curvature as Figure 11-46*B*, one must consider the velocities and direction of flow for which the fitting is being designed. For low velocities (5 feet per second or less) the internal shape has very little effect upon efficiency, regardless of flow direction. For higher velocities (a majority of cases) several factors must be considered. For example, the square pipes should first be sized as for round pipes (vortices develop in the corners thus increasing resistance and reducing capacity). The direction of flow is a factor at high velocity because a given fitting will develop different internal turbulences when flow direction is reversed. As a general rule, any feature which contributes to full flow is desirable and less turbulence occurs when surfaces are curved in making transitions of shape, size, or direction. The flow patterns within elbows, tees, Y branches, and other fittings are complicated, to say the least. Any change in velocity, shape, angular impact, etc., further complicates the problem. At present, much research is being done using high-speed photography and high-velocity smoke to check patterns within fittings.

As an example of the development of one branch of a fitting, refer to Figure 11-47. Notice that, as in the problem of Figure

Fig. 11-47

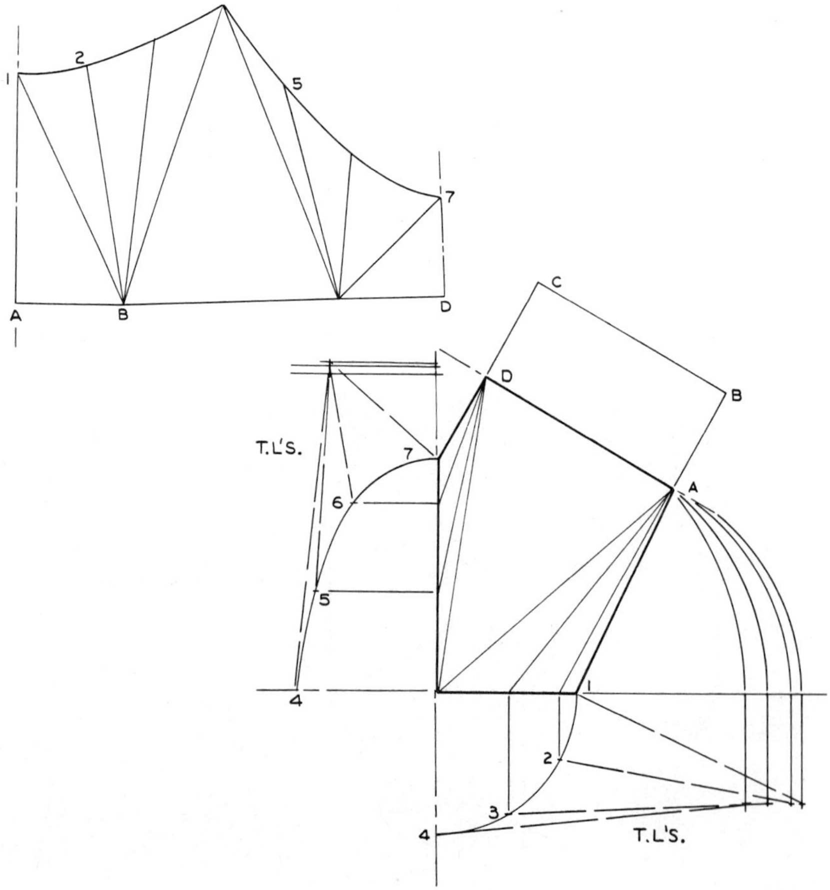

TRIANGULATION 239

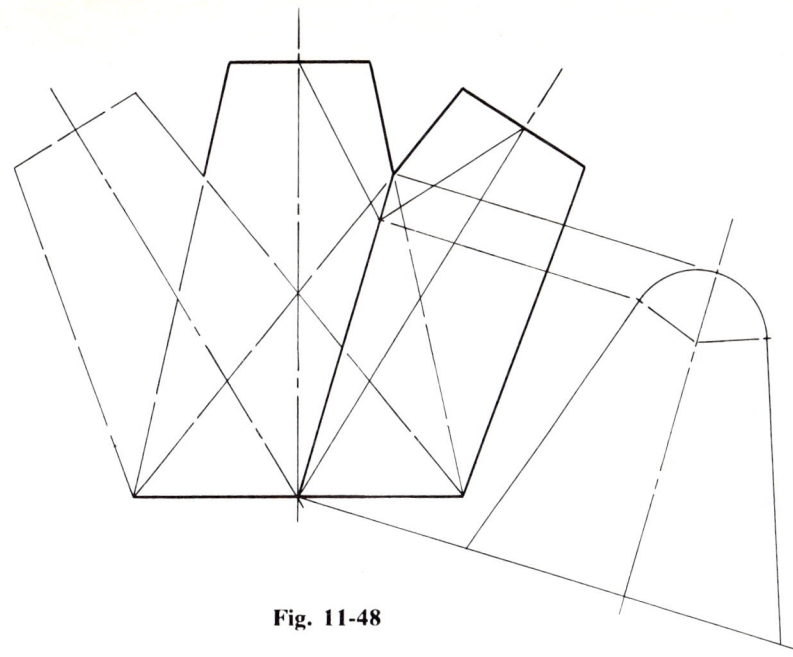

Fig. 11-48

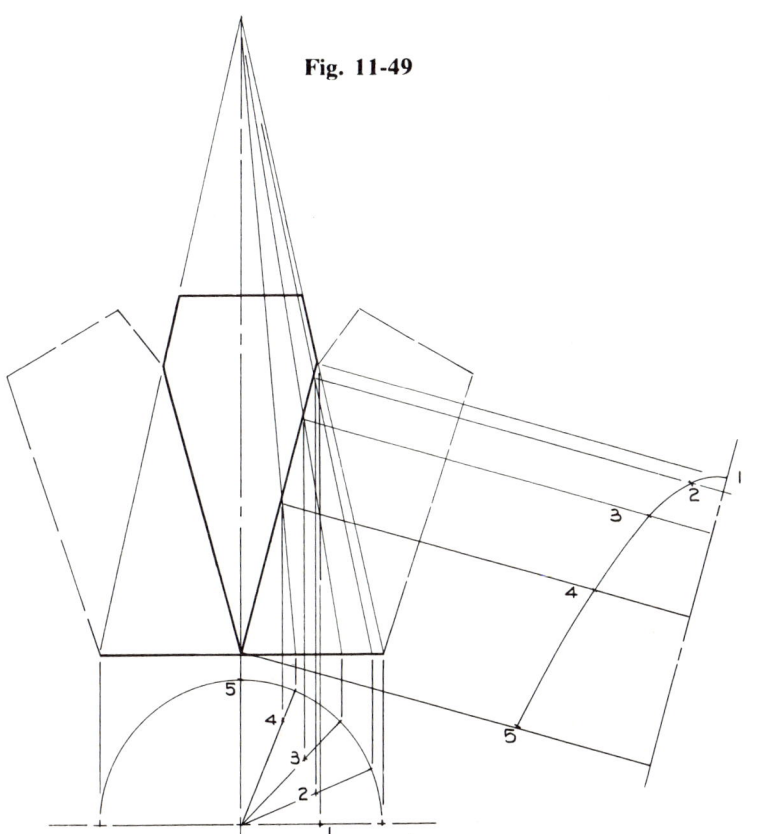

Fig. 11-49

11-41, the front and the partial views of the openings are the only views required to determine the true length of each line for the pattern.

11-21 Three-prong single-plane reducer

The three-pronged Y branch may be designed, as shown in Figure 11-48, using a curved joint section arbitrarily established (as for previous problems). The fitting may also be constructed with the center prong considered a part of a right circular cone as shown in Figure 11-49. If the center prong is to be considered a right circular cone, the true shape of the joint junction is thereby fixed and must be plotted. Regardless of joint shape, two patterns are required for the fitting, one for the center pipe and one for a side branch. Because of similarity with previous solutions, development procedure is not illustrated. Additional pronged intersections are studied in Chapter 12.

11-22 Clustered branches

Y branches may be designed with the branches clustered symmetrically. Any number of branches could be clustered, but a reasonable limit of four is mentioned as being practical relative to fabrication. In the problems previously illustrated (Figures 11-36 and 11-41) the joint plane appears edgewise in both top and front views. Thus a single plane is required to join two pipes. When three pipes symmetrically spaced are joined, three vertical planes are required. The planes appear as lines in the top view spaced 120 degrees apart, and as curves in the front view (see Figure 11-50). When four pipes are involved, two vertical planes

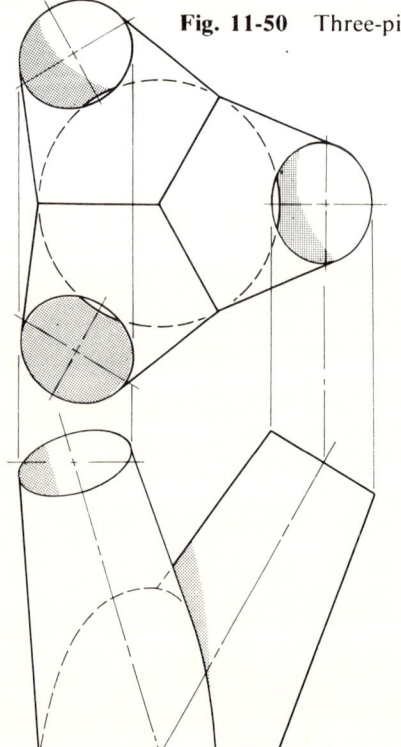

Fig. 11-50 Three-pipe Y branch.

at 90 degrees are required. The planes again appear as straight lines in the top view, but appear as curves in the front view (see Figure 11-51). Since the pipes of any given fitting are all alike, only one branch need be drawn and the pattern developed. The drawing and pattern for one branch of a three-pipe cluster is illustrated in Figure 11-52. Notice that the shape of the joint is first established in the front view on the joint plane *AB* (where it is true size and shape). The required curve points are projected to the edge view of the plane in the top view, then are rotated to the joint plane *BC*. After rotation, the points are returned to the front view to establish the front view of the joint curve. The usual numbering system is employed to trace the surface of the branch. Four true-length diagrams are established. The two diagrams to the left concern lines for numbers 1 through 11. The two diagrams to the right of the front view pertain to lines numbered from 11 to 18. (Two diagrams are used, in each case, to separate the odd-to-even and even-to-odd, increasingly numbered lines.) A half-pattern for one branch

Fig. 11-51 Four-pipe Y branch.

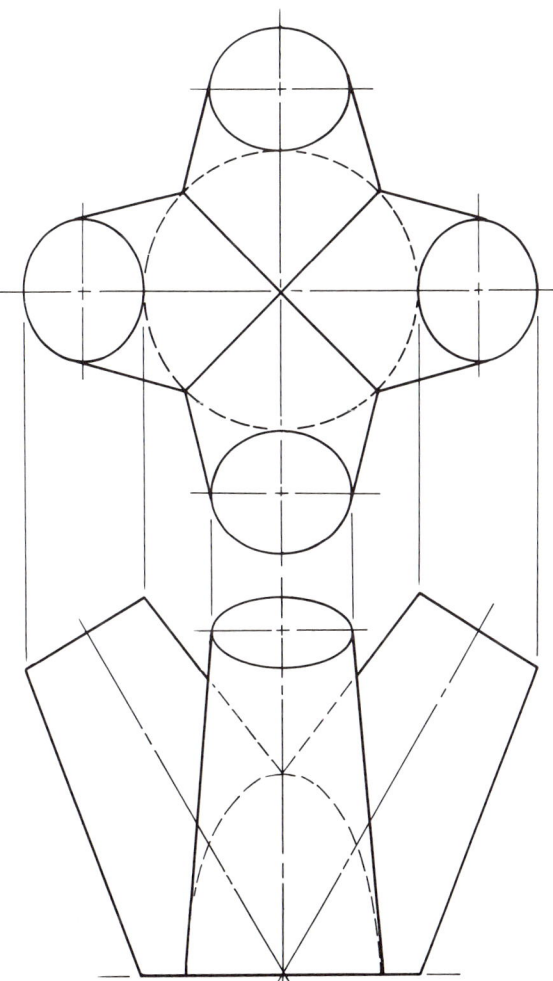

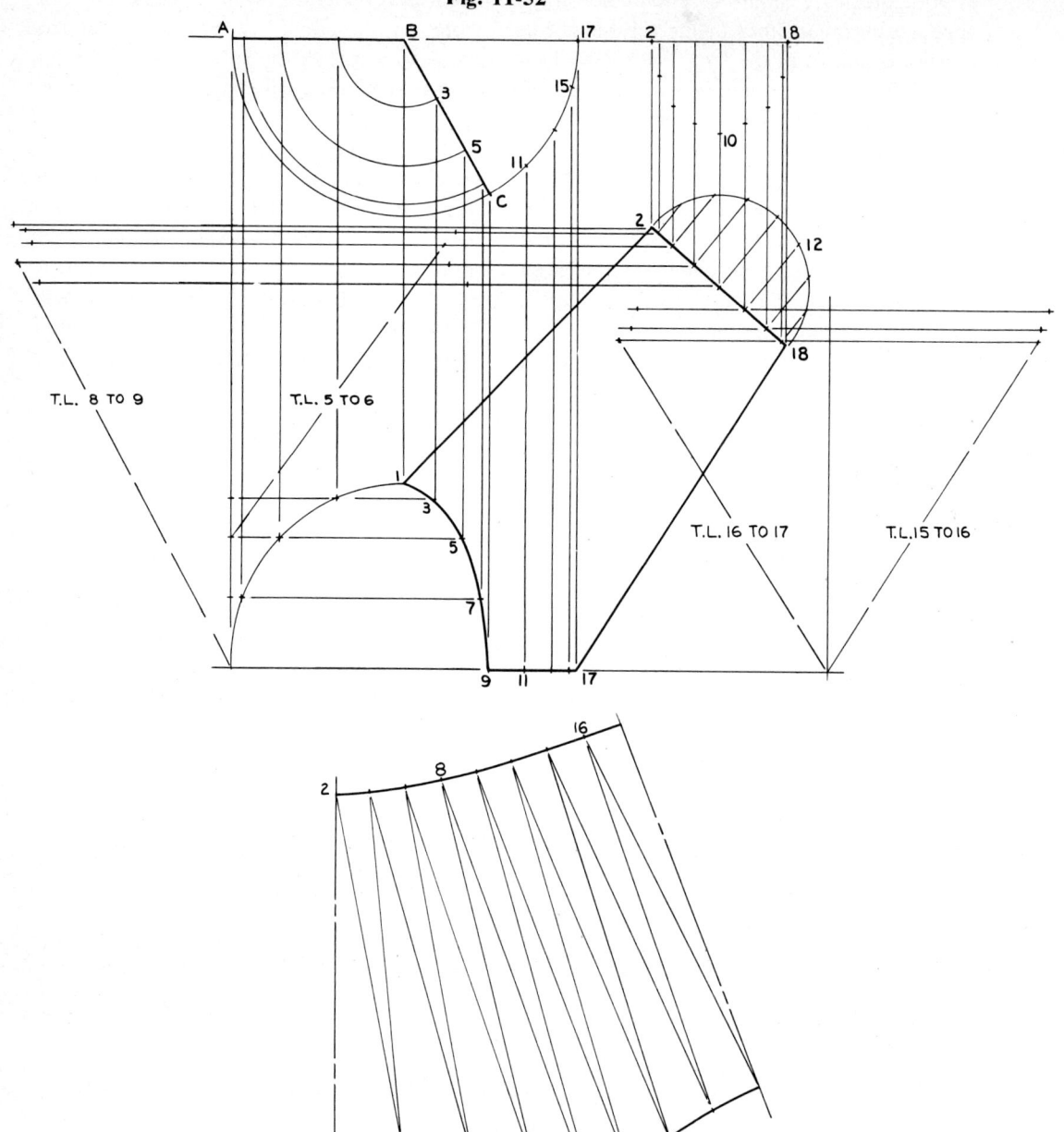

Fig. 11-52

TRIANGULATION

Fig. 11-53

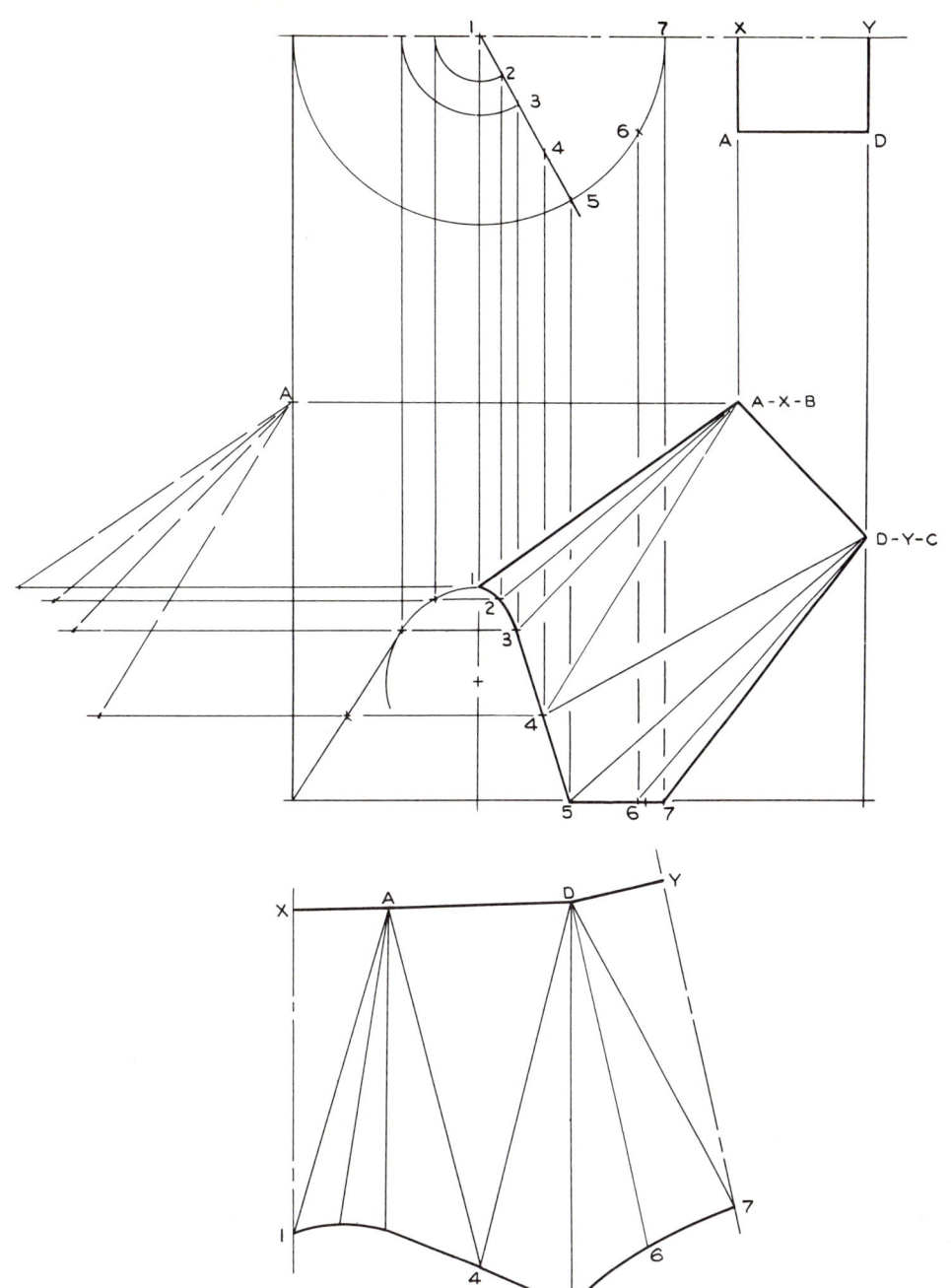

of the fitting is shown developed below the front view of Figure 11-52.

The drawing and pattern of Figure 11-53 illustrates a three-pipe cluster in which the necks are square. Figure 11-54 represents one treatment of a three-way square-to-square branch.

Using the principles set forth in this chapter, the student should be able to deal with additional branches as required.

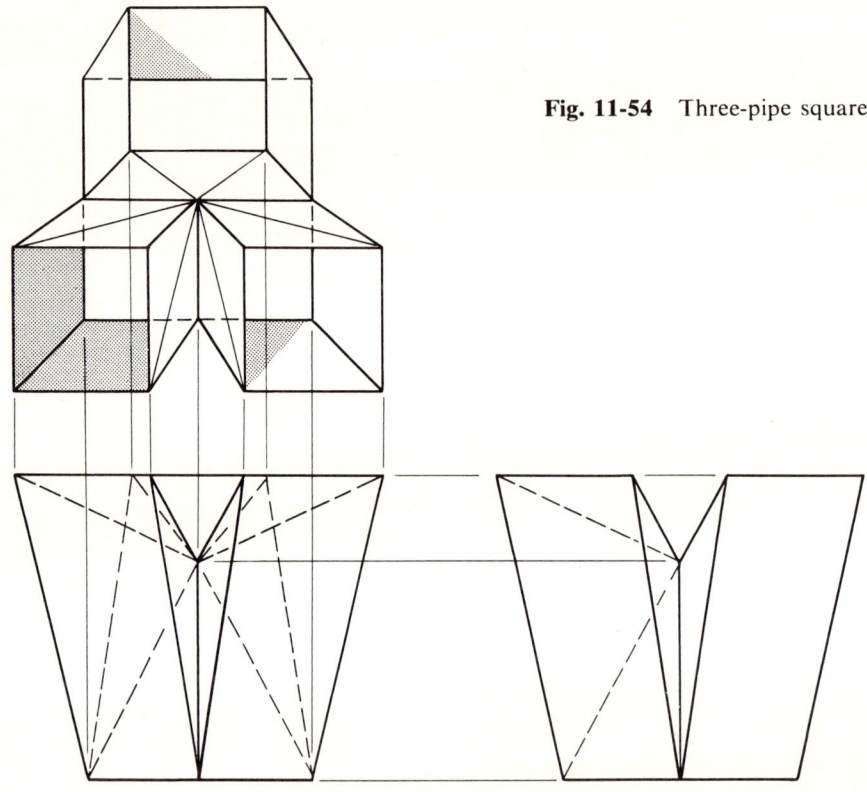

Fig. 11-54   Three-pipe square cluster.

EXERCISES

1. Develop a pattern for a twist fitting to meet the following specifications: 3 in [76 mm] square-to-square, 45-degree twist, 1 in [25 mm] off center, 3½-in [89-mm] rise.
2. Develop a pattern for a fitting to join the pipe openings shown in Figure 11-55.
3. Develop patterns for the square-to-round transition of the ventilator shown in Figure 11-56. Specifications: roof rise equals one-fourth span; square duct size equals 24 in [610 mm]; round opening equals 18 in [457 mm] diameter; total height of square-to-round fitting equals 14 in [356 mm].
4. Develop a pattern for a square-to-octagon transition to meet the following specifications: 3 in [76 mm] square, 3½ in [89 mm] octagon flat diameter, 1 in [25 mm] off center, 3½-in [89-mm] rise.
5. Develop a pattern for the transition indicated in Figure 11-57 to join the round pipe with an opening which uses A as a top view.
6. Same directions for problem 5, except use B as the top view.
7. Same directions as for problem 5, except use C as the top view.
8. Develop a pattern for the offset transition

TRIANGULATION

DIMENSIONED IN MILLIMETERS

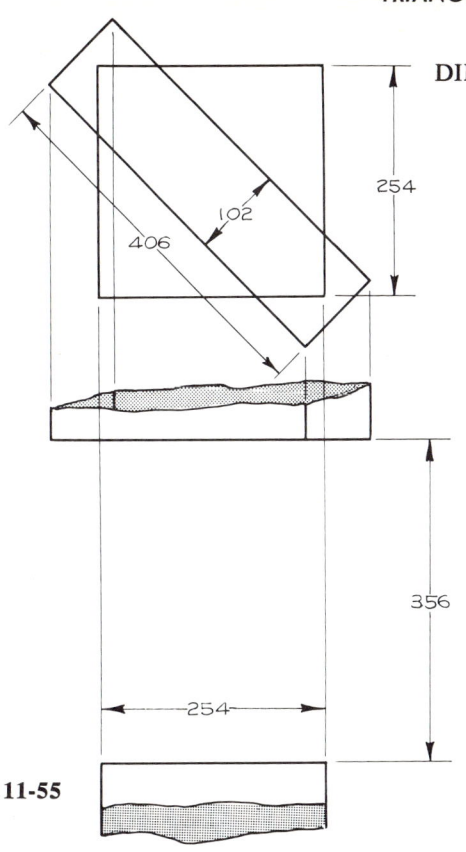

Fig. 11-55

Fig. 11-56 Square-to-round ventilator transition. (*Burt Manufacturing Co., Akron, Ohio.*)

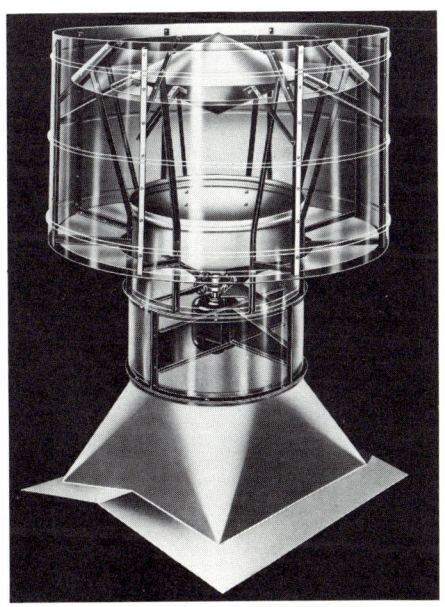

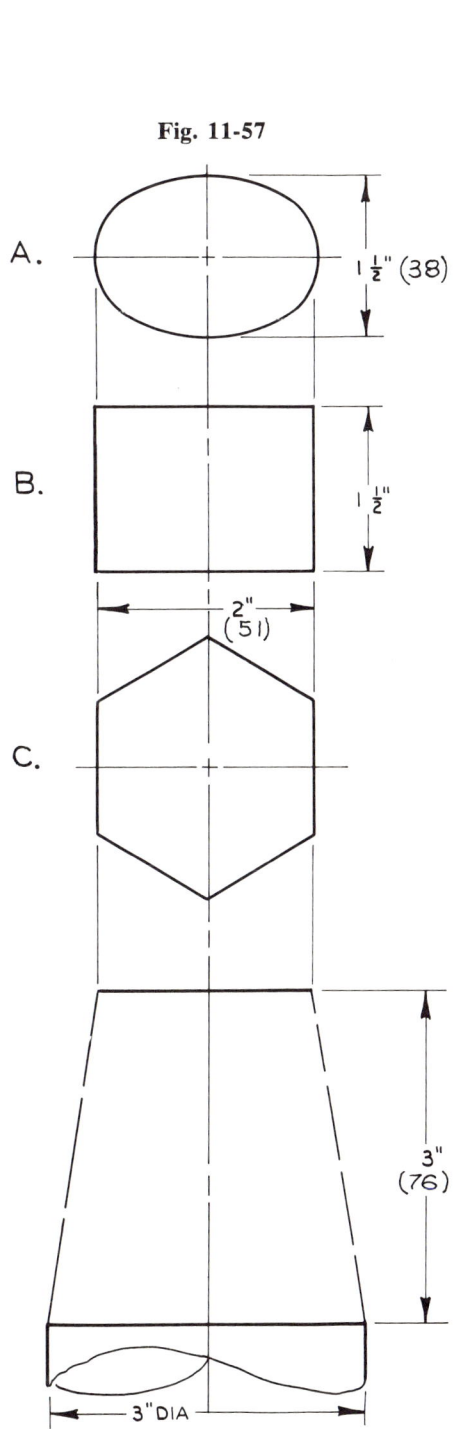

Fig. 11-57

246 SHEET METAL DRAFTING

indicated in Figure 11-58 to join the round pipe with an opening which uses *A* as the top view.

9. The directions same as for problem 8 except use *B* as the top view.
10. Same directions as for problem 8 except use *C* as the top view.
11. Develop a pattern for the transitional elbow in Figure 11-59 so as to join the rectangular duct with an opening which uses *A* as the right-side view.
12. Same directions as for problem 11, except use side view *B*.
13. Same directions as for problem 11, except use side view *C*.
14. Develop the patterns for a three-piece 75-degree transitional elbow to meet specifications as follows: 3-in [76-mm] square base to a 2-in [51-mm] square inclined opening; center-line radius equals 3½-in [89 mm].
15. Develop the patterns for a three-piece 90-degree transitional elbow to meet the following specifications: 3-in [76-mm] square base to a 2-in [51-mm] round opening; center-line radius equals 4 in [102 mm].
16. Develop cheek, throat, and heel patterns for a curved transitional duct elbow to connect a 1½- × 2-in [38- × 51-mm] horizontal duct with a 3-in [76 mm] square vertical duct. Use a 2-in [51-mm] throat radius.

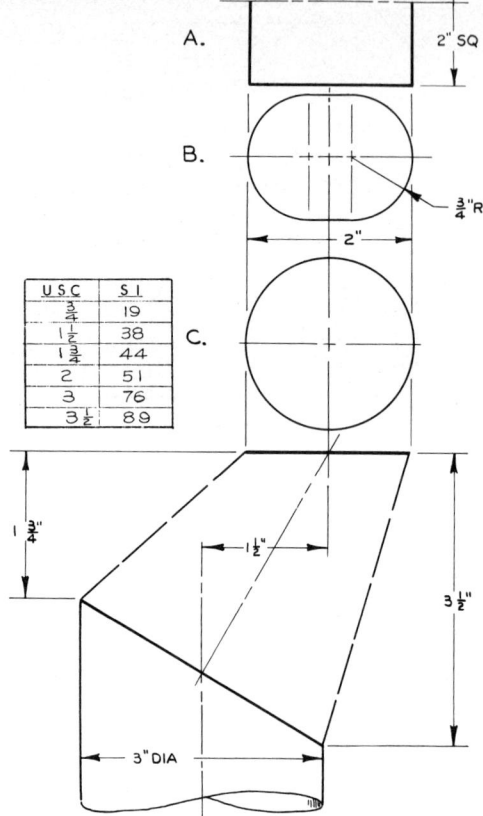

Fig. 11-58

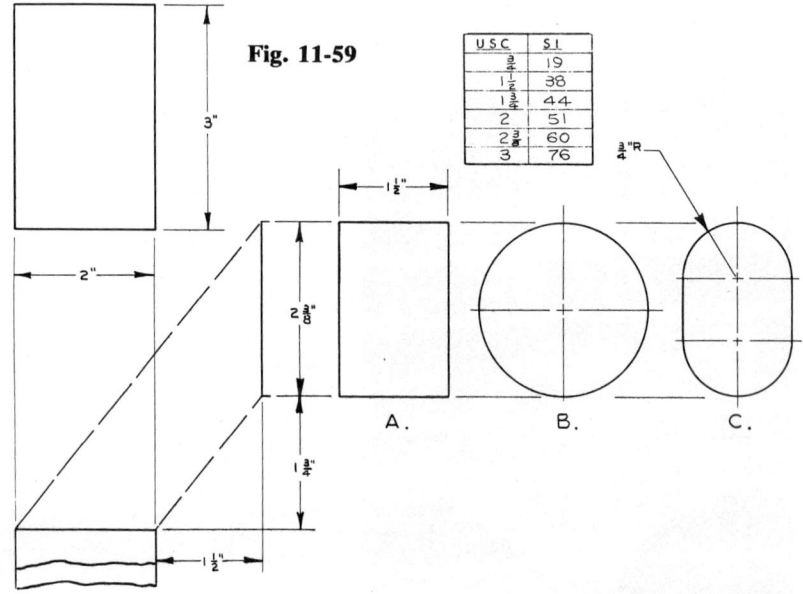

Fig. 11-59

TRIANGULATION

17. Develop a pattern for one branch of the symmetrical fitting shown in Figure 11-60. All openings are cylindrical.
18. Develop a pattern for one branch of the symmetrical fitting shown in Figure 11-61. Use a joint section with top curve of 1-in [25-mm] radius. Consider that all openings are cylindrical.
19. Develop a pattern for one branch of the symmetrical fitting shown in Figure 11-61. Use an elliptical joint section. Consider all openings cylindrical.
20. Develop a half-pattern for the fitting shown in Figure 11-62. Place the seams at the top and throats of the side pipes.

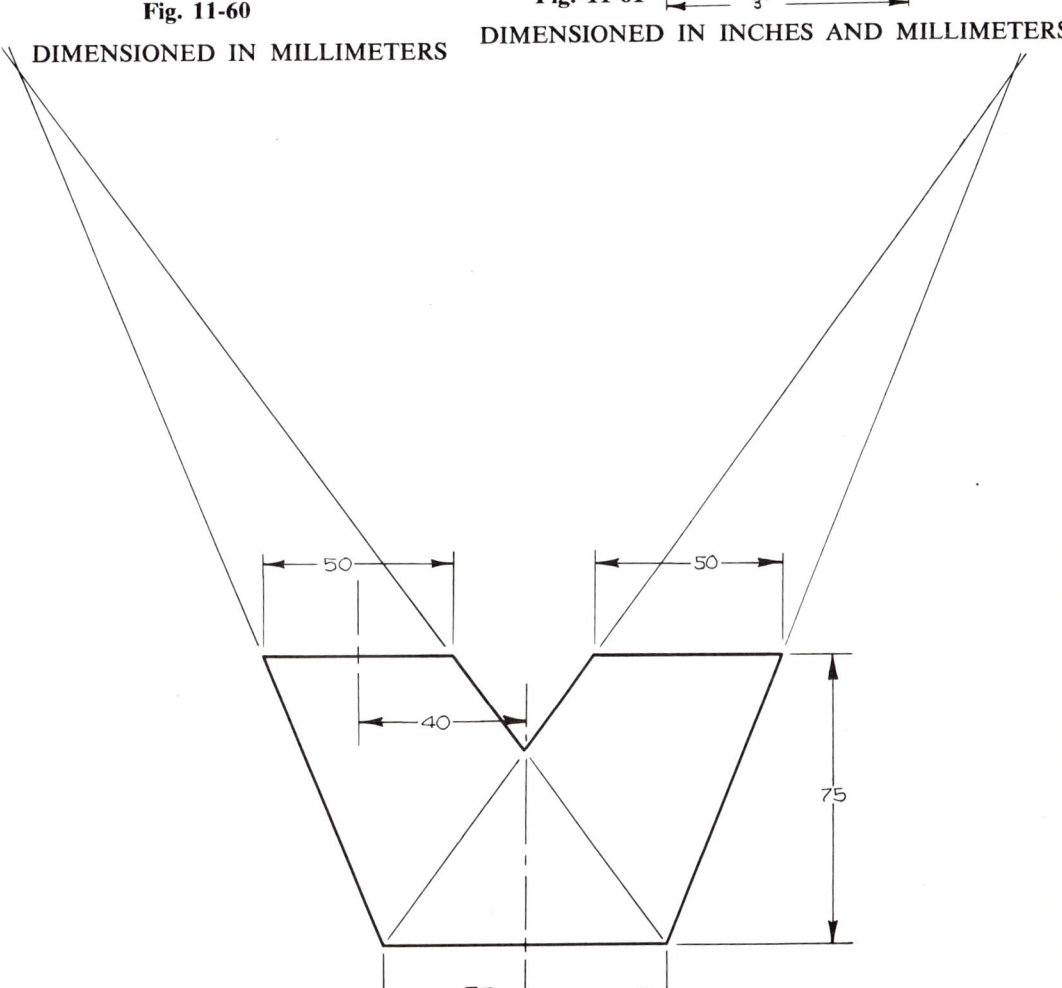

Fig. 11-60
DIMENSIONED IN MILLIMETERS

Fig. 11-61
DIMENSIONED IN INCHES AND MILLIMETERS

21. Develop patterns for the branches of the fitting shown in Figure 11-63. Use a joint section with top curve of 1¼-in [32-mm] radius. Consider all openings cylindrical.
22. Develop a pattern for one branch of the fitting shown in Figure 11-60. Consider the 75-mm main as cylindrical and the branch necks as square.
23. Develop a pattern for one branch of the fitting shown in Figure 11-61. Consider the 3-in [76-mm] main as a square duct and the branch necks cylindrical. Use an elliptical joint section.
24. Develop a pattern for one branch of the fitting shown in Figure 11-61. Consider the 3-inch main as a square duct and the branch necks cylindrical. Use an elliptical joint section.
25. Develop patterns for the branches of the fitting shown in Figure 11-63. Consider the 3-in [76-mm] main and left branch as cylindrical. The right branch to be square. (The joint section measures 2 × 3 in [51 × 76 mm] rectangular.)
26. Develop a pattern for one branch of a three-way cluster. Consider all pipes cylindrical and use data as given for one branch of Figure 11-61.

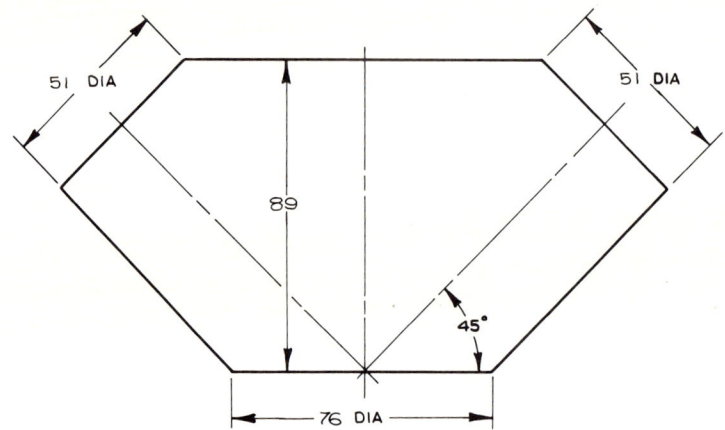

Fig. 11-62 DIMENSIONED IN MILLIMETERS

Fig. 11-63 DIMENSIONED IN INCHES AND MILLIMETERS

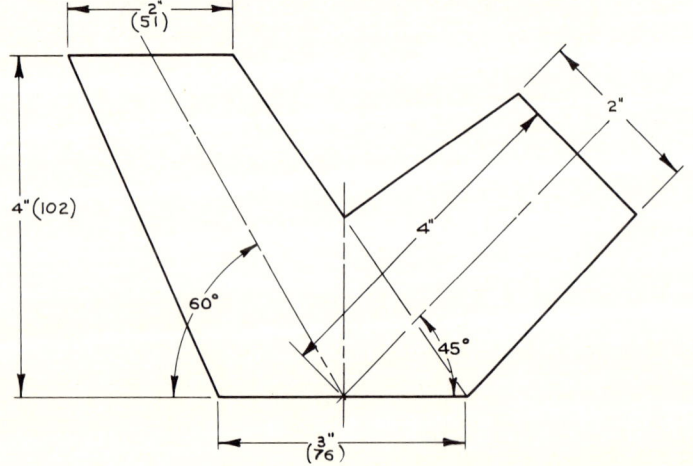

chapter 12

COMBINATION PROBLEMS

The more complicated sheet metal fittings are not of a single surface type but are combinations of cylinders, prisms, cones, pyramids, and warped surfaces. Therefore, the problems to be studied in this chapter are the result of the intersection of two or more basic geometric forms, or are a combination of forms separately arranged as an installation. Since parallel-line procedures are used to develop cylinders and prisms, radial-line procedures for cones and pyramids, and triangulation for all types of ruled surfaces, it is logical to assume that one or more of these methods will be required in the solution of combination problems. Because of the flexibility of parallel-line, radial-line, and triangulation methods, more than one method might be applied to accomplish the same result for many problems. Therefore, a careful analysis of the problem is in order before attempting development. Then those methods which are easiest, most direct, most accurate, and best suited to the problem at hand should be used.

12-1 Reducer tee

The reducer tee in Figure 12-1 consists of the intersection of a cylinder and a right circular cone. The problem is similar to a straight tee, and the methods for finding the line of intersection may be applied in the same manner. Elements are placed on the cone base and numbered. The elements are projected to the front view and transferred, by reference, to a partial side view of the cylinder. The side view establishes the piercing point for each element. The piercing points are projected to their respective element in the front view. The cylinder and cone are developed using methods previously discussed in Chapters 9 and 10. In developing the cone frustum, notice that the piercing points for each element are rotated to the true-length position. The pattern for the hole in the cylinder is developed, using the spaces as taken from the side view (see section 9-20).

12-2 Offset pitched reducer tee

A reducer tee with the additional features of offset and pitch is shown in Figure 12-2. The method of determining the line of intersection is the same as for the previous problem. Notice, however, that each numbered element, because of the offset and pitch, has an individual piercing point which must be carefully projected to plot the curve in the front view. Since the general procedures for this problem are the same as have previously been performed, only partial views, where necessary, and a started pattern are shown in Figure 12-2.

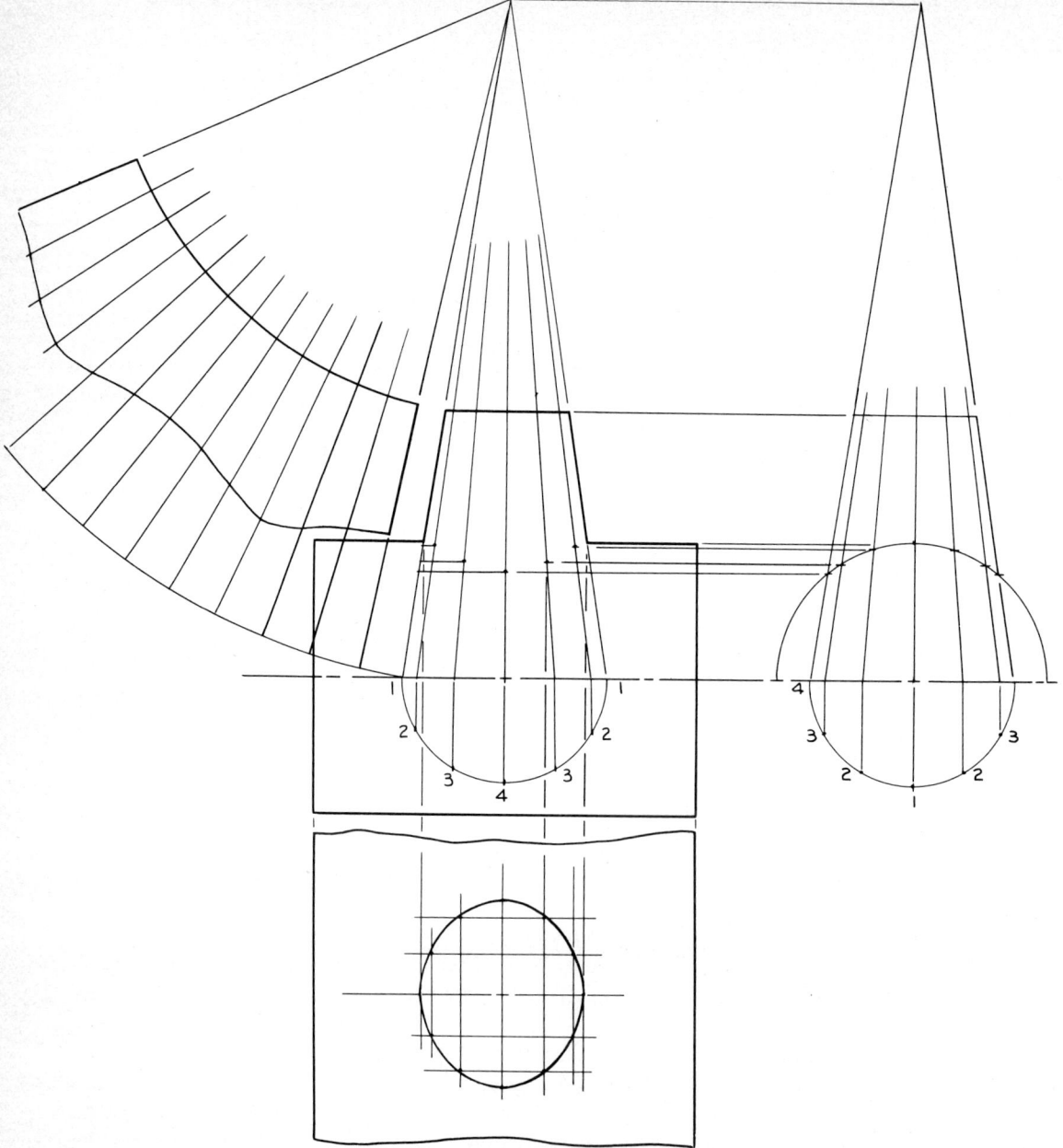

Fig. 12-1 Tee with conical take-off.

12-3 Three-piece reducing elbow

A reducing elbow in which all of the reduction takes place in the center piece of the fitting is shown in Figure 12-3. The reducer section is part of a right circular cone. The elbow is constructed with the same spacing as is used for straight cylindrical elbows. The center lines are established from specifications. The tangent-sphere method is used,

COMBINATION PROBLEMS

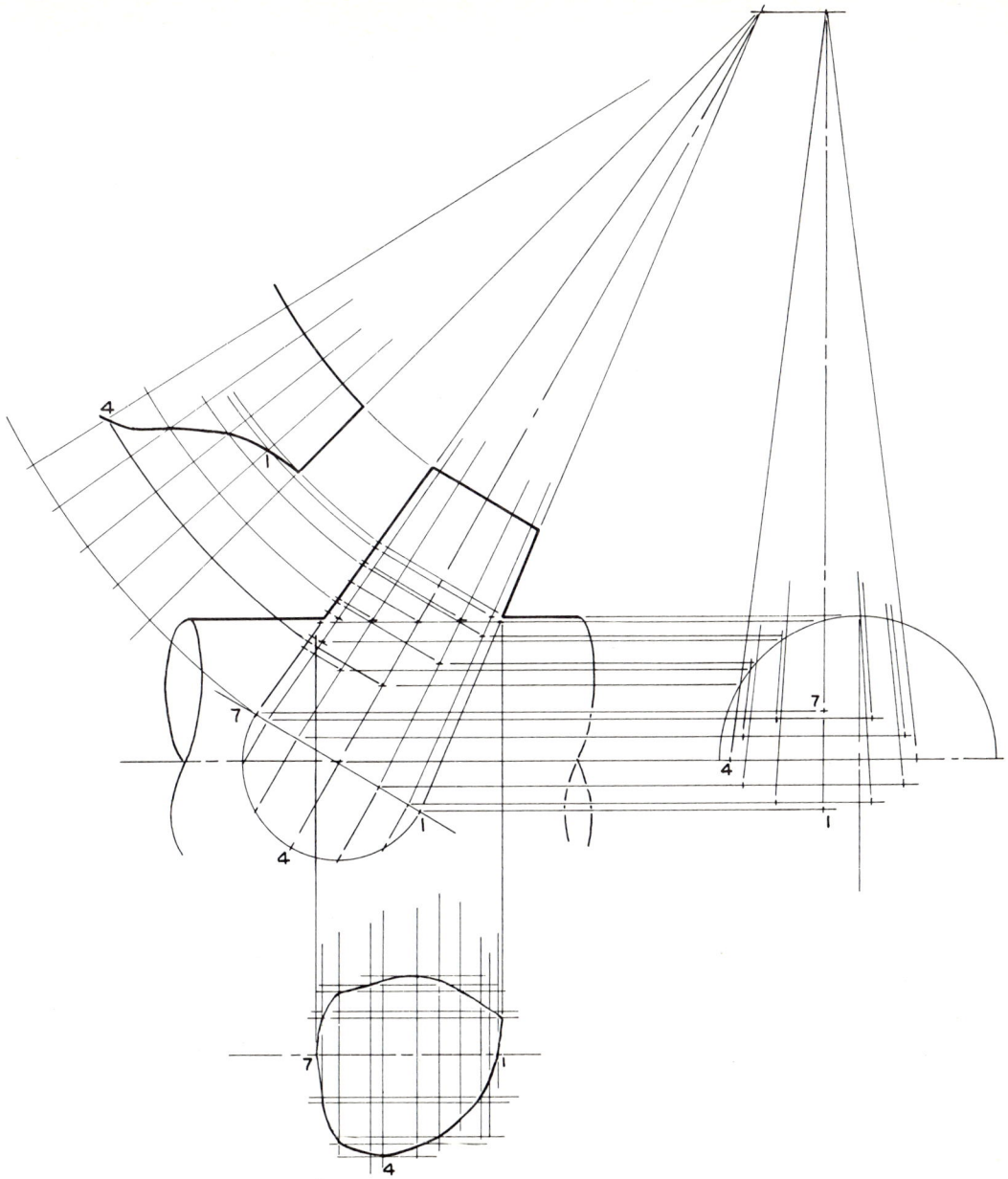

Fig. 12-2 Pitched tee with conical take-off.

as shown in Figure 12-3, to locate the miter lines (see Figures 9-26 and 10-25) and to represent a center piece whose cross section is circular. The contour elements of the cone are extended to intersect at the apex *A*. A right circular base is established *perpendicular to the cone axis*. The cone is developed, using the radial methods, which are explained thoroughly in Chapter 10. The cylindrical end pieces are developed, using the parallel-line methods, which are explained in Chapter 9.

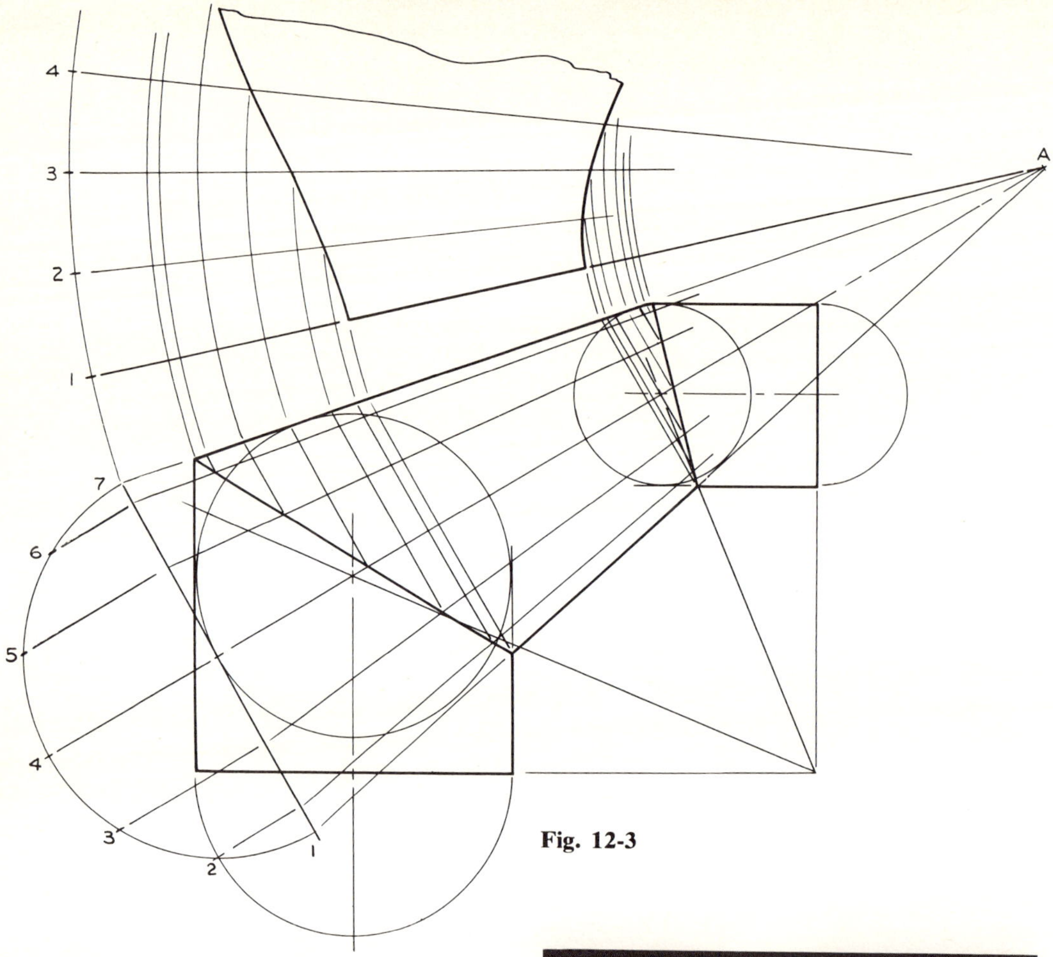

Fig. 12-3

12-4 Cone section for refinery reactor

The industrial application shown in Figure 12-4 is similar to the previous problems of this chapter in that the intersections are those of cone and cylinders. The cone frustum may be developed as for similar prob-

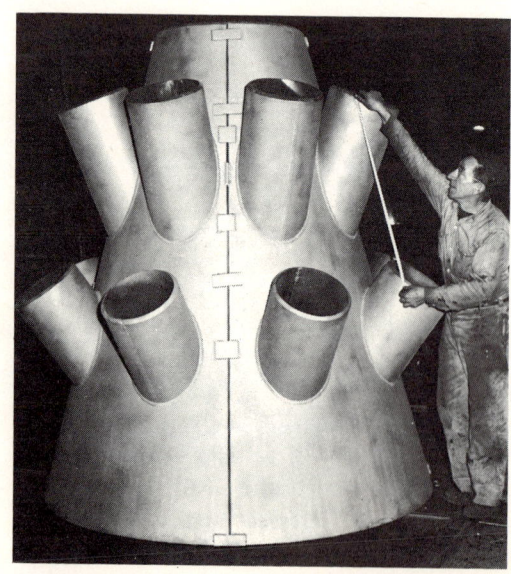

Fig. 12-4 Cone section for refinery reactor. (*Missouri Boiler and Sheet Iron Works, St. Louis, Mo.*)

lems of Chapter 10, the only additional new feature being the location and cutting of the proper shaped holes. The cylinders may also be developed by using principles previously explained. First, however, the line of intersection between the cone and cylinder must be established accurately. This is done as illustrated in Figure 12-5. Note that partial views of the cone base are attached to the front view and are divided into a number of closely spaced elements. Partial auxiliary views are drawn which will present each cylinder edgewise (as a circle). When the elements are transferred to the auxiliary view, the piercing points are thus located. The points of the curve are projected to the front view and the curve established.

Figure 12-6 shows the development of the holes in the cone plate. The procedure is a radial-line technique. The development of cylinder patterns is not illustrated because of the similarity with previous problems.

Fig. 12-5

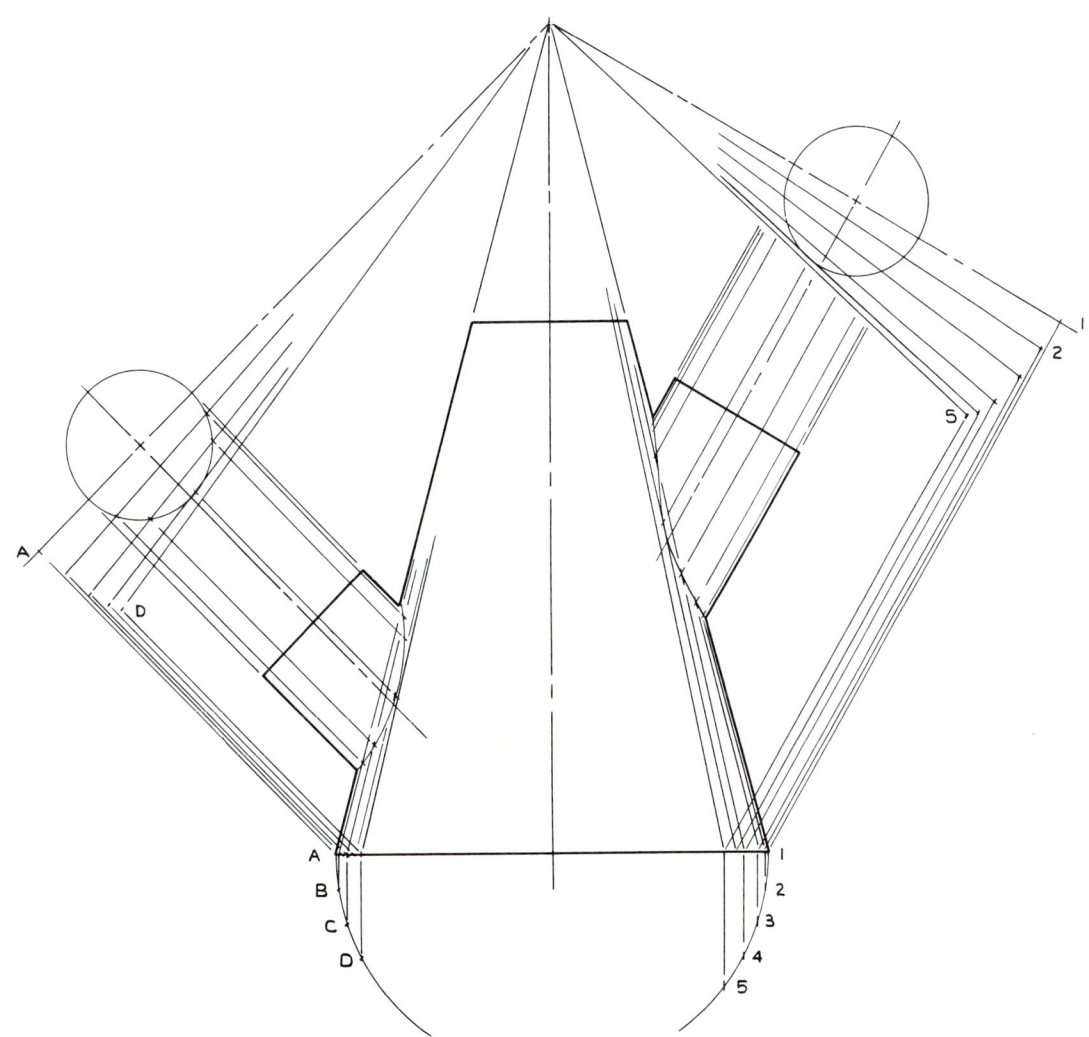

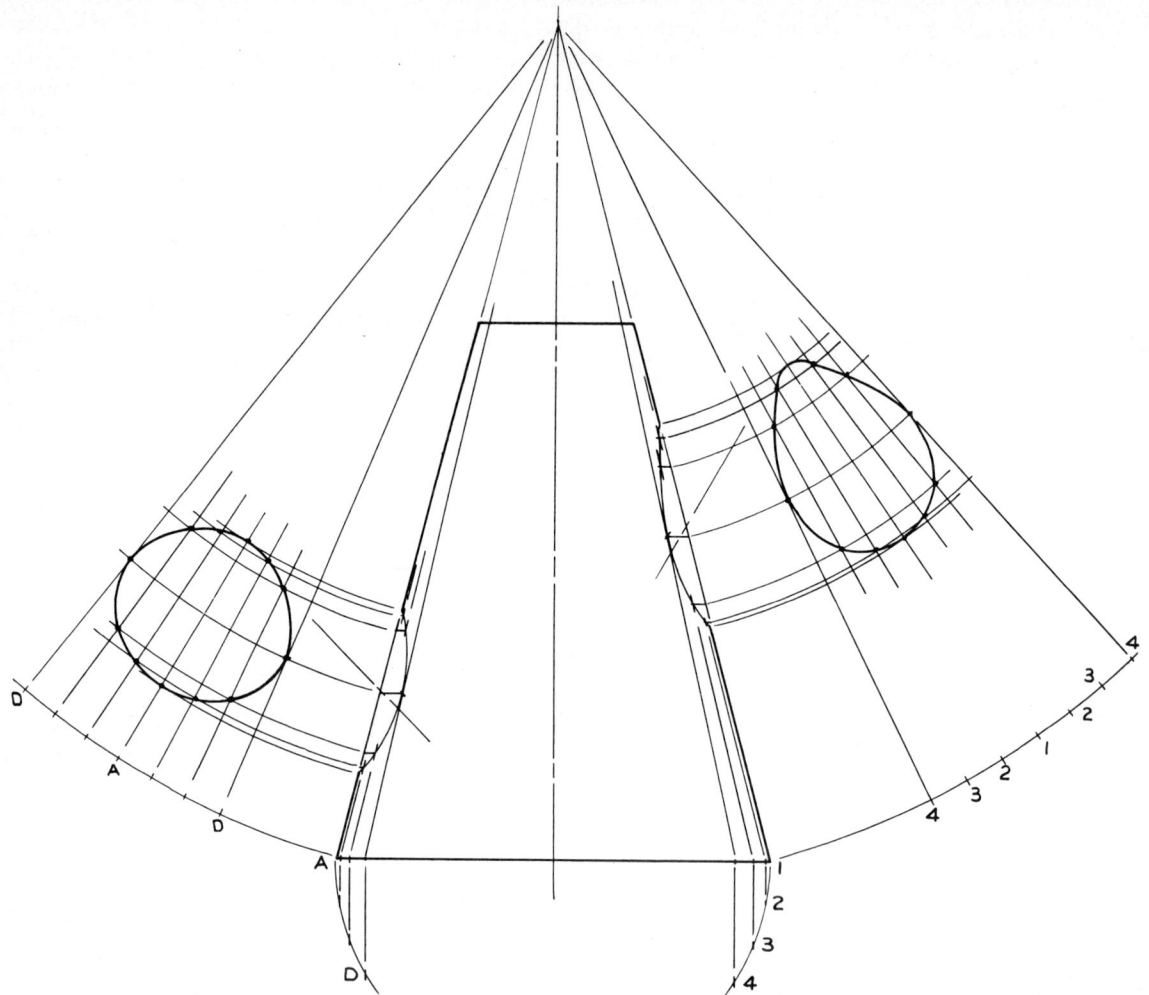

Fig. 12-6

12-5 Cone frustum with horizontal take-off

The problem is shown in Figure 12-7 as the intersection of a vertical cone and a horizontal pipe. The cutting-plane method is used to determine the line of intersection. Cutting planes 1, 2, 3, 4, etc., produce elements on the pipe. Each cutting plane, being perpendicular to the axis of the right circular cone, also produces a circular trace on the cone (as numbered in the top view of Figure 12-7). The intersection of the circular trace and its related elements locates points for the line of intersection. Patterns for the pipe and for the cone frustum are developed as shown in Figure 12-8. To develop the pattern hole: (1) Draw the trace lines on the pattern surface. (2) Construct an element on the cone in the top view as an arbitrary reference line. Place the element on the pattern at the desired location for beginning the hole. (3) Using a convenient setting, step off and transfer the curved distance from the reference line to each point (as taken from the top view and transferred to each trace line on the pattern—see sample dimension D). (4) You will then complete the pattern of the hole by connecting the points.

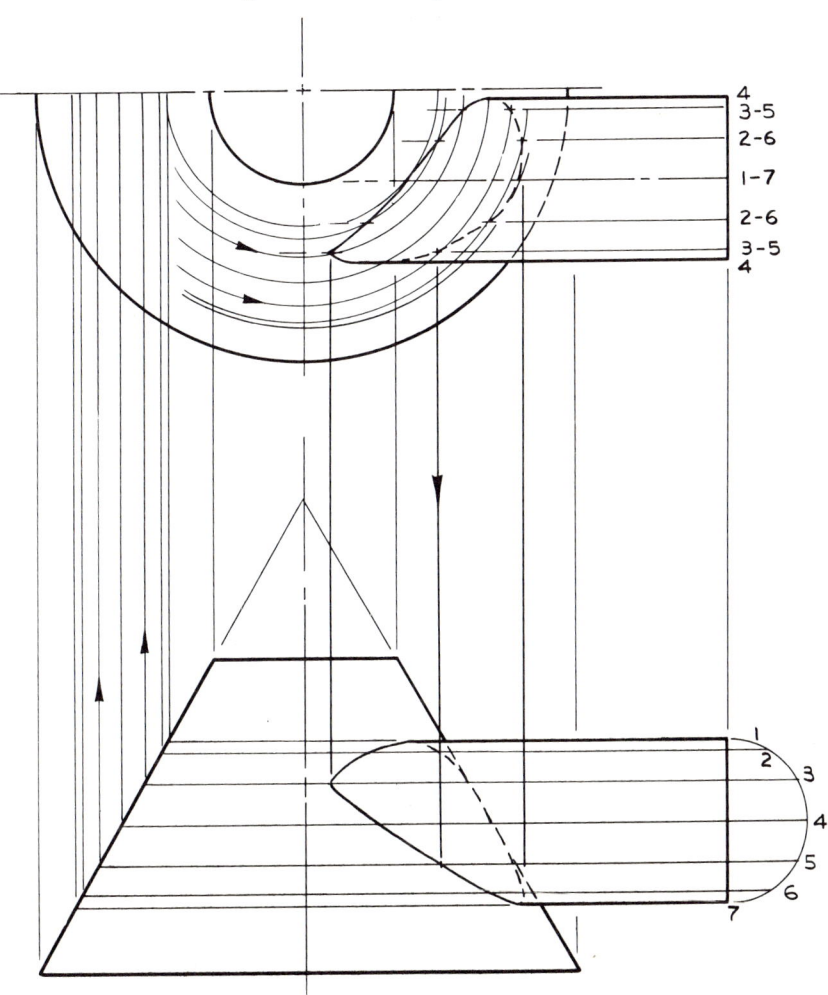

Fig. 12-7 Centrifugal connection.

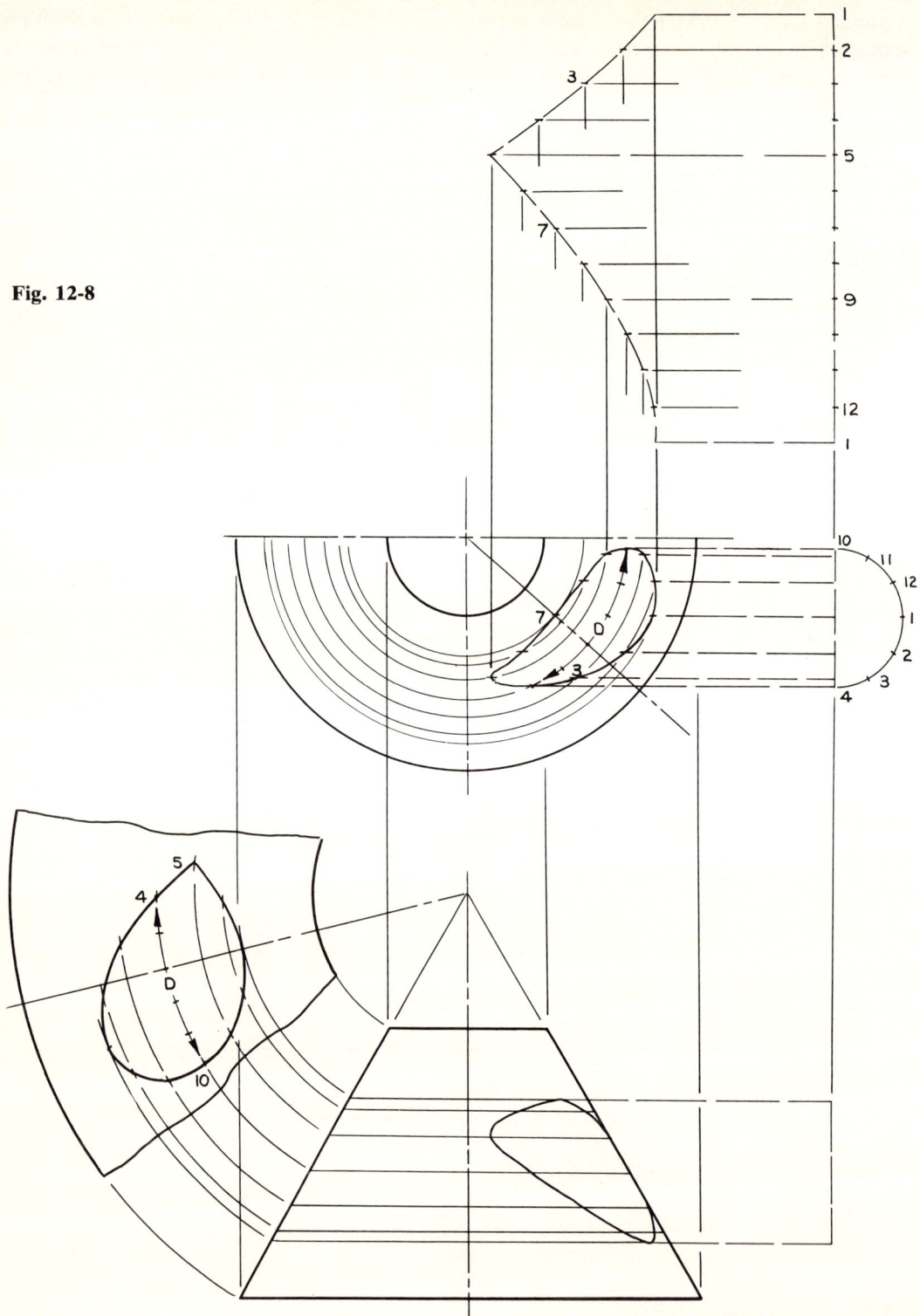

Fig. 12-8

COMBINATION PROBLEMS

12-6 Cone frustum with vertical take-off

The cutting-plane method, used in the same manner as for the previous problem, determines the line of intersection as shown in Figure 12-9. Patterns for the pipe, cone, and hole are developed as for those shown in Figure 12-8.

12-7 Pyramidal vent hood with vertical take-off

The lightweight plastic hood shown in Figure 12-10 is made of Hetron (R)92 in order to surmount a tough corrosion prob-lem involving warm chemical gases. The hood is in the shape of a right pyramid intersecting a vertical stack. Top and front views are drawn as shown in Figure 12-11. By means of elements uniformly spaced on the front surface of the pyramid, the piercing points necessary to establish line of intersection are determined (for that face only). Patterns are then developed as shown, using principles previously applied, for one quarter of the pipe and for one panel of the pyramid.

12-8 Square hopper and cylinder

Materials are often transferred by means of conveyors in which hoppers empty into

Fig. 12-9 Offset stack.

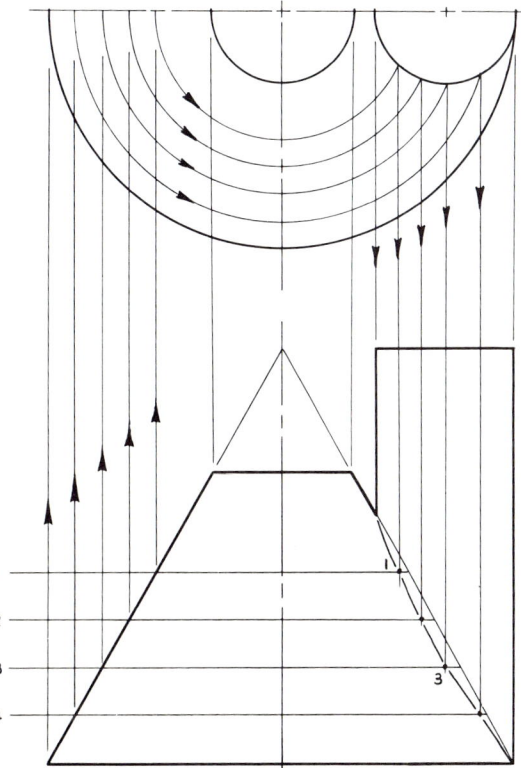

258   SHEET METAL DRAFTING

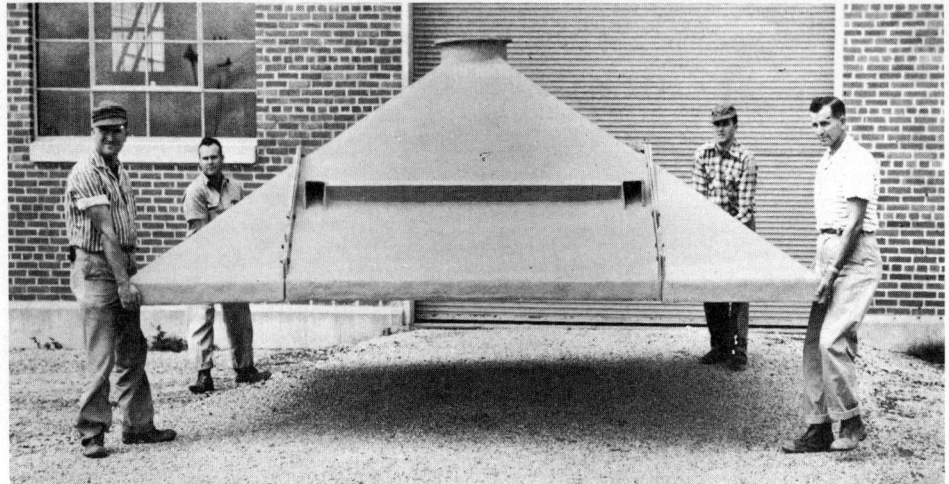

Fig. 12-10 Plastic hood made of Hetron (R) 92. (*Hooker Electrochemical Co., Niagara Falls, N.Y.*)

Fig. 12-11

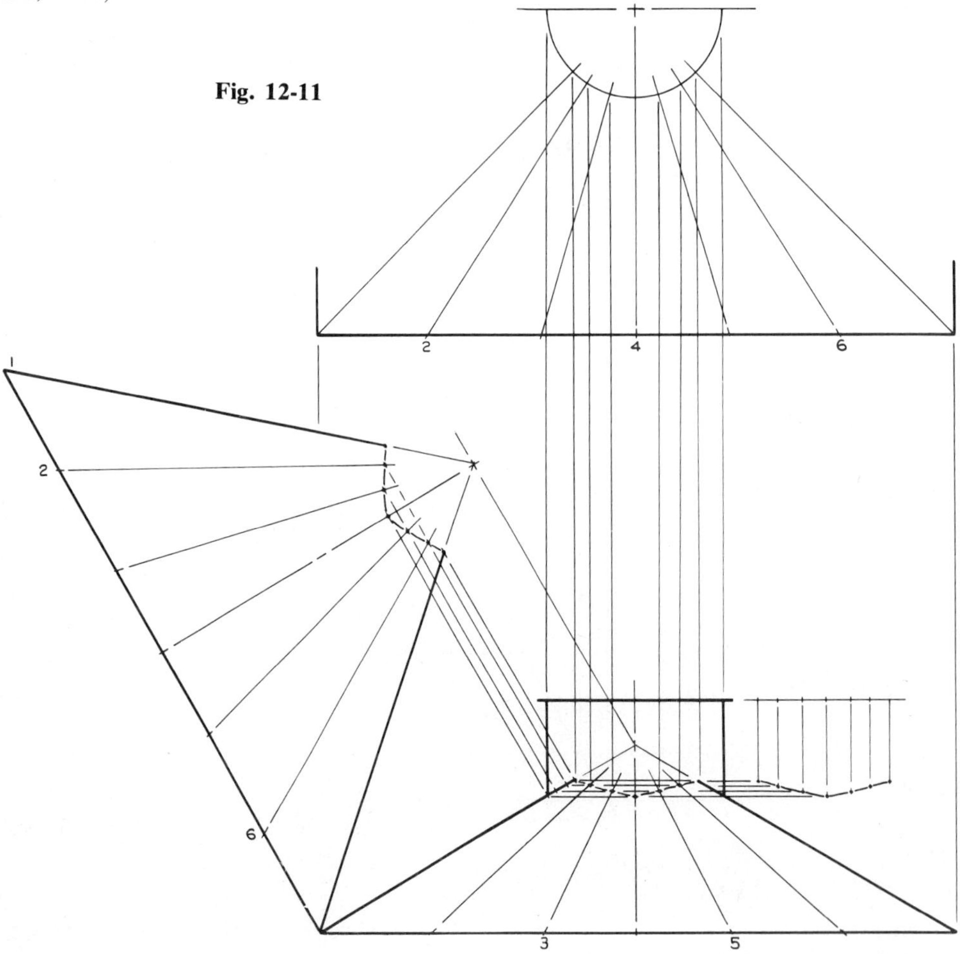

COMBINATION PROBLEMS

Fig. 12-12

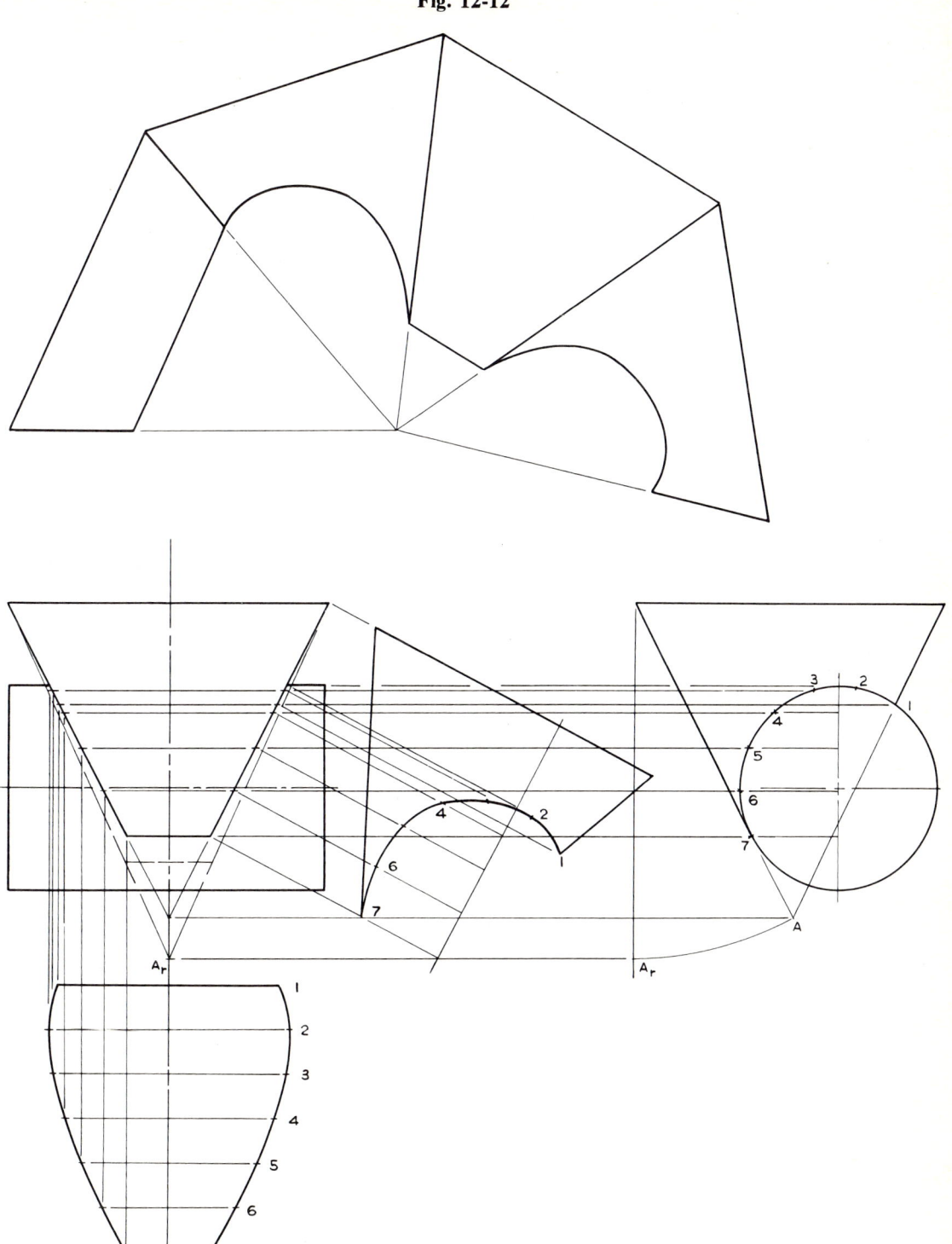

260 SHEET METAL DRAFTING

round pipes. The intersection problem of Figure 12-12 represents a square hopper which is off center in order to feed to the lower side of the conveyor screw operating within the cylinder. The front and right-side views represent a good shape description of the project. It is required to develop patterns for the hopper panels and the hole in the cylinder.

The hole pattern is a parallel-line problem. Seven equally spaced elements appear as points in the side view and as true-length lines in the front view. The stretchout for the hole is projected below the front view, as shown.

Since the hopper is a pyramid (with apex at A) it could be developed as a complete pyramid pattern (see Figure 10-56). The elliptical curve could then be plotted on the side panels and the back and front panels trimmed to form the hopper pattern. A faster and more direct method, however, would be as shown in Figure 12-12 in which the front and back panels are rotated until parallel to the front plane of projection. Thus they become true size and shape in the front view (see Figure 8-5). Since both side panels are the same, an auxiliary view of one of them is constructed. The four panels may then be put together (by means of triangulation) to form a complete hopper pattern, as shown.

12-9 Cone and vertical duct intersection

A fitting composed of cone and duct, as shown in Figure 12-13, produces a line of intersection which is a hyperbolic curve on each side of the duct. To plot the curve for one side, draw elements (1 and 7) which pass through the duct corners in the top view. Project the elements to the front view. Project the corners to the front view, there-

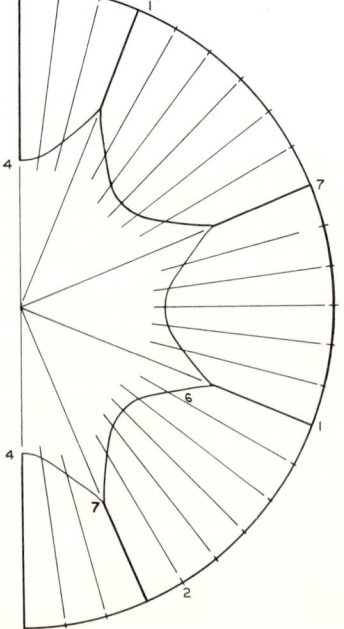

Fig. 12-13

COMBINATION PROBLEMS   261

upon locating the piercing points. Additional elements (2 through 6) uniformly spaced on the cone locate sufficient points for completing the curve. Develop patterns for the cone and duct in the usual manner. Notice that the elements of the curve must be rotated to become contour elements before the piercing points can be placed on the pattern. Because elements 1 through 7 represent one quarter of the cone and one side of the duct, the elements may be repeated to develop a full pattern for each part of the fitting, as shown in Figure 12-13.

12-10 Cone and horizontal duct intersection

The intersection shown in Figure 12-14 occurs quite often in the fabrication of separators and fans. A top and a front view are drawn from specifications. The intersection in the front view is a hyperbolic curve, determined in a manner similar to the last problem. Equally spaced elements (1 through 6) on the cone surface in the top view pierce the front plane of the square duct. The piercing points are projected to their respective elements in the front view to establish the curve.

Develop a pattern for the entire cone frustum in the manner described in Chapter 10. To develop the hole in the pattern, begin with six equally spaced elements (1 through 6 as taken from the top view). The curve points are rotated in the front view in the usual manner and swung into the pattern as shown. The curved distance from 6 to 7 is determined in the top view by means of step-offs and the distance is transferred to the base line of the pattern to locate point 7. Point 10 lies on the same cone element as 7. Point 11 lies on the same cone element as 1.

Fig. 12-14

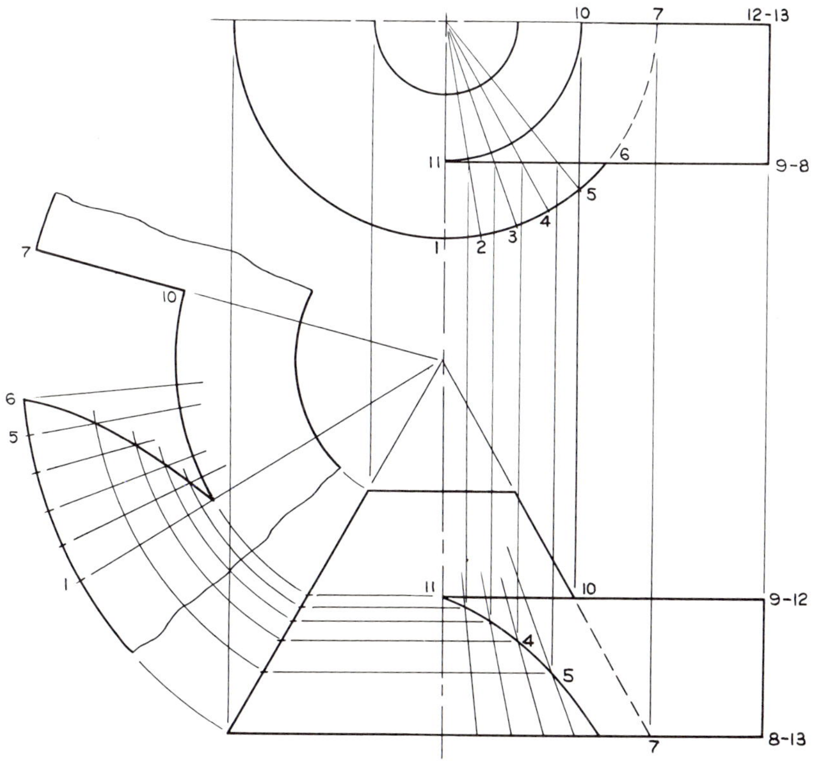

The patterns for the four sides of the duct may be taken from the given views. Top panel 10-11-9-12 and bottom panel 6-7-8-13 are true size and shape in the top view. Front panel 6-8-9-11 and back panel 7-10-12-13 are true size in the front view. They may be transferred, by triangulation, to form a complete pattern.

12-11 Reducer take-off

To draw the front view of a reducer take-off, as shown in Figure 12-15, construct the center lines from specifications. Using the specified pipe diameters, construct spheres at the intersection of the center lines. By drawing lines tangent to the spheres, the outline of the warped cone is established and the miter line between the cone and the small pipe may be drawn. Because the reducer is to be a warped surface, the miter lines between the reducer and the large cylinder are located by choice to cut through center as shown (rather than to use the spheres for locating them).

The miter planes cut the large cylinder at two different angles. In each case, however, the miter sections are semielliptical. Since the semimajor and minor diameters are known, the trammel method is used to draw a partial view of the semiellipses, attached to the front view as shown in Figure 12-16. Likewise, the miter section at the small end of the cone must be an ellipse in order to match the cut of the small cylinder. Attach a half-view, using the trammel method as before. The small elliptic curve is divided into a convenient number of equal spaces (8 in Figure 12-16) and the entire joint curves of the large end are spaced so as to equal the number of spaces at the small end. As in the problems of Chapter 11, a numbering system is used which will help in the triangulation of the zigzag arrangement of lines. True-length diagrams are constructed as before, using the distance between points (from the

Fig. 12-15

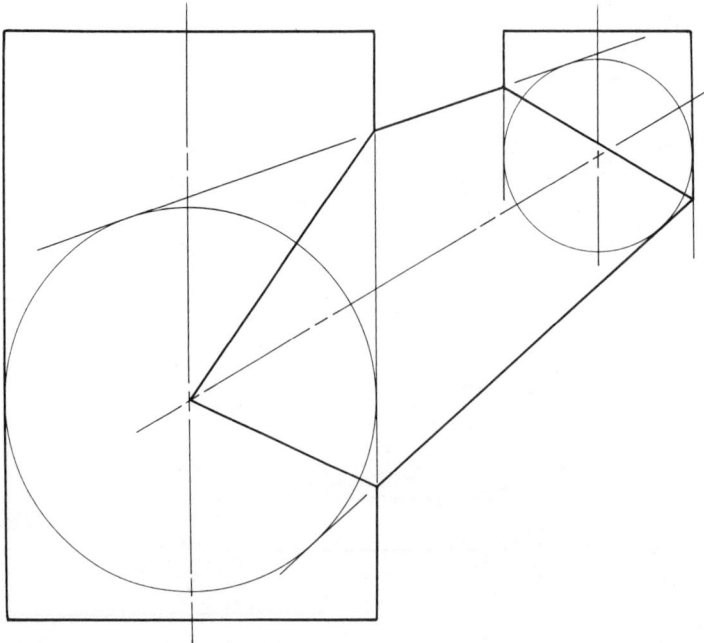

COMBINATION PROBLEMS 263

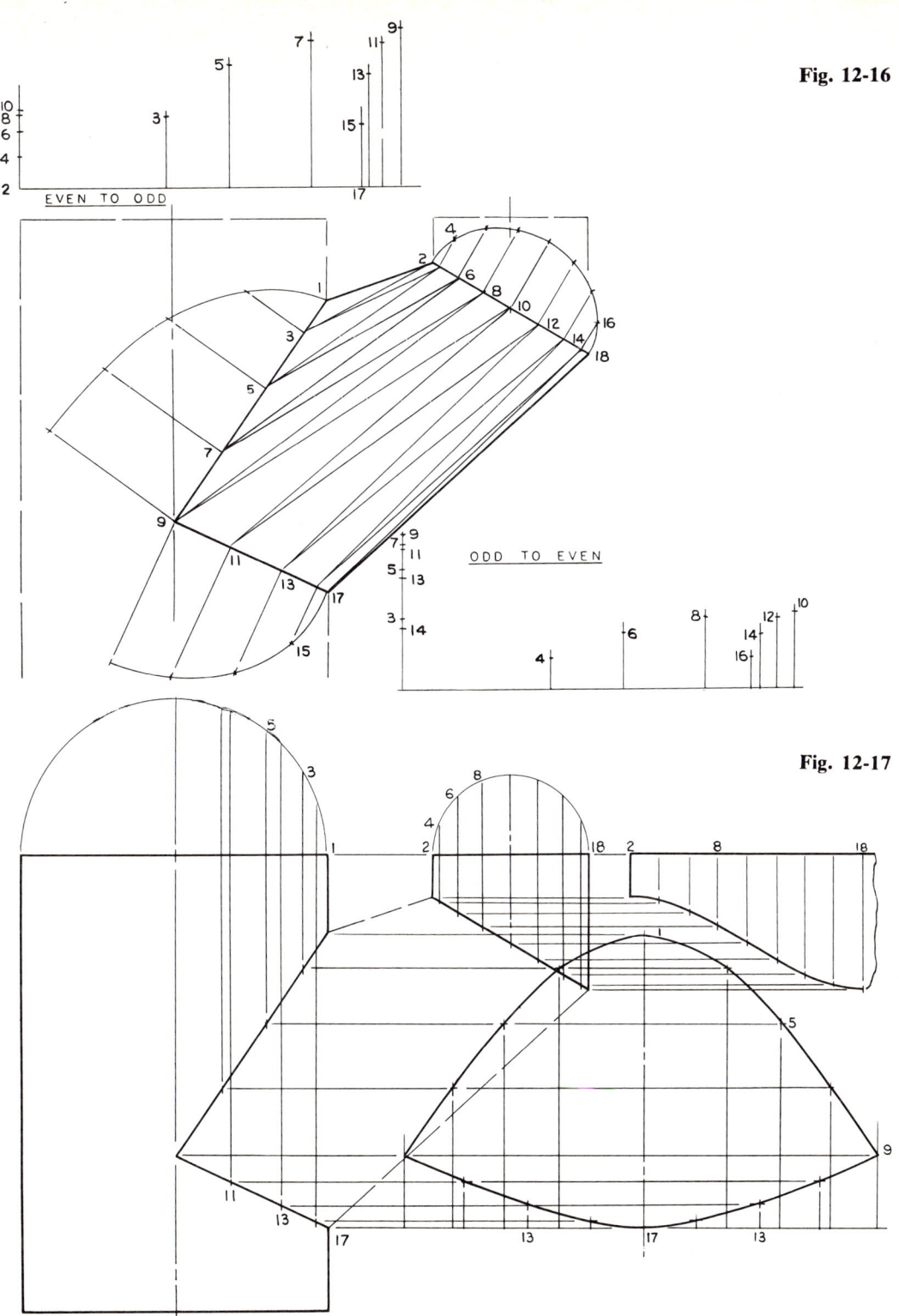

Fig. 12-16

Fig. 12-17

front view) as a horizontal measurement, and the distances from center (as taken from partial views) as vertical measurements (see Figure 12-16). The pattern for the warped surface may then be triangulated in the usual manner. Partial top views are used to develop the cylindrical features, as shown in Figure 12-17.

12-12 Warped cone elbow

The cyclone separator shown in Figure 12-18 combines the usual development of cones and cylinders with an interesting reducer elbow in the form of a warped cone.

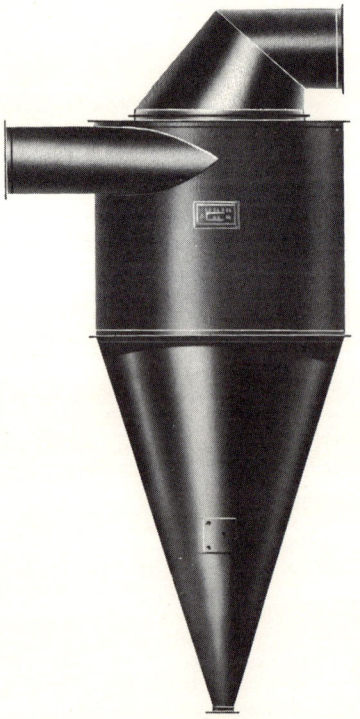

Fig. 12-18 Cyclone separator. (*Northern Blower Co., Cleveland, Ohio.*)

The elbow is drawn as shown in Figure 12-19. Begin by constructing the center lines from specifications. At the intersection of the cone and horizontal cylinder center lines, draw a sphere with a diameter equal to that of the cylinder. Draw the cone and cylinder outline tangent to the sphere, thus locating the miter line. The joint section between the cone and the horizontal cylinder is an ellipse in order to match the cut of the cylinder. The base of the cone is a circle; therefore, partial views are attached which verify the shape of the joints. To complete the pattern of the warped cone, use the same numbering and triangulation procedures as are explained for previous problems.

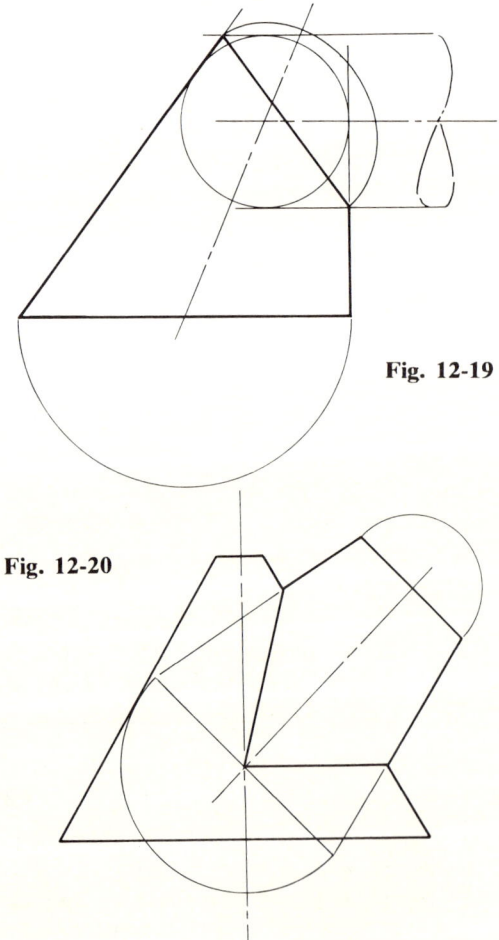

Fig. 12-19

Fig. 12-20

COMBINATION PROBLEMS 265

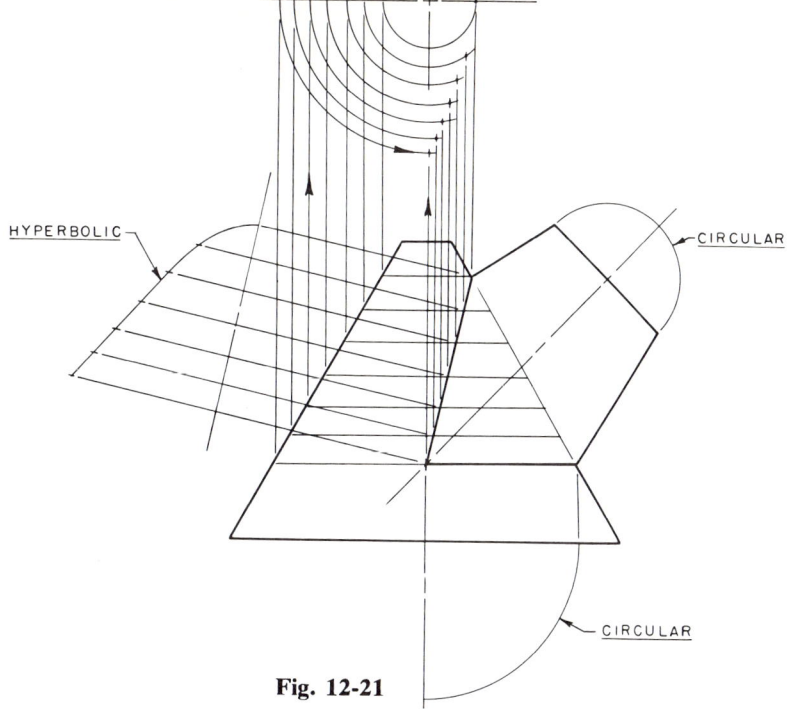

Fig. 12-21

12-13 Right circular cone and warped cone intersections

Many interesting and variable designs are possible when a warped cone is made to intersect a right circular cone, as shown in Figure 12-20. The variable factors are the intersection angle, the taper and length of each cone, and the inlet-outlet sizes. The feature noteworthy of repeated discussion at this time is *the importance* of determining the true shape of the miter sections. For the fitting as established in Figure 12-20, the true shape of each curve is determined by projection principles of Figure 12-21. Each curve is identified as being circular, elliptic, parabolic, or hyperbolic. The procedures for constructing such curves are discussed fully in section 10-8. Once the partial views representing the miter sections have been drawn, the remaining development is routine, based on earlier examples.

In concluding this discussion of sheet metal development, may the student and the person in industry be encouraged to solve as many and as varied problems as time will allow. It is with practice that one gains the confidence and the experience necessary for results and advancement.

EXERCISES

1. Design and develop patterns for a 90-degree tee to meet the following specifications: main pipe, 3-in [76-mm] diameter, 5-in [127-mm] length; taper branch pipe from main diameter to a 2-in [51-mm] diameter in a 5-in [127-mm] total fitting height.
2. Use the same measurements as for problem 1 but place the branch opening off center

an amount sufficient to make pipe vertically tangent on the back side of the fitting (flat on one side).

3. Develop patterns for the elbow in Figure 12-22.
4. Develop patterns for the offset transition in Figure 12-22.
5. Use the same measurements as for problem 1, but make the tee a 45-degree fitting (pitched). The length of the branch pipe from the intersection of center lines to the branch opening is 3½ in [89 mm].
6. Use the same directions as for problem 5, but make the 45-degree branch with the contour element on the back side in the same vertical plane as the back contour element of the main pipe (flat on back side).
7. Design and develop a 75-degree three-piece reducer elbow. The end sections are cylinders of 3- and 2-in [76- and 51-mm] diameter, respectively. The size reduction is accomplished in the center section. Make the center-line radius 3 in [76 mm].
8. Develop a pattern for the cone frustum of the collector as shown in Figure 12-23.
9. Develop patterns for one horizontal takeoff and hole of the collector shown in Figure 12-23. Use data as shown in Figure 12-24.

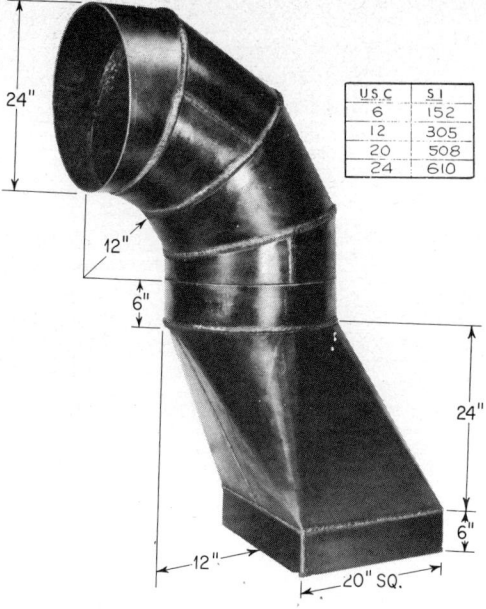

U.S.C	S I
6	152
12	305
20	508
24	610

Fig. 12-22 (*Naylor Pipe Co., Chicago, Ill.*)

Fig. 12-23 (*Prater Pulverizer Co., Chicago, Ill.*)
DIMENSIONED IN MILLIMETERS

COMBINATION PROBLEMS

DIMENSIONED IN MILLIMETERS

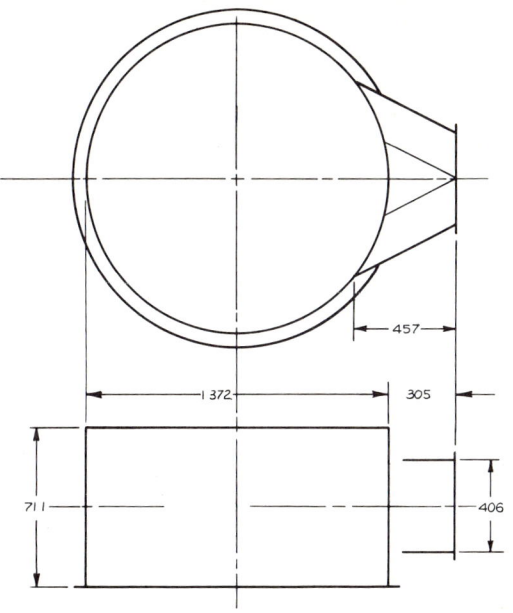

Fig. 12-24

Fig. 12-25

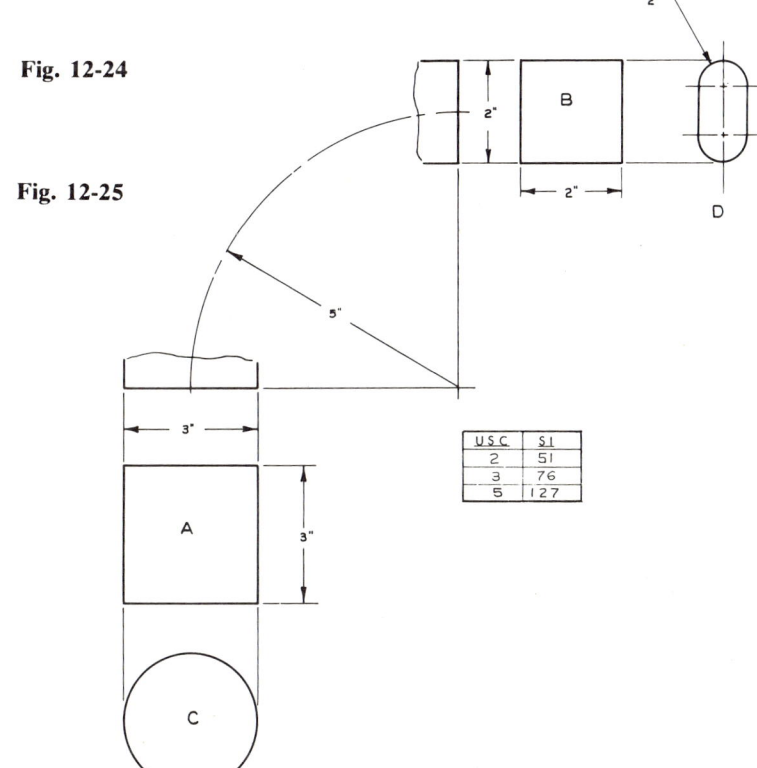

10. Develop a 90-degree three-piece transitional elbow as shown in Figure 12-25. Use bottom view *A* and side view *B* (all transition to take place in the center section).
11. Develop patterns as indicated in Figure 12-25, using bottom view *A* and side view *D*.
12. Same directions as for problem 11, except use views *C* and *B*.
13. Use views *C* and *D* of Figure 12-25.
14. The rectangular manifold shown in Figure 12-26 has an opening which measures 9 by 24 in [229 by 610 mm]. Following the center lines as specified in Figure 12-27, draw a top and a front view of the installation. Use a five-piece elbow of 18-in [457-mm] diameter and make the square-to-round transition 20 in [508 mm] high.
15. As assigned, develop patterns for the fittings of the layout of problem 14.
16. Develop patterns for the pipe, and intersection opening, as shown in Figure 12-28. Use pipe shape *A*.

USC	SI
2	51
3	76
5	127

268 SHEET METAL DRAFTING

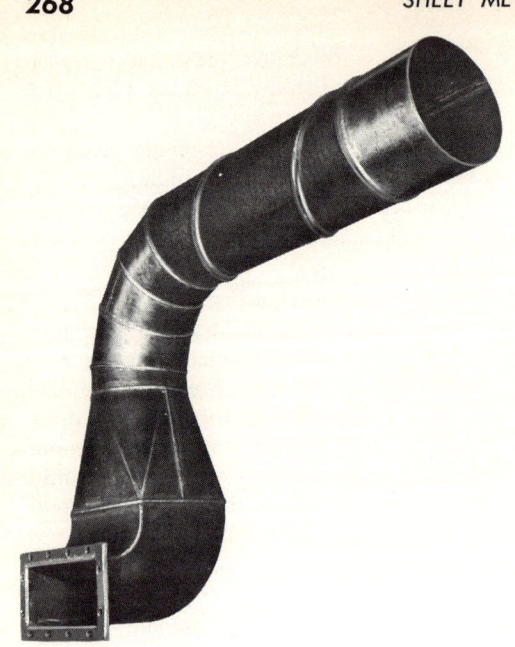

Fig. 12-26 (*Naylor Pipe Co., Chicago, Ill.*)

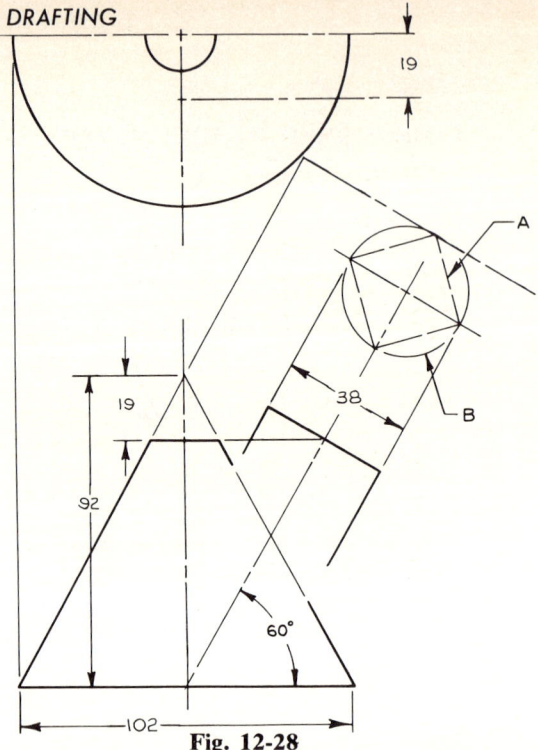

Fig. 12-28
DIMENSIONED IN MILLIMETERS

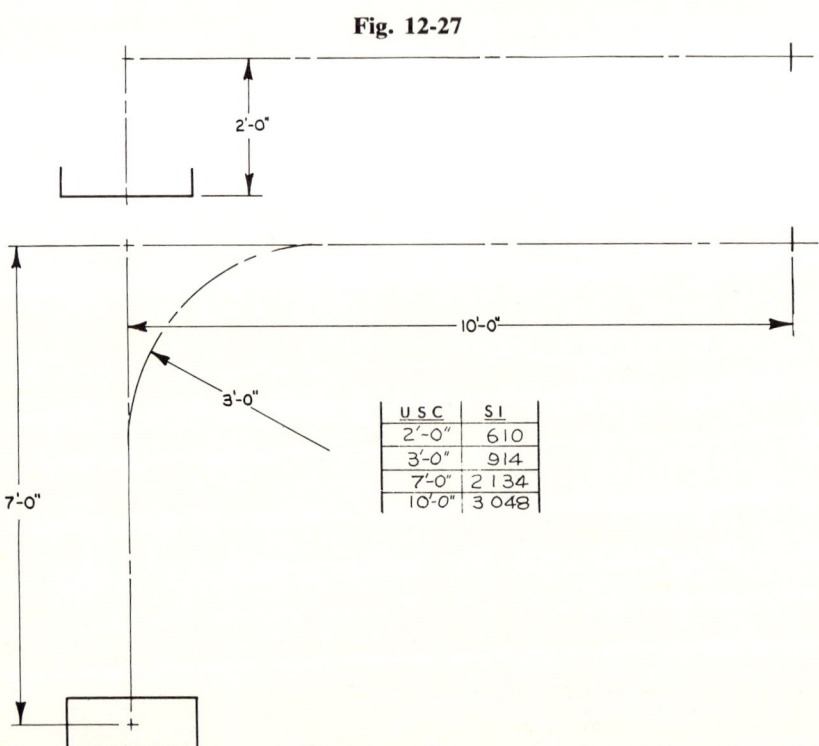

Fig. 12-27

USC	SI
2'-0"	610
3'-0"	914
7'-0"	2134
10'-0"	3048

COMBINATION PROBLEMS 269

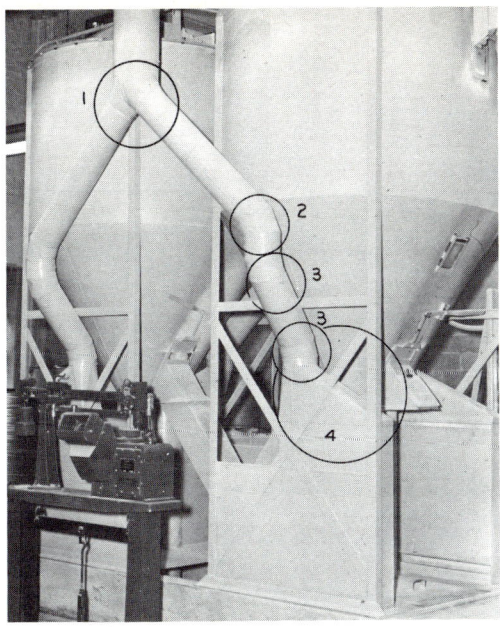

Fig. 12-29 (*Prater Pulverizer Co., Chicago, Ill.*)

17. Develop patterns for the pipe, and intersection opening, using shape *B* of Figure 12-28.
18. Develop patterns for the assigned fittings of Figure 12-29. Use data as follows:
    1. 45-degree Y branch, 14- and 10-in [356- and 254-mm-] diameter pipe, 3-in [76-mm] throat on each branch, 7-in [178-mm] miter line.
    2. 45-degree three-piece elbow, 10-in [254-mm] diameter and throat radius.
    3. Two-piece 30-degree elbow, 10-in [254-mm] diameter and throat radius.
    4. Transition, 10-in [254-mm] diameter to a 6- by 36-in [152- by 914-mm] duct, vertical on one side as shown, height of fitting 18 in [457-mm].
19. Develop patterns for the pipe and intersection opening as shown in Figure 12-30. Use pipe shape *A*.
20. Develop patterns for the pipe and intersection opening using shape *B* of Figure 12-30.

Fig. 12-30
DIMENSIONED IN MILLIMETERS

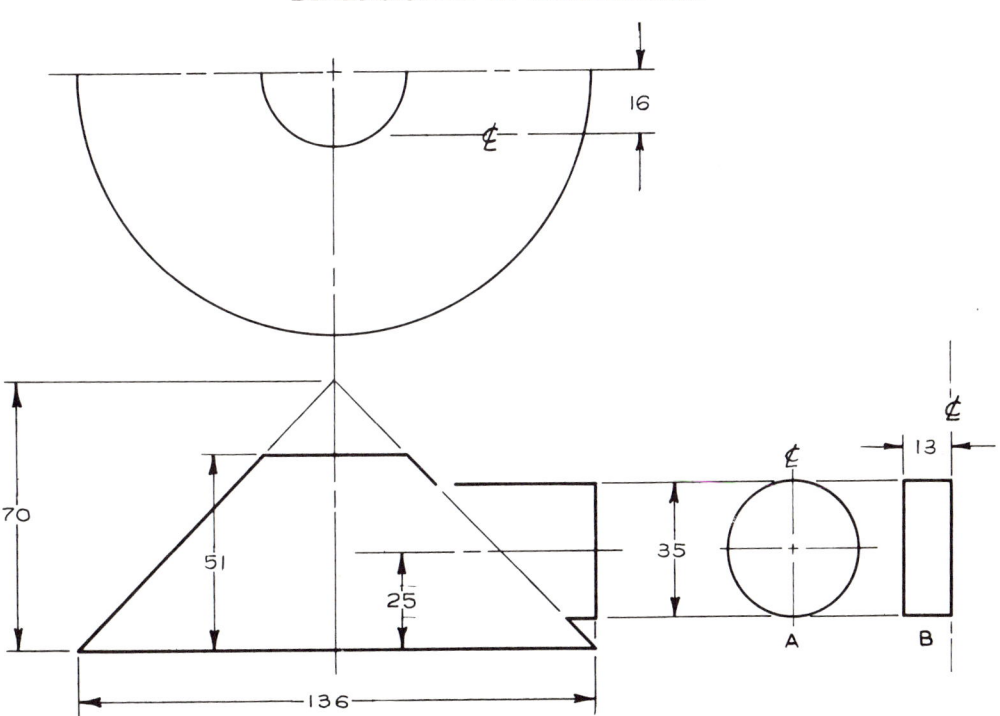

270 SHEET METAL DRAFTING

21. Develop patterns for the pipe and intersection opening as shown in Figure 12-31. Use pipe shape *A*.
22. Develop patterns for the pipe and intersection opening using shape *B* of Figure 12-31.
23. Develop patterns as assigned for the fittings of Figure 12-32.
24. Develop patterns for the fitting shown in Figure 12-33.
25. Develop patterns for the five-piece 60-degree elbow to be used to ventilate the 1,500-gallon no. 2 kettle, as shown in Figure 12-34. Elbow is made of 20-gauge [1 mm], type 316 Allegheny stainless steel. Diameter and throat radius are each 18 in [457 mm].

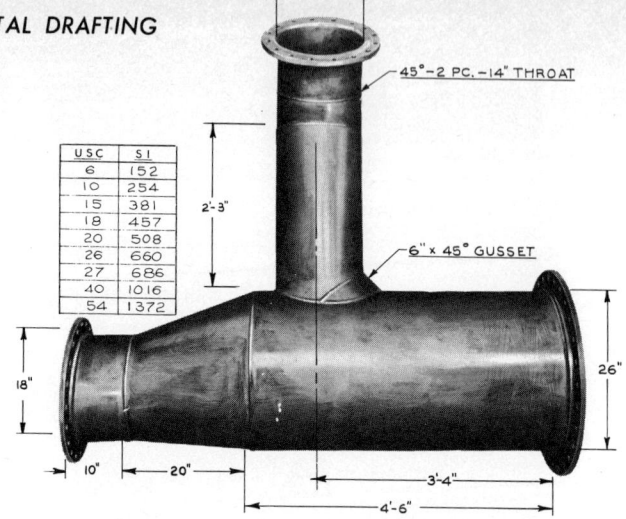

USC	SI
6	152
10	254
15	381
18	457
20	508
26	660
27	686
40	1016
54	1372

Fig. 12-32 (*Naylor Pipe Co., Chicago, Ill.*)

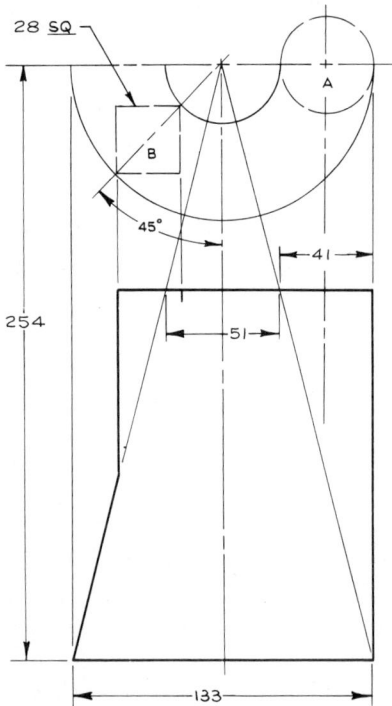

DIMENSIONED IN MILLIMETERS
Fig. 12-31

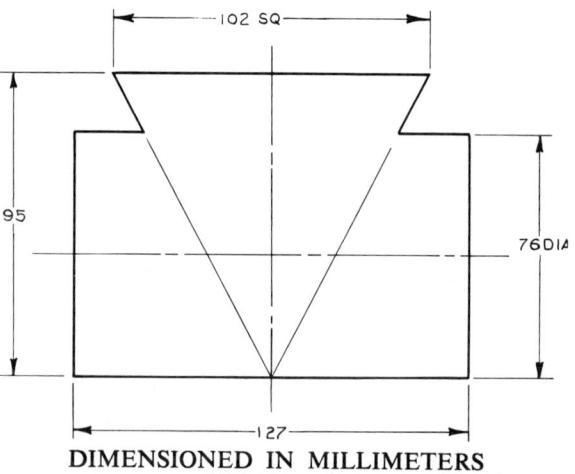

DIMENSIONED IN MILLIMETERS
Fig. 12-33

Fig. 12-34 Kettle ventilation in cannery. (*Lincoln Electric Co., Cleveland, Ohio.*) (*Red Wing Co., Fredonia, N.Y.*)

26. Develop patterns for connecting the elbow of problem 25 with the 32-in [813-mm] trunk.

    Make connection at 45 degrees on center with a branch throat length of 6 in [152 mm].

27. Develop patterns for the conical kettle hoods and covers shown in Figure 12-34. Diameter at base is 7 feet [2 134 mm], diameter at stack 18 in [457 mm], altitude at frustum 24 in [610 mm]. Width of door at base 3 ft [914 mm], true length of door side, 30 in [762 mm].

28. Develop patterns for the five-piece 45-degree elbow used above the no. 1 kettle of Figure 12-34. Diameter is 18 in [457 mm] and throat radius is 24 in [610 mm].

29. Based upon the hood dimensions of problem 27, and ignoring the slight curvature of the hood surface, construct an ellipse suitable as a pattern for the entrance of the 5-in [127-mm] steam pipe. The outside diameter of the steam pipe is 5½ in [140-mm]. Make pattern ¼ in [6 mm] undersize all around for flanging.

30. Develop patterns for the seven-piece elbow used in the main pipe of Figure 12-34. Diameter and throat radius are each 32 in [813 mm]. Intersect the top and second pieces with an 18-in-[457-mm-] diameter pipe centered on a radial line through the miter of the two pieces. The branch is tangent to the elbow and at 45 degrees to the plane of the elbow. Intersect the bottom pieces with the 32-in [813-mm] fan pipe.

31. The first aluminum concrete mixer ever

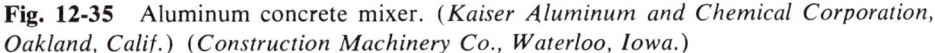

Fig. 12-35 Aluminum concrete mixer. (*Kaiser Aluminum and Chemical Corporation, Oakland, Calif.*) (*Construction Machinery Co., Waterloo, Iowa.*)

272   SHEET METAL DRAFTING

constructed is shown in Figure 12-35. It was built by Construction Machinery Company, Waterloo, Iowa, in cooperation with Kaiser Aluminum and Chemical Corporation, to undergo extensive testing. Develop patterns for the mixer as assigned. Use the measurements indicated in Figures 12-36 and 12-37.

Fig. 12-36

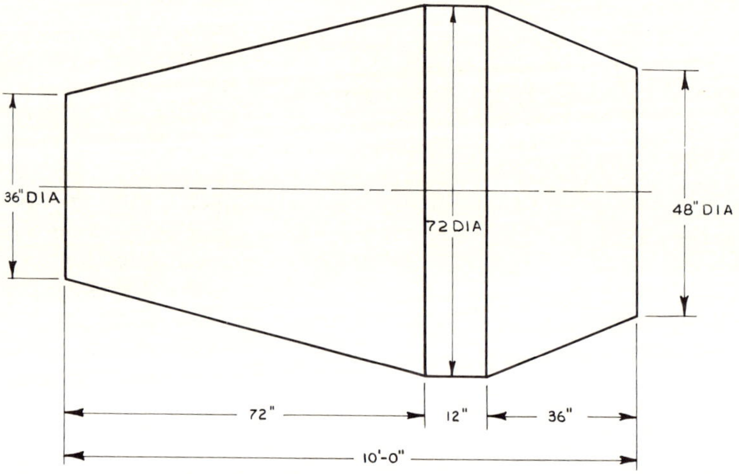

USC	SI	USC	SI
1	25	12	305
1½	38	17	432
2	51	22	559
2½	64	24	610
3	76	28	711
3½	89	38	965
4	102	42	1 067
5	127	48	1 219
7	178	72	1 829
9	229	10'-0"	3 408
10	254		

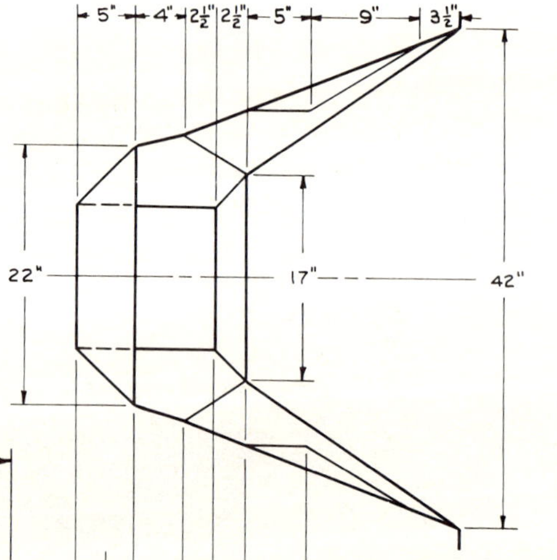

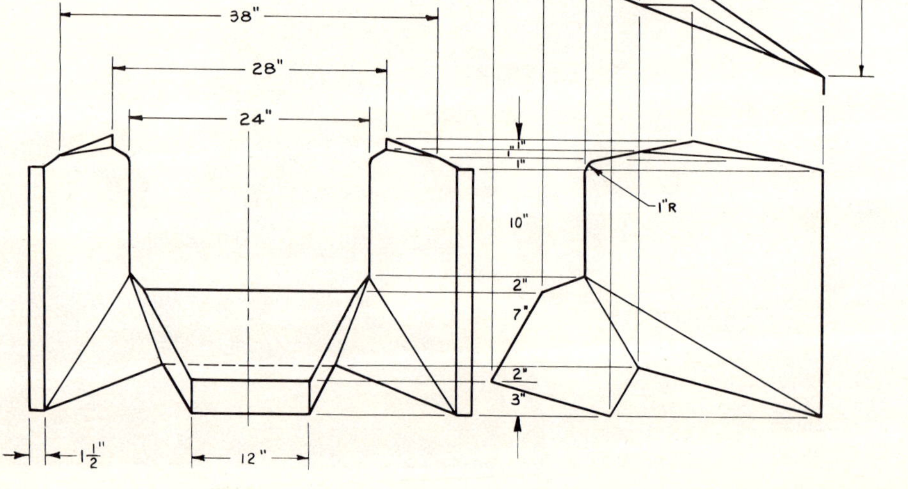

COMBINATION PROBLEMS

Fig. 12-37

DIMENSIONED IN MILLIMETERS

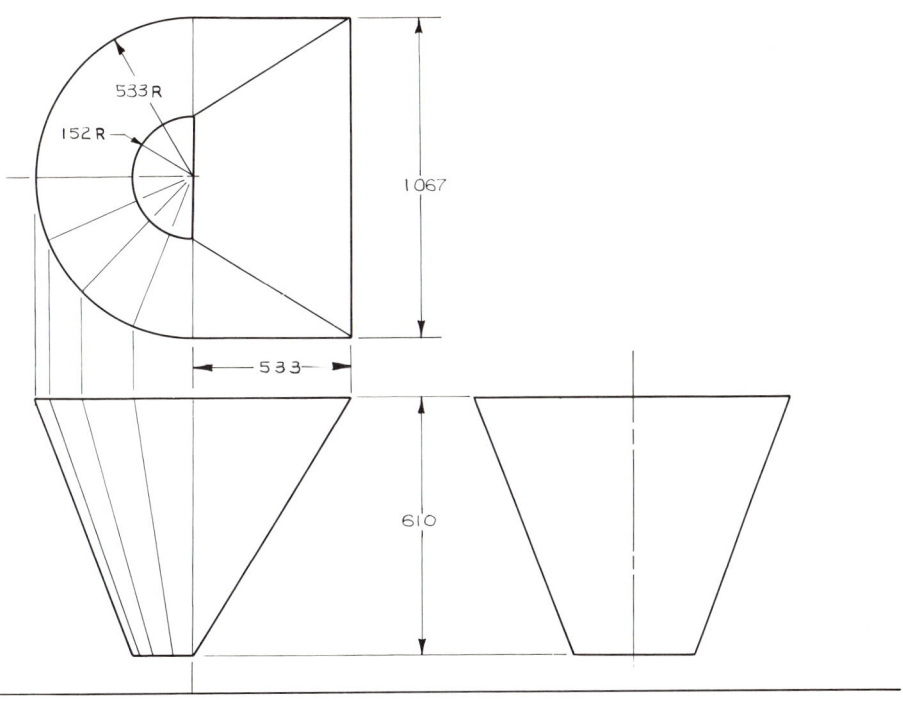

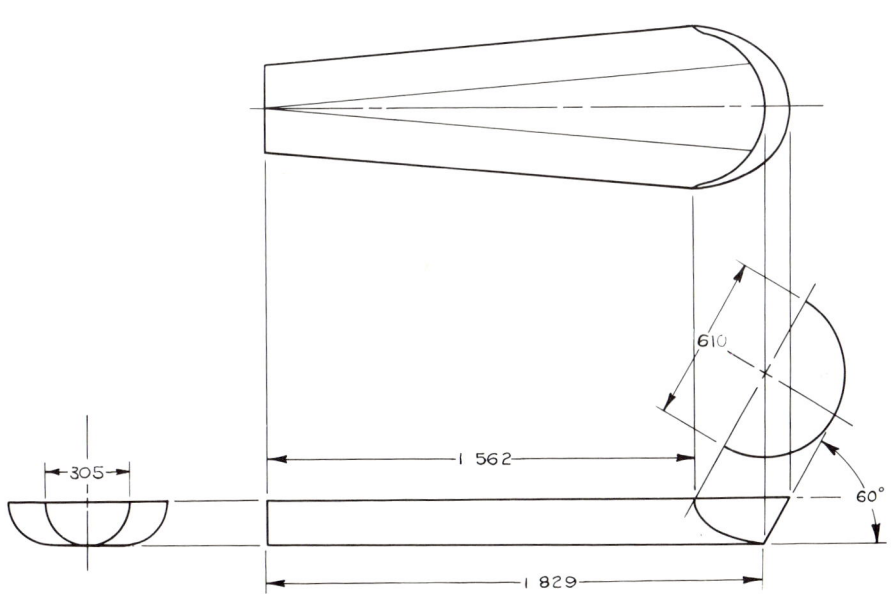

chapter 13

DIMENSIONING

The chapters thus far have dealt primarily with the description of the *shape* of the sheet metal object and with the shape of the *pattern* required to make it. Used only incidentally in illustrations and the specifications of problems, *but of equal importance,* is a *size description* of the sheet metal object. A two-view drawing may give the complete shape description of a cylinder, as shown in Figure 13-1A, however, unless size description, such as diameter and length dimensions, and material specifications are given, different workers could interpret the drawing to represent anything from a small satellite cylinder to the body of a missile, a boiler, a tank, or what have you. When the size and material specifications are added, as in Figure 13-1B, the drawing becomes a

Fig. 13-1 Sheet metal working drawing.
DIMENSIONED IN MILLIMETERS

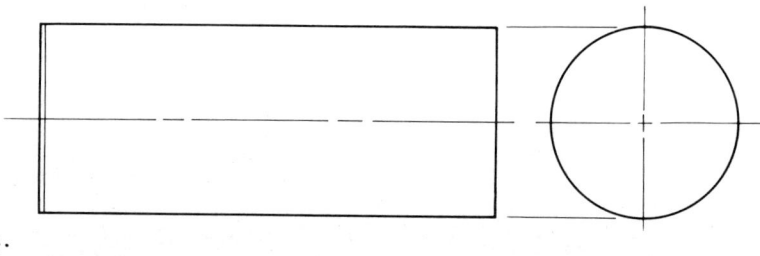

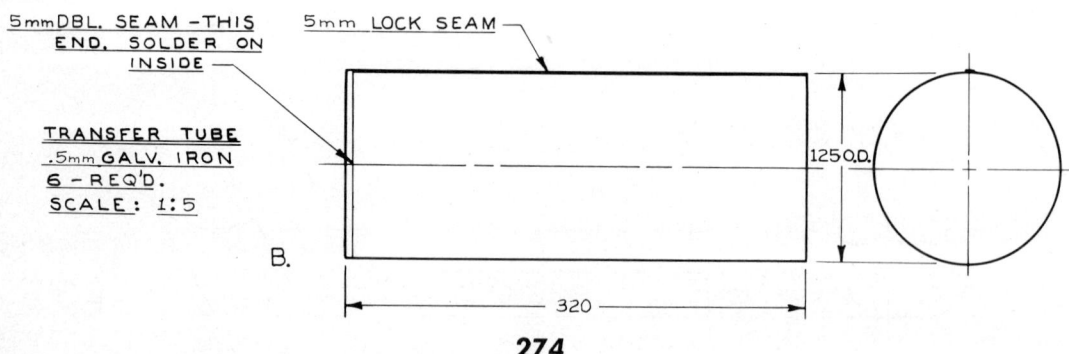

DIMENSIONING

working drawing, which is suitable for use in the shop.

In the study of sheet metal dimensioning, the student must learn three basic skills. First are the *techniques* used in dimensioning. Second, the student must learn to *choose* or *select* those dimensions which are needed in order to make the object. Third, the student must place the chosen dimensions according to accepted standards or in logical locations for the convenience of the workers using them.

DIMENSIONING TECHNIQUES

13-1 Lines

Three types of lines are used in dimensioning a sheet metal drawing. One of these, *center lines,* is first discussed in sections 3-12 and 3-13. Center lines are a necessary part of the shape description and, when used to locate certain features, are also a part of the size description.

Extension lines are thin black lines which "extend" away from the object in order to remove the dimension from the view. In this manner, the view remains unobstructed and easier to read. See Figure 13-2. Notice that

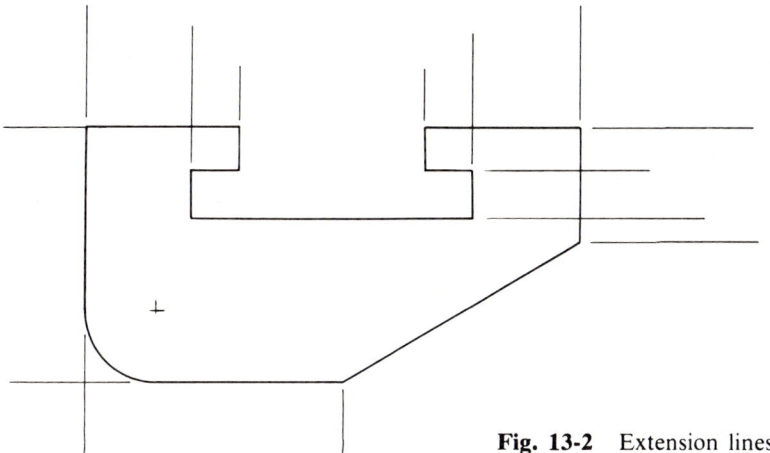

Fig. 13-2 Extension lines.

Fig. 13-3 Dimension lines.

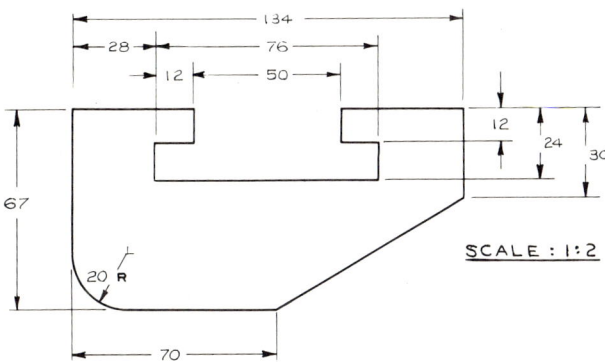

DIMENSIONED IN MILLIMETERS

SCALE : 1:2

the extension line has a gap of 1/16 inch [2 millimeters] between it and the object. Observe also that, when an extension line crosses an object line, in order to remove the dimension of an interior feature (Figure 13-2), a gap is used at the end of the extension line only. Extension lines may cross other extension lines. This usually occurs when the dimensions are perpendicular to each other. However, in the case of several parallel dimensions, the longer dimensions are placed outside the shorter ones so that extension lines need not cross dimension lines.

Dimension lines are thin black lines which are used to indicate, by means of arrowheads, the beginning and ending of the dimension involved and the magnitude of the dimension in the form of a figure. They are placed, as shown in Figure 13-3, with the arrows touching extension lines. Dimension lines are spaced from 3/8 to 1/2 in [10 to 13 mm] away from the object, and 3/8 in [10 mm] from one another when several are parallel. In mechanical engineering practice, the dimension line has an opening or break midway between arrows for the placement of the dimension. See Figure 13-3. In structural drafting and when using SI units, the dimension line is made continuous and the dimension value is placed above the line. See Figure 13-4. Both systems are used in sheet metal dimensioning, depending upon whether the drawing is made for stamping or other machine operations, made for an industry closely allied to structural work, or made to the preferences of the engineering department and chief draftsman. It would seem most efficient to make the line continuous as it takes less time than to plan a gap in it.

13-2 Arrowheads

Dimension lines, as shown in Figures 13-3 and 13-4, are initiated and terminated with *arrowheads*. For most drawings, arrowheads are made about 1/8 in [3 mm] long. On large sheet metal, auto, ship, or aircraft drawings, they may be preferred slightly larger. The proportions, however, remain constant as illustrated in Figure 13-5. Notice that the width of the arrowhead is about one-

Fig. 13-4   DIMENSIONED IN MILLIMETERS

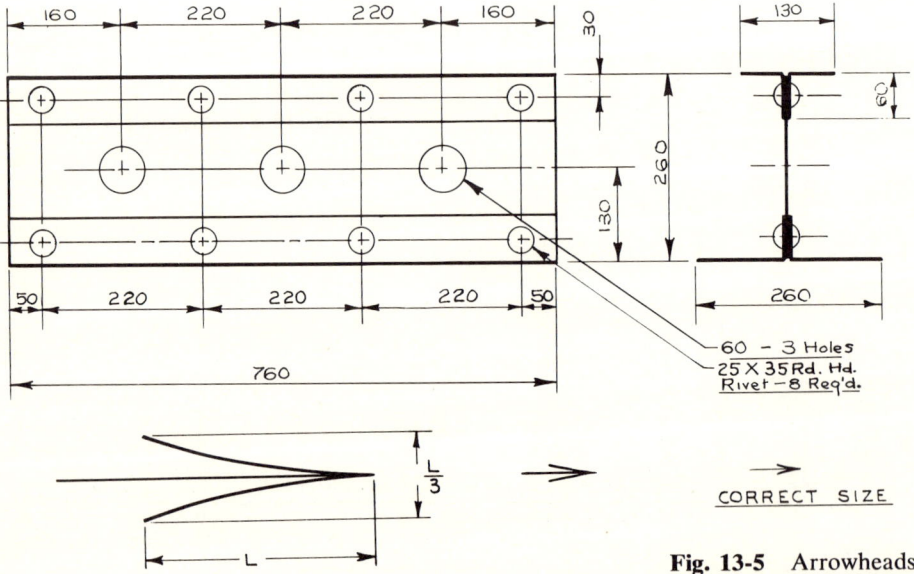

Fig. 13-5 Arrowheads.

third its length. Usually the arrowhead is made in two strokes. The order and direction of the strokes will vary with draftsmen. The important thing is that the arrowhead be made in the same way each time, and that it have proper size and proportions. Most arrowheads, made in two strokes with pencil, are preferred open at the stem (Figure 13-5). Arrowheads made in ink, however, have a tendency to "close" about the stem. (Notice the arrowheads used in all text figures. They are closed and about $\frac{1}{16}$ in [2 mm] long, hence very narrow.)

13-3 Figures

The magnitude of the dimension is shown by means of figures placed in the gap of the dimension line (in the structural system and in the SI system, above the dimension line). Numbers, fractions, and the necessary guide lines are discussed in section 4-4. Avoid crowding dimensions into small spaces. Figure 13-3 illustrates small spaces properly dimensioned. Note that the arrows are the first item to be placed outside the dimensioned area and that both the arrows and the dimension may be outside if necessary. When several dimensions are parallel, the numbers may be placed in staggered gaps. An advantage of the SI system is that no planning for the figures is necessary when putting the dimension lines on the drawing.

Two systems for the reading of dimensions are as follows: (1) the aligned system as shown in Figure 13-6, and (2) the unidirectional system of Figure 13-7. The unidirectional system was introduced by the automotive industry and has been adopted by the aircraft and shipbuilding industries. It is ideally suited to large drawings and therefore is recommended for sheet metal drawings.

In the past, sheet metal drawings were di-

Fig. 13-6 Aligned dimensions.
DIMENSIONED IN MILLIMETERS

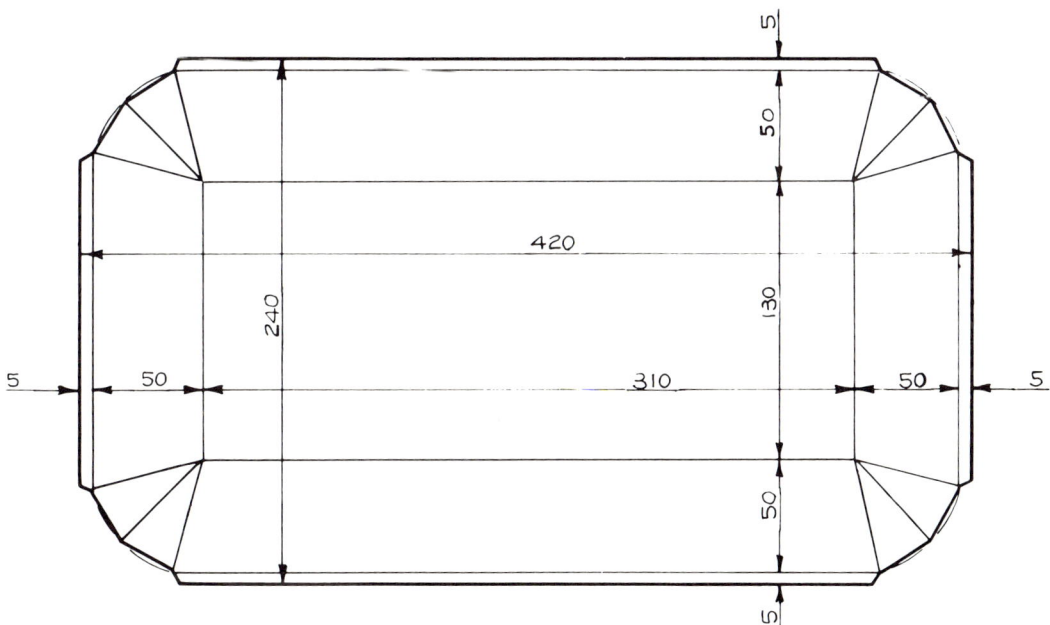

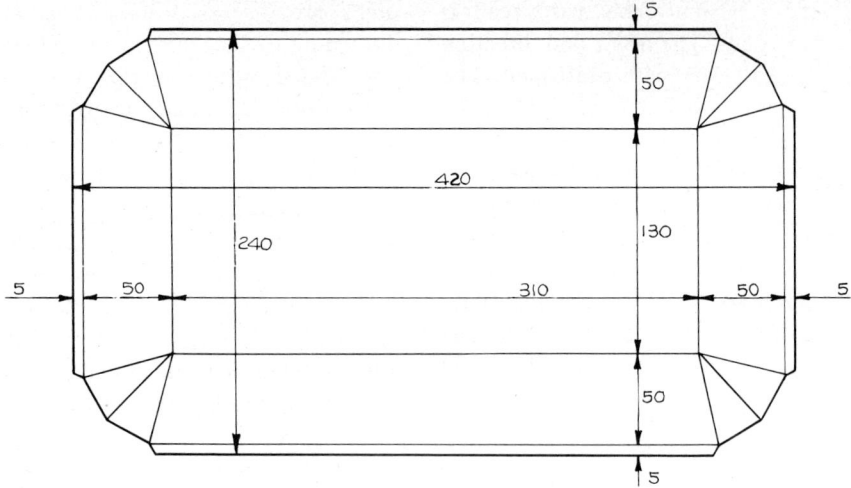

DIMENSIONED IN MILLIMETERS

Fig. 13-7 Unidirectional dimensions.

mensioned in inches. The inch marks (″) were placed above and to the right of the dimension. Inch marks were not required if all dimensions were in inches but, nevertheless, were a convenience in readability. For large objects, over 72 in, measurements were given in feet and inches (7′-6″) for the convenience of the worker, who usually carried a 6-foot scale. A scale which reads in both inches and millimeters is most useful to a worker. The majority of sheet metal measurements are of a magnitude that can be expressed in millimeters, e.g., 51 mm (2″), 254 mm (10″) etc. If the measurement is equivalent to 1 meter (39.3″) or greater, then the SI measurement is shown as 1 254 mm, 3 048 mm, etc., in which the first digit represents meters and the remainder millimeters.

13-4 Notes and leaders

While size description is given by means of dimensions, many features are better described by means of a note. *General notes* are placed in a convenient location on the drawing and are carefully lettered to be *read from the bottom of the drawing.* Typical

Fig. 13-8 Leaders.

DIMENSIONED IN MILLIMETERS

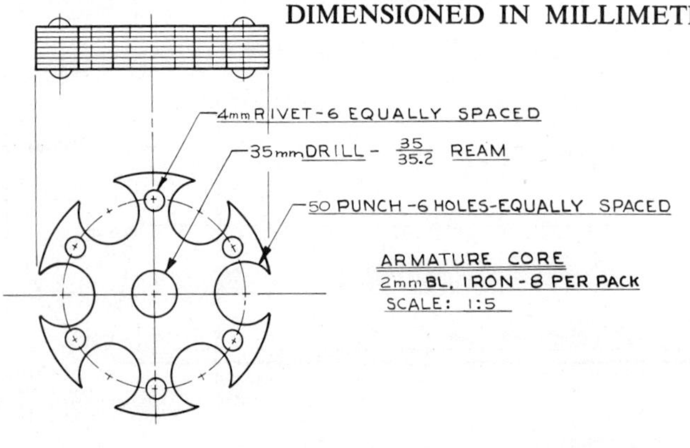

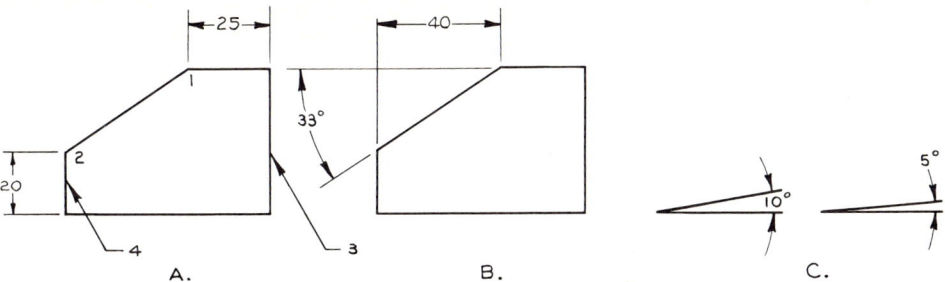

DIMENSIONED IN MILLIMETERS

Fig. 13-9 Dimensioning angles.

examples, such as ".5 mm Galv. Iron—6 reqd; Scale; etc.," are shown in Figure 13-1*B*. *Specific notes* have a thin black line leading from the note to the feature involved. The line is called a *leader* and is initiated with a horizontal line (about ¼ in [5 mm]) and is terminated with an arrowhead. Leaders to circular features should, when possible, be radial lines (aimed at the center point). The leader should be a *straight* line, placed at an angle (preferably 60 degrees upward to the right or downward to the left). The drawing appearance is enhanced if all leaders can be arranged parallel. The leader *must start at the beginning or at the end of the note,* never at the beginning or end of an intermediate line of the note (see Figure 13-8). Those features which are usually described by note

and leader are discussed later when considering dimension placement.

13-5 Angles

Angles are dimensioned as shown in Figure 13-9. Example *A* illustrates the coordinate method of dimensioning so as to locate the end points of the angular line from functional edges or surfaces. Corner 1 may be measured from edge 3 or 4, depending upon which edge is used as a reference for other dimensions. Notice that corner 2 is measured from the base edge, but could be measured from the top edge if that edge is more important or functional. Example *B* illustrates angular dimensioning by degrees. The degree method is not as accurate because the

Fig. 13-10 Dimensioning arcs.
DIMENSIONED IN MILLIMETERS

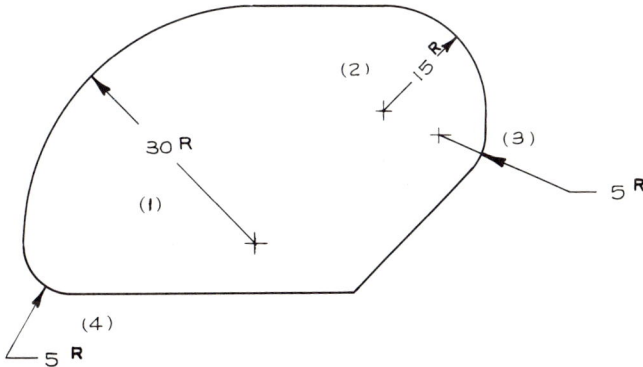

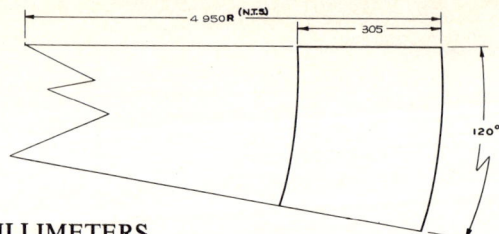

Fig. 13-11

DIMENSIONED IN MILLIMETERS

amount of error increases with angle leg length. Example C indicates the method for dimensioning degrees in small spaces. Notice that crowding is thus avoided.

13-6 Arcs

An arc is dimensioned by giving the radius in the view in which it appears true size. The dimension value is followed by the letter *R*. In Figure 13-10, four arcs are illustrated. Example 1 has the arc center indicated and the arc is large enough for the dimension and arrow to be placed within the circular feature. Example 2 is the same as 1 except an aligned dimension is shown. Example 3 illustrates the dimensioning of a small radius. Example 4 is the same as 3 except that the center point is not required. Since sheet metal drawings are usually large, it may be necessary to dimension to the center of an arc when the center falls outside the drawing sheet or material layout. Such a situation is illustrated in Figure 13-11.

CHOOSING SUITABLE DIMENSIONS

13-7 Size dimensions

Dimensions which tell "how big the object is" are commonly referred to as *size dimensions*. Certain basic principles apply to the selection of dimensions for various shapes. For example, square, rectangular, and angular pieces of flat stock have the two basic dimensions of width and length; the thickness is usually given by means of a note or other specification. Flat disks have only one basic dimension, diameter; the thickness is specified as before. If the flat stock is square or rectangular but has rounded corners, the radius of the rounded corner is given. See Figures 13-3 and 13-12. In other words, a rectangular strip with rounded corners would have over-all dimensions and the corner radius shown. In this case, it is not necessary to locate the radius because it is tangent to a square corner. Flat stock with *rounded ends*, however, would be dimensioned as shown in Figure 13-13. Notice that only the center-to-center distance (150 mm) and the radius (35 mm) are given because these are the only dimensions needed to lay out and cut the blanks. If a die is to be made for blanking, then the width of the blank 70 mm would also be shown. Observe that an over-all dimension is *not* shown.

Prisms are dimensioned by giving the width and length measurements on one view and the thickness dimension on a related view.

Fig. 13-12

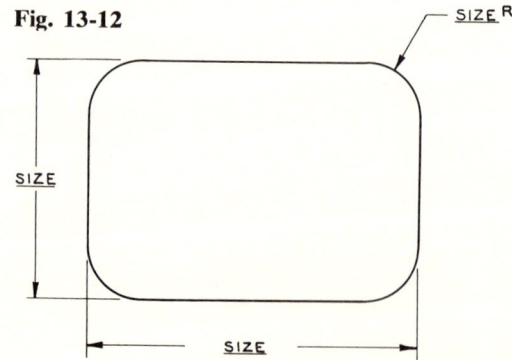

DIMENSIONING

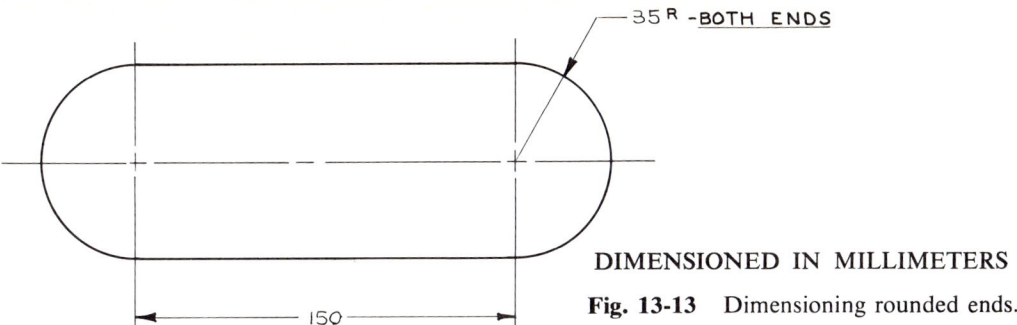

DIMENSIONED IN MILLIMETERS

Fig. 13-13 Dimensioning rounded ends.

Cylinders are dimensioned, as shown in Figure 13-1, by giving the diameter and length on the *noncircular view*.

Cones are dimensioned by giving the diameter of the base in the noncircular view and the altitude in a view where it is true length (Figure 13-14A). A cone frustum is dimensioned by giving the base diameter, the frustum diameter, and the altitude of the frustum, all in the noncircular view as shown in Figure 13-14B. An alternative method sometimes required on dimensioning a cone frustum is to give an angular value either between the cone base and the elements or between elements, instead of the frustum diameter or the frustum altitude (Figure 13-14C).

Pyramids are dimensioned by giving the measurements for the base in the view where the base appears true size and the altitude in a view where it is true length. See Figure 13-15A. In case the pyramid is nonsymmetrical or in case the apex requires location, dimensions should be given as shown in Figure 13-15B. Pyramid frustums are dimensioned as shown in Figure 13-15C or D.

Fig. 13-14 Dimensioning cones.

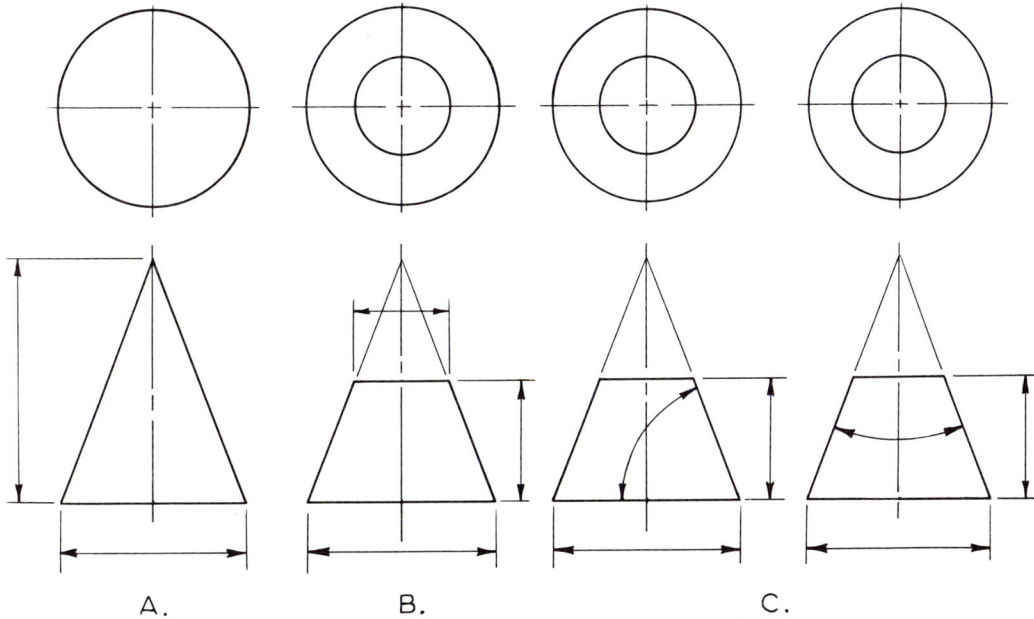

A.   B.   C.

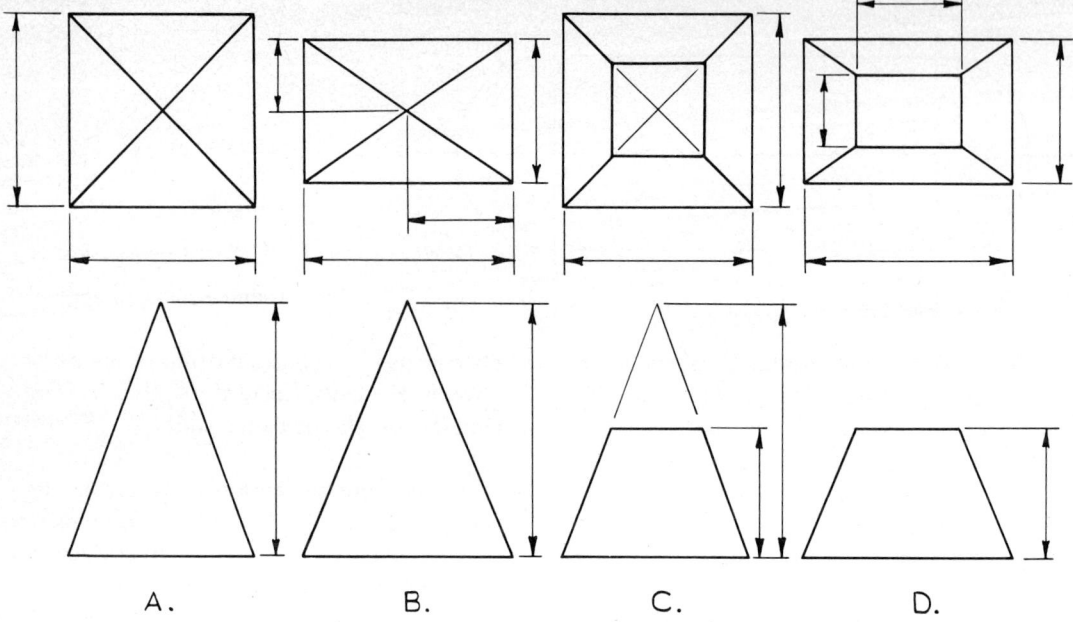

Fig. 13-15 Dimensioning pyramids.

13-8 Location dimensions

Since each feature being dimensioned for size is also related, in a functional manner, to other features of the object, it becomes necessary to locate them properly. Dimensions which relate the hole, notch, depression, center, or other geometric shape to the main features of the object are called *location dimensions*. See Figure 13-4 for examples of location dimensions on flat stock. Prisms are located by means of dimensions to their faces. If one face is coincident with a surface of the larger object, no location dimension is needed in that direction. For example, in Figure 13-16, the "box within

Fig. 13-16

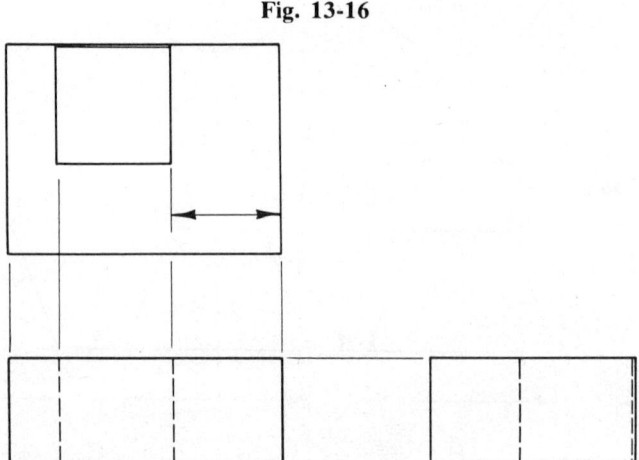

DIMENSIONING

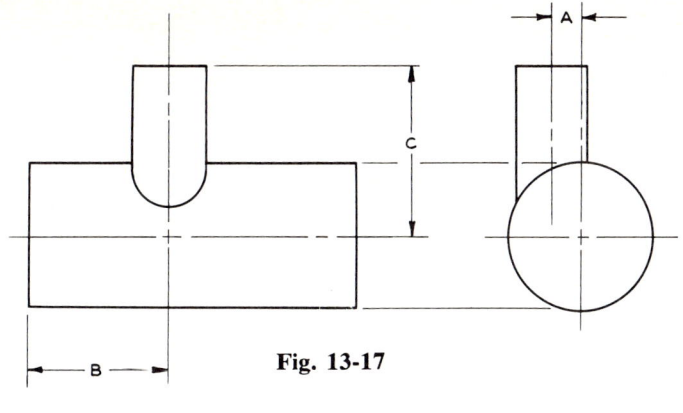

Fig. 13-17

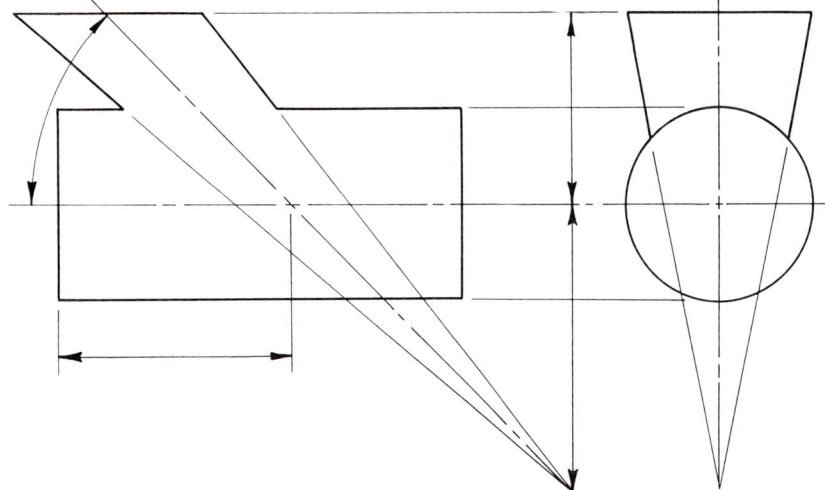

Fig. 13-18

Fig. 13-19

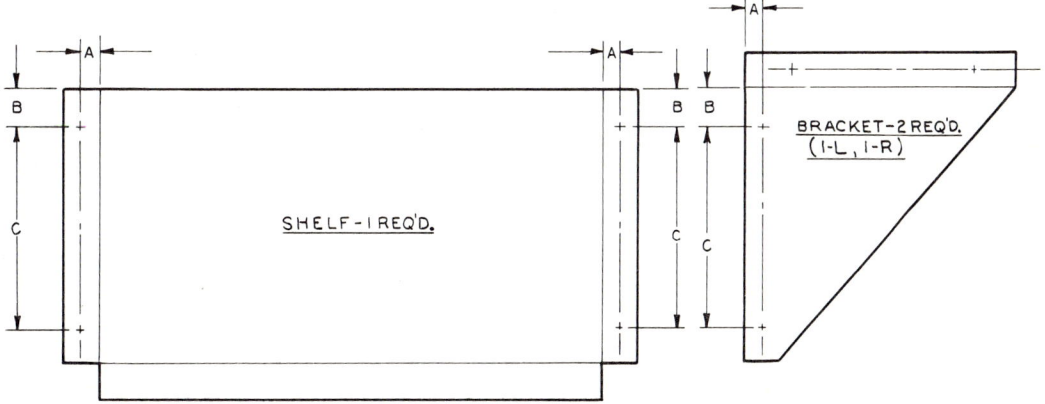

a box" is completely located with the one dimension shown. Cylinders are located by means of their centers and one face. In Figure 13-17 the center-line distance A and the center-to-face distances B and C relate the cylinders to each other. Since a hole may be considered a cylinder (a negative one) many location dimensions are used to locate the center lines of holes. Such dimensions are always placed in the view where the holes appear as circles. Likewise, cones and pyramids are located by means of the centers and bases. The dimensions are placed on a view where they show true length and where they will be of most value to the worker (see Figure 13-18). As has probably been observed by now, location dimensions are used between center lines, highly functional edges, or combinations of both. A hole, for example, is located by center lines from other center lines, or from a reference surface or edge. If the hole is to mate with a feature on another piece, the mating dimensions correlate the features and should be used in dimensioning the other piece. See the shelf and bracket of Figure 13-19.

13-9 Curves and profiles

Since they are easily dimensioned and located, curves should be designed by using circular arcs whenever possible. See Figure 13-20A. If the curves are not parts of circles, they may be accurately plotted by giving a series of offset measurements, uniformly spaced at a number of measuring stations, as shown in Figure 13-20B.

The profile of a bent strip is dimensioned as shown in Figure 13-21. Notice that all dimensions are made to one surface of the

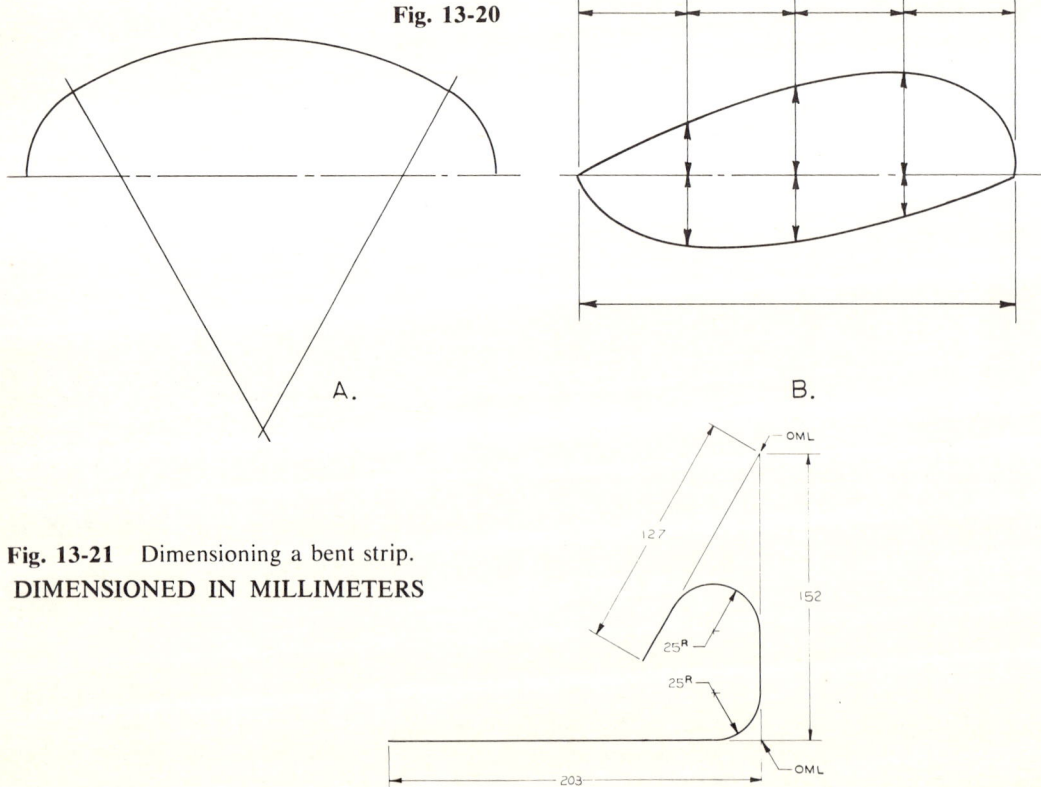

Fig. 13-20

Fig. 13-21 Dimensioning a bent strip.
DIMENSIONED IN MILLIMETERS

DIMENSIONED IN MILLIMETERS

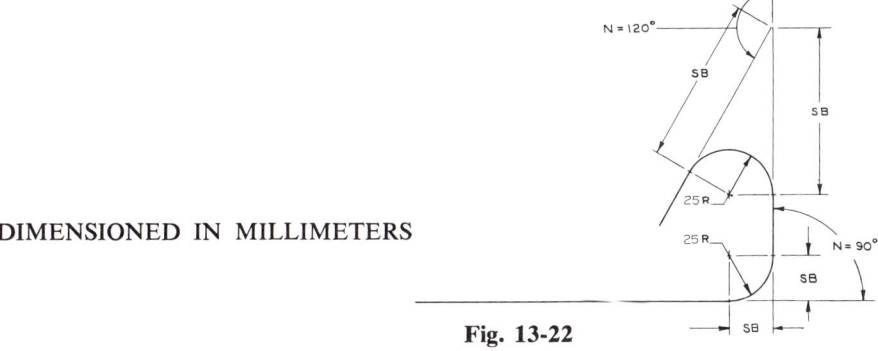

Fig. 13-22

strip and that dimensions are given from outside mold line OML to outside mold line, and to the ends of the strip. In such a profile, the sum of the dimensions [482 mm] will be greater than the actual stretchout (developed) length of the strip required.

To compute the stretchout or developed length from the dimensions shown in Figure 13-21, refer to Figure 13-22 and proceed as follows: (1) Calculate the *setback* SB for each leg of each bend. Use the formula: SB = radius $R \times \tan N/2$. Thus, at the 90-degree bend, SB = 25 tan 45°, or SB = 25 × 1 = 25; therefore, in the case of a right angle, the setback distance is equal to the radius (SB = R). At the 120-degree bend, SB = 25 × tan 60°, or 25 × 1.73 = 43 mm. (2) The distance around curves may be calculated, using the formula

Curved distance CD =
$$\frac{\text{radius } R \times \text{angle of bend } N}{57.3}$$

(see section 5-9)

Hence, for the 90-degree bend, CD = 39 mm, and for the 120-degree curve, CD = 52 mm. (3) Add the legs (482 mm) and subtract the setback total (136 mm). Thus CD = 482 − 136 = 346 mm. (4) Add the total distance around the bends (39 + 52 = 91); therefore, the total length of the strip required is found to be 346 + 91 = 437 mm.

The above computations are based upon relatively thin metal, in which the bend characteristics for both the inside and outside surfaces, for all practical purposes, are uniform. If the metal has considerable thickness, additional factors must be considered in determining stretchout or developed length.

13-10 Bend allowance

When forming heavy sheet metal, allowance must be made for the fact that the metal on the outside of the bend tends to stretch while the metal on the inside of the bend compresses. If both forces were equal, the line between them, called the *neutral axis*, would remain at the thickness center. When this is true, the neutral axis may be used to calculate the stretchout. Unfortunately, the tendency to stretch is greater than the tendency to compress; hence, the neutral axis is moved inward on the bend. Factors which affect the shift of neutral axis are the ductility, hardness, and yield strength properties of the metal, and the bend radius in proportion to the bend angle. For any radius greater than twice the metal thickness, the

DIMENSIONED IN MILLIMETERS

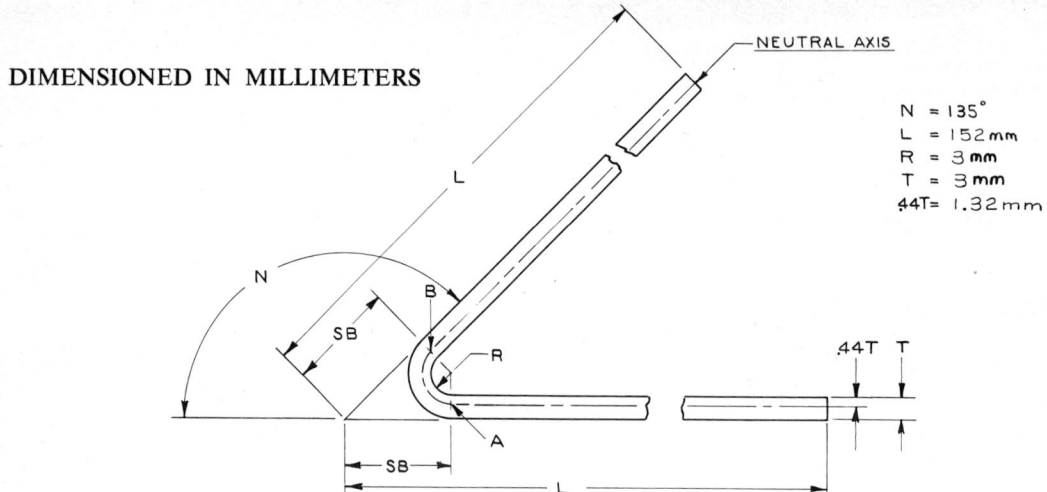

Fig. 13-23

neutral axis may be assumed to remain at the center of the piece. The aircraft industries have established that, for bend radii of two thicknesses or less, the neutral axis should be assumed at 0.44 of the metal thickness, *measured from the inside radius*. They have also arrived at an empirical formula for computing the amount of material needed for a bend. Where R = inside bend radius, T = thickness of metal, and N = angle of bend in degrees, BA (bend allowance) = $(0.01743R + 0.0078T)N$.

Several examples should illustrate the calculations necessary to determine stretchout or developed length in heavy materials. As the first example, assume that a cylinder of ¼-in [6-mm] plate is required to have a 38-in [965-mm] inside diameter. Since the bend radius is many times greater than the material thickness, the neutral axis may be assumed in the center of the stock. Thus the *neutral diameter* is equal to $38 + T$, or 38¼ in [965 + T, or 971 mm]. The length of strip needed is $38.25 \times \pi$ or 120.1 in [$971 \times \pi$, or 3 049 mm]. If only the inside diameter were used in the computations, the plate strip would be 1 in [25 mm] short. If the outside diameter were used, the strip would be ¾ in [19 mm] long.

As a second example, for a bend as shown in Figure 13-23, assume the neutral axis to be $0.44T$ from the inside edge. Note that to determine SB, *using the outside surface*, the formula should read: SB = $(R + T)$ tan $N/2$. Thus, $(3 + 3) \times 2.41 = 14$ mm. To calculate BA, use the formula

$$\mathrm{BA} = \frac{(R + 0.44T)N}{57.3} = \frac{4.32 \times 135}{57.3} = 10 \text{ mm}$$

The developed length is determined, using:

Length = 2L − 2SB + BA = 304 − 28 + 10 = 286 mm

As a third example, compare the empirical formula results with the BA of 10 mm previously computed. Thus BA = [(0.01743 × 3) + (0.0078 × 3)] 135 = (.05229 + .0234) 135 = .07569 × 135 = 10.2 mm. The difference in methods is only: .2 mm.

The problems of bend allowance have been troubling plate layout workers for many years. Fortunately much data from experimental bending has been tabulated and bend allowance tables are available.

PLACEMENT OF DIMENSIONS

13-11 Prisms, cones, pyramids

Prisms are dimensioned by placing two of the three dimensions required in one view. Dimensions which are shown by both views (such as 1 and 2 of Figure 13-24) are placed between the views, but are applied (by means of extension lines) to only one view. The required dimensions for cones are placed as shown in Figure 13-14 and the dimensions for pyramids are placed as shown in Figure 13-15.

13-12 Cylinders and holes

Positive cylinders are dimensioned in the *noncircular* view. Holes (negative cylinders) are usually dimensioned by means of a note applied with leader to the circular view. The note, besides giving size, may also specify the method of producing the hole. If possible, the dimensions which locate the hole should be given in the circular view, since that is the view to which the layout worker would naturally refer. Holes spaced on a circular center line about a center are dimensioned by giving the size (as a note) in the circular view and by giving the *diameter* of the center line in the same view. The wording "equally spaced" may be used to locate the holes on the center line. If unequal spacing is involved, angular or offset measurements are used to locate the holes (see Figure 13-8).

13-13 The contour principle

There is a best place for the location of each dimension on a drawing. One of the most valuable basic rules in determining the best placement is known as the *contour principle*. The contour principle may be expressed as follows: Since several views are required to completely describe the shape of an object, some views describe only certain features. Whereas the best views are determined and constructed first, place the dimensions describing size and location of a certain feature in the view in which the feature best shows. The object shown in Figure 13-25 is dimensioned according to the contour principle.

Fig. 13-24

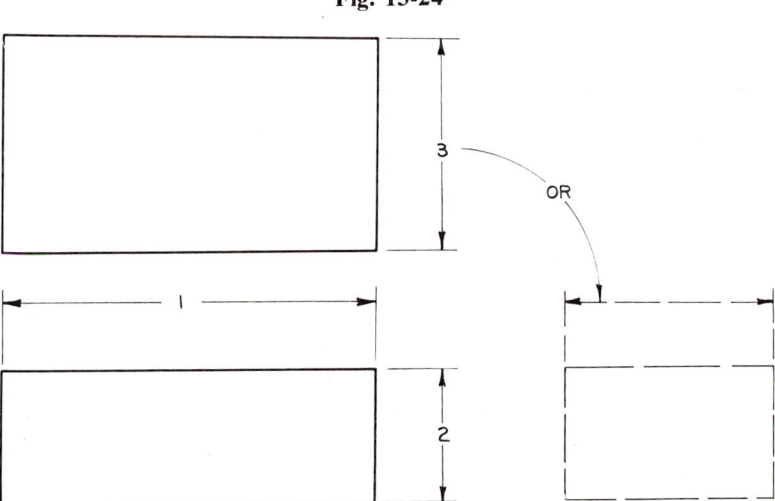

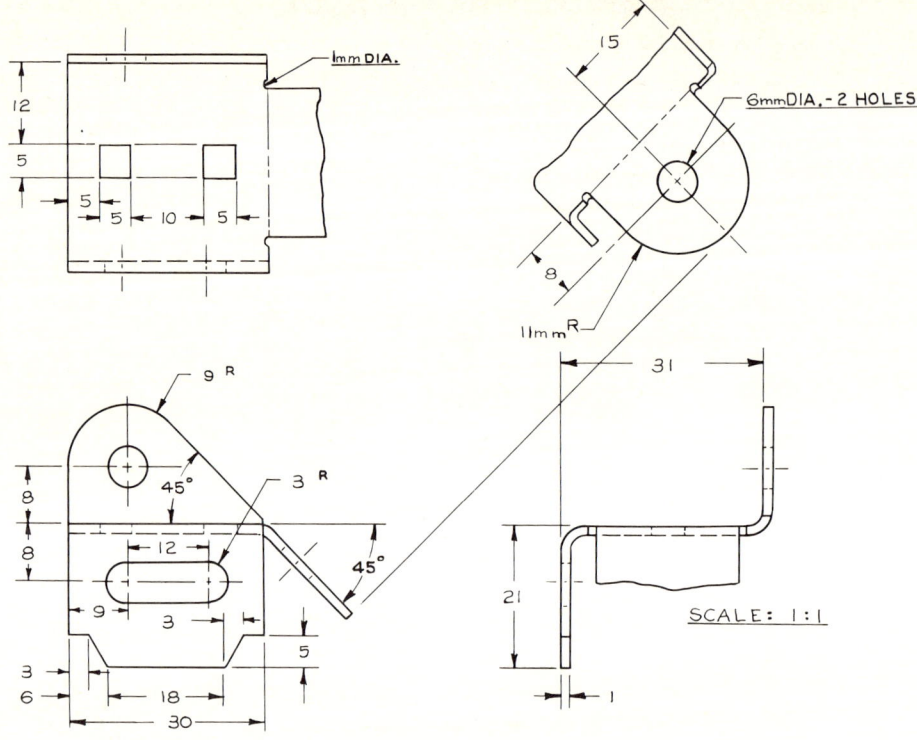

DIMENSIONED IN MILLIMETERS

Fig. 13-25 Dimensioning according to the contour principle.

13-14 Clarity

Dimensions may be placed according to basic rules and yet the drawing may be vague or difficult to read. This occurs because a certain amount of good judgment is necessary as to exact placements. The following statements will illustrate points to keep in mind in order to make the dimensioned drawing clear and easy to read.

a. Dimensions should always be placed off the view unless clearness and directness are gained by placing them on the view. However, most drawings will usually have some dimensions which must be placed on the view.

b. Each dimension should be given only once and the same information should not be repeated in a different form.

c. Dimensions should be given to visible features and hidden line dimensioning should be avoided.

d. All lettering should be done horizontally so as to be readable from the bottom of the drawing. Guide lines should be used for all lettering. Dimension figures should be legibly made.

e. Give all exact information to the worker, allowing no necessity for calculations, scaling, assumptions, centering, etc.

Lastly, keep in mind at all times, when dimensioning a drawing, the necessity for clearness. If any changes or deviations in dimensioning will add to the clearness of the drawing, obviously such changes are acceptable.

13-15 Assignments in dimensioning

Student assignments in dimensioning practice may be taken from the previous chapters 9, 10, 11, and 12. It is suggested that the instructor assign problems, which have been previously drawn and developed, as dimensioning problems to be put in good form for actual use in the shop. Many dimensioning problems may be taken from the illustrations of the previous chapters by establishing a scale and using it to transfer them to the drawing.

APPENDIX

Equivalents

mm-SI	Decimal-Inch	Fraction-Inch	mm-SI	Decimal-Inch	Fraction-Inch
.397	.0156	1/64	9.128	.3594	23/64
.794	.0313	1/32	9.525	.3750	3/8
1.000	.0394		9.922	.3906	25/64
1.191	.0469	3/64	10.000	.3937	
1.588	.0625	1/16	10.319	.4063	13/32
1.984	.0781	5/64	10.716	.4219	27/64
2.000	.0787		11.000	.4331	
2.381	.0938	3/32	11.113	.4375	7/16
2.778	.1094	7/64	11.509	.4531	29/64
3.000	.1181		11.906	.4688	15/32
3.175	.1250	1/8	12.000	.4724	
3.572	.1406	9/64	12.303	.4844	31/64
3.969	.1563	5/32	12.700	.5000	1/2
4.000	.1575		13.000	.5118	
4.366	.1719	11/64	13.097	.5156	33/64
4.763	.1875	3/16	13.494	.5313	17/32
5.000	.1969		13.891	.5469	35/64
5.159	.2031	13/64	14.000	.5512	
5.556	.2188	7/32	14.288	.5625	9/16
5.953	.2344	15/64	14.684	.5781	37/64
6.000	.2362		15.000	.5906	
6.350	.2500	1/4	15.081	.5938	19/32
6.747	.2656	17/64	15.478	.6094	39/64
7.000	.2756		15.875	.6250	5/8
7.144	.2813	9/32	16.000	.6299	
7.541	.2969	19/64	16.272	.6406	41/64
7.938	.3125	5/16	16.669	.6563	21/32
8.000	.3150		17.000	.6693	
8.334	.3281	21/64	17.066	.6719	43/64
8.731	.3438	11/32	17.463	.6875	11/16
9.000	.3543		17.859	.7031	45/64
18.000	.7087		21.828	.8594	55/64
18.256	.7188	23/32	22.000	.8661	
18.653	.7344	47/64	22.225	.8750	7/8
19.000	.7480		22.622	.8906	57/64
19.050	.7500	3/4	23.000	.9055	
19.447	.7656	49/64	23.019	.9063	29/32
19.844	.7813	25/32	23.416	.9219	59/64
20.000	.7874		23.813	.9375	15/16
20.241	.7969	51/64	24.000	.9449	
20.638	.8125	13/16	24.209	.9531	61/64
21.000	.8268		24.606	.9688	31/32
21.034	.8281	53/64	25.003	.9844	63/64
21.431	.8438	27/32	25.400	1.0000	1

Wire and Sheet-metal Gages *Dimensions in decimal parts of an inch*

No. of gage	American or Brown and Sharpe[a]	Washburn & Moen or American Steel & Wire Co.[b]	Birmingham or Stubs iron wire[c]	Music wire[d]	Imperial wire gage[e]	U.S. Std. for plate[f]
0000000	.....	0.4900	.....	.....	0.5000	0.5000
000000	0.5800	0.4615	.....	0.004	0.4640	0.4688
00000	0.5165	0.4305	0.500	0.005	0.4320	0.4375
0000	0.4600	0.3938	0.454	0.006	0.4000	0.4063
000	0.4096	0.3625	0.425	0.007	0.3720	0.3750
00	0.3648	0.3310	0.380	0.008	0.3480	0.3438
0	0.3249	0.3065	0.340	0.009	0.3240	0.3125
1	0.2893	0.2830	0.300	0.010	0.3000	0.2813
2	0.2576	0.2625	0.284	0.011	0.2760	0.2656
3	0.2294	0.2437	0.259	0.012	0.2520	0.2500
4	0.2043	0.2253	0.238	0.013	0.2320	0.2344
5	0.1819	0.2070	0.220	0.014	0.2120	0.2188
6	0.1620	0.1920	0.203	0.016	0.1920	0.2031
7	0.1443	0.1770	0.180	0.018	0.1760	0.1875
8	0.1285	0.1620	0.165	0.020	0.1600	0.1719
9	0.1144	0.1483	0.148	0.022	0.1440	0.1563
10	0.1019	0.1350	0.134	0.024	0.1280	0.1406
11	0.0907	0.1205	0.120	0.026	0.1160	0.1250
12	0.0808	0.1055	0.109	0.029	0.1040	0.1094
13	0.0720	0.0915	0.095	0.031	0.0920	0.0938
14	0.0641	0.0800	0.083	0.033	0.0800	0.0781
15	0.0571	0.0720	0.072	0.035	0.0720	0.0703
16	0.0508	0.0625	0.065	0.037	0.0640	0.0625
17	0.0453	0.0540	0.058	0.039	0.0560	0.0563
18	0.0403	0.0475	0.049	0.041	0.0480	0.0500
19	0.0359	0.0410	0.042	0.043	0.0400	0.0438
20	0.0320	0.0348	0.035	0.045	0.0360	0.0375
21	0.0285	0.0317	0.032	0.047	0.0320	0.0344
22	0.0253	0.0286	0.028	0.049	0.0280	0.0313
23	0.0226	0.0258	0.025	0.051	0.0240	0.0281
24	0.0201	0.0230	0.022	0.055	0.0220	0.0250
25	0.0179	0.0204	0.020	0.059	0.0200	0.0219
26	0.0159	0.0181	0.018	0.063	0.0180	0.0188
27	0.0142	0.0173	0.016	0.067	0.0164	0.0172
28	0.0126	0.0162	0.014	0.071	0.0148	0.0156
29	0.0113	0.0150	0.013	0.075	0.0136	0.0141
30	0.0100	0.0140	0.012	0.080	0.0124	0.0125
31	0.0089	0.0132	0.010	0.085	0.0116	0.0109
32	0.0080	0.0128	0.009	0.090	0.0108	0.0102
33	0.0071	0.0118	0.008	0.095	0.0100	0.0094
34	0.0063	0.0104	0.007	0.100	0.0092	0.0086
35	0.0056	0.0095	0.005	0.106	0.0084	0.0078
36	0.0050	0.0090	0.004	0.112	0.0076	0.0070
37	0.0045	0.0085	.....	0.118	0.0068	0.0066
38	0.0040	0.0080	.....	0.124	0.0060	0.0063
39	0.0035	0.0075	.....	0.130	0.0052	
40	0.0031	0.0070	.....	0.138	0.0048	

[a] Recognized standard in the United States for wire and sheet metal of copper and other metals except steel and iron.

[b] Recognized standard for steel and iron wire. Called the "U.S. steel wire gage."

[c] Formerly much used, now nearly obsolete.

[d] American Steel & Wire Co.'s music wire gage. Recommended by U.S. Bureau of Standards.

[e] Official British Standard.

[f] Legalized U.S. Standard for iron and steel plate, although plate is now always specified by its thickness in decimals of an inch.

Preferred thicknesses for uncoated thin flat metals (under 0.250 in.): ASA B32—1941 gives recommended sizes for sheets.

Trigonometric Functions

Angle	Sine		Cosine		Tangent		Cotangent		Angle
	Nat.	Log.	Nat.	Log.	Nat.	Log.	Nat.	Log.	
0° 00′	.0000	∞	1.0000	0.0000	.0000	∞	∞	∞	90° 00′
10	.0029	7.4637	1.0000	0000	.0029	7.4637	343.77	2.5363	50
20	.0058	7648	1.0000	0000	.0058	7648	171.89	2352	40
30	.0087	9408	1.0000	0000	.0087	9409	114.59	0591	30
40	.0116	8.0658	.9999	0000	.0116	8.0658	85.940	1.9342	20
50	.0145	1627	.9999	0000	.0145	1627	68.750	8373	10
1° 00′	.0175	8.2419	.9998	9.9999	.0175	8.2419	57.290	1.7581	89° 00′
10	.0204	3088	.9998	9999	.0204	3089	49.104	6911	50
20	.0233	3668	.9997	9999	.0233	3669	42.964	6331	40
30	.0262	4179	.9997	9999	.0262	4181	38.188	5819	30
40	.0291	4637	.9996	9998	.0291	4638	34.368	5362	20
50	.0320	5050	.9995	9998	.0320	5053	31.242	4947	10
2° 00′	.0349	8.5428	.9994	9.9997	.0349	8.5431	28.636	1.4569	88° 00′
10	.0378	5776	.9993	9997	.0378	5779	26.432	4221	50
20	.0407	6097	.9992	9996	.0407	6101	24.542	3899	40
30	.0436	6397	.9990	9996	.0437	6401	22.904	3599	30
40	.0465	6677	.9989	9995	.0466	6682	21.470	3318	20
50	.0494	6940	.9988	9995	.0495	6945	20.206	3055	10
3° 00′	.0523	8.7188	.9986	9.9994	.0524	8.7194	19.081	1.2806	87° 00′
10	.0552	7423	.9985	9993	.0553	7429	18.075	2571	50
20	.0581	7645	.9983	9993	.0582	7652	17.169	2348	40
30	.0610	7857	.9981	9992	.0612	7865	16.350	2135	30
40	.0640	8059	.9980	9991	.0641	8067	15.605	1933	20
50	.0669	8251	.9978	9990	.0670	8261	14.924	1739	10
4° 00′	.0698	8.8436	.9976	9.9989	.0699	8.8446	14.301	1.1554	86° 00′
10	.0727	8613	.9974	9989	.0729	8624	13.727	1376	50
20	.0756	8783	.9971	9988	.0758	8795	13.197	1205	40
30	.0785	8946	.9969	9987	.0787	8960	12.706	1040	30
40	.0814	9104	.9967	9986	.0816	9118	12.251	0882	20
50	.0843	9256	.9964	9985	.0846	9272	11.826	0728	10
5° 00′	.0872	8.9403	.9962	9.9983	.0875	8.9420	11.430	1.0580	85° 00′
10	.0901	9545	.9959	9982	.0904	9563	11.059	0437	50
20	.0929	9682	.9957	9981	.0934	9701	10.712	0299	40
30	.0958	9816	.9954	9980	.0963	9836	10.385	0164	30
40	.0987	9945	.9951	9979	.0992	9966	10.078	0034	20
50	.1016	9.0070	.9948	9977	.1022	9.0093	9.7882	0.9907	10
6° 00′	.1045	9.0192	.9945	9.9976	.1051	9.0216	9.5144	0.9784	84° 00′
10	.1074	0311	.9942	9975	.1080	0336	9.2553	9664	50
20	.1103	0426	.9939	9973	.1110	0453	9.0098	9547	40
30	.1132	0539	.9936	9972	.1139	0567	8.7769	9433	30
40	.1161	0648	.9932	9971	.1169	0678	8.5555	9322	20
50	.1190	0755	.9929	9969	.1198	0786	8.3450	9214	10
7° 00′	.1219	9.0859	.9925	9.9968	.1228	9.0891	8.1443	0.9109	83° 00′
10	.1248	0961	.9922	9966	.1257	0995	7.9530	9005	50
20	.1276	1060	.9918	9964	.1287	1096	7.7704	8904	40
	Nat.	Log.	Nat.	Log.	Nat.	Log.	Nat.	Log.	
Angle	Cosine		Sine		Cotangent		Tangent		Angle

Trigonometric Functions (Cont.)

Angle	Sine		Cosine		Tangent		Cotangent		Angle
	Nat.	Log.	Nat.	Log.	Nat.	Log.	Nat.	Log.	
30	.1305	1157	.9914	9963	.1317	1194	7.5958	8806	30
40	.1334	1252	.9911	9961	.1346	1291	7.4287	8709	20
50	.1363	1345	.9907	9959	.1376	1385	7.2687	8615	10
8° 00′	.1392	9.1436	.9903	9.9958	.1405	9.1478	7.1154	0.8522	82° 00′
10	.1421	1525	.9899	9956	.1435	1569	6.9682	8431	50
20	.1449	1612	.9894	9954	.1465	1658	6.8269	8342	40
30	.1478	1697	.9890	9952	.1495	1745	6.6912	8255	30
40	.1507	1781	.9886	9950	.1524	1831	6.5606	8169	20
50	.1536	1863	.9881	9948	.1554	1915	6.4348	8085	10
9° 00′	.1564	9.1943	.9877	9.9946	.1584	9.1997	6.3138	0.8003	81° 00′
10	.1593	2022	.9872	9944	.1614	2078	6.1970	7922	50
20	.1622	2100	.9868	9942	.1644	2158	6.0844	7842	40
30	.1650	2176	.9863	9940	.1673	2236	5.9758	7764	30
40	.1679	2251	.9858	9938	.1703	2313	5.8708	7687	20
50	.1708	2324	.9853	9936	.1733	2389	5.7694	7611	10
10° 00′	.1736	9.2397	.9848	9.9934	.1763	9.2463	5.6713	0.7537	80° 00′
10	.1765	2468	.9843	9931	.1793	2536	5.5764	7464	50
20	.1794	2538	.9838	9929	.1823	2609	5.4845	7391	40
30	.1822	2606	.9833	9927	.1853	2680	5.3955	7320	30
40	.1851	2674	.9827	9924	.1883	2750	5.3093	7250	20
50	.1880	2740	.9822	9922	.1914	2819	5.2257	7181	10
11° 00′	.1908	9.2806	.9816	9.9919	.1944	9.2887	5.1446	0.7113	79° 00′
10	.1937	2870	.9811	9917	.1974	2953	5.0658	7047	50
20	.1965	2934	.9805	9914	.2004	3020	4.9894	6980	40
30	.1994	2997	.9799	9912	.2035	3085	4.9152	6915	30
40	.2022	3058	.9793	9909	.2065	3149	4.8430	6851	20
50	.2051	3119	.9787	9907	.2095	3212	4.7729	6788	10
12° 00′	.2079	9.3179	.9781	9.9904	.2126	9.3275	4.7046	0.6725	78° 00′
10	.2108	3238	.9775	9901	.2156	3336	4.6382	6664	50
20	.2136	3296	.9769	9899	.2186	3397	4.5736	6603	40
30	.2164	3353	.9763	9896	.2217	3458	4.5107	6542	30
40	.2193	3410	.9757	9893	.2247	3517	4.4494	6483	20
50	.2221	3466	.9750	9890	.2278	3576	4.3897	6424	10
13° 00′	.2250	9.3521	.9744	9.9887	.2309	9.3634	4.3315	0.6366	77° 00′
10	.2278	3575	.9737	9884	.2339	3691	4.2747	6309	50
20	.2306	3629	.9730	9881	.2370	3748	4.2193	6252	40
30	.2334	3682	.9724	9878	.2401	3804	4.1653	6196	30
40	.2363	3734	.9717	9875	.2432	3859	4.1126	6141	20
50	.2391	3786	.9710	9872	.2462	3914	4.0611	6086	10
14° 00′	.2419	9.3837	.9703	9.9869	.2493	9.3968	4.0108	0.6032	76° 00′
10	.2447	3887	.9696	9866	.2524	4021	3.9617	5979	50
20	.2476	3937	.9689	9863	.2555	4074	3.9136	5926	40
30	.2504	3986	.9681	9859	.2586	4127	3.8667	5873	30
40	.2532	4035	.9674	9856	.2617	4178	3.8208	5822	20
50	.2560	4083	.9667	9853	.2648	4230	3.7760	5770	10
	Nat.	Log.	Nat.	Log.	Nat.	Log.	Nat.	Log.	
Angle	Cosine		Sine		Cotangent		Tangent		Angle

Trigonometric Functions (Cont.)

Angle	Sine		Cosine		Tangent		Cotangent		Angle
	Nat.	Log.	Nat.	Log.	Nat.	Log.	Nat.	Log.	
15° 00'	.2588	9.4130	.9659	9.9849	.2679	9.4281	3.7321	0.5719	75° 00'
10	.2616	4177	.9652	9846	.2711	4331	3.6891	5669	50
20	.2644	4223	.9644	9843	.2742	4381	3.6470	5619	40
30	.2672	4269	.9636	9839	.2773	4430	3.6059	5570	30
40	.2700	4314	.9628	9836	.2805	4479	3.5656	5521	20
50	.2728	4359	.9621	9832	.2836	4527	3.5261	5473	10
16° 00'	.2756	9.4403	.9613	9.9828	.2867	9.4575	3.4874	0.5425	74° 00'
10	.2784	4447	.9605	9825	.2899	4622	3.4495	5378	50
20	.2812	4491	.9596	9821	.2931	4669	3.4124	5331	40
30	.2840	4533	.9588	9817	.2962	4716	3.3759	5284	30
40	.2868	4576	.9580	9814	.2994	4762	3.3402	5238	20
50	.2896	4618	.9572	9810	.3026	4808	3.3052	5192	10
17° 00'	.2924	9.4659	.9563	9.9806	.3057	9.4853	3.2709	0.5147	73° 00'
10	.2952	4700	.9555	9802	.3089	4898	3.2371	5102	50
20	.2979	4741	.9546	9798	.3121	4943	3.2041	5057	40
30	.3007	4781	.9537	9794	.3153	4987	3.1716	5013	30
40	.3035	4821	.9528	9790	.3185	5031	3.1397	4969	20
50	.3062	4861	.9520	9786	.3217	5075	3.1084	4925	10
18° 00'	.3090	9.4900	.9511	9.9782	.3249	9.5118	3.0777	0.4882	72° 00'
10	.3118	4939	.9502	9778	.3281	5161	3.0475	4839	50
20	.3145	4977	.9492	9774	.3314	5203	3.0178	4797	40
30	.3173	5015	.9483	9770	.3346	5245	2.9887	4755	30
40	.3201	5052	.9474	9765	.3378	5287	2.9600	4713	20
50	.3228	5090	.9465	9761	.3411	5329	2.9319	4671	10
19° 00'	.3256	9.5126	.9455	9.9757	.3443	9.5370	2.9042	0.4630	71° 00'
10	.3283	5163	.9446	9752	.3476	5411	2.8770	4589	50
20	.3311	5199	.9436	9748	.3508	5451	2.8502	4549	40
30	.3338	5235	.9426	9743	.3541	5491	2.8239	4509	30
40	.3365	5270	.9417	9739	.3574	5531	2.7980	4469	20
50	.3393	5306	.9407	9734	.3607	5571	2.7725	4429	10
20° 00'	.3420	9.5341	.9397	9.9730	.3640	9.5611	2.7475	0.4389	70° 00'
10	.3448	5375	.9387	9725	.3673	5650	2.7228	4350	50
20	.3475	5409	.9377	9721	.3706	5689	2.6985	4311	40
30	.3502	5443	.9367	9716	.3739	5727	2.6746	4273	30
40	.3529	5477	.9356	9711	.3772	5766	2.6511	4234	20
50	.3557	5510	.9346	9706	.3805	5804	2.6279	4196	10
21° 00'	.3584	9.5543	.9336	9.9702	.3839	9.5842	2.6051	0.4158	69° 00'
10	.3611	5576	.9325	9697	.3872	5879	2.5826	4121	50
20	.3638	5609	.9315	9692	.3906	5917	2.5605	4083	40
30	.3665	5641	.9304	9687	.3939	5954	2.5386	4046	30
40	.3692	5673	.9293	9682	.3973	5991	2.5172	4009	20
50	.3719	5704	.9283	9677	.4006	6028	2.4960	3972	10
22° 00'	.3746	9.5736	.9272	9.9672	.4040	9.6064	2.4751	0.3936	68° 00'
10	.3773	5767	.9261	9667	.4074	6100	2.4545	3900	50
20	.3800	5798	.9250	9661	.4108	6136	2.4342	3864	40
	Nat.	Log.	Nat.	Log.	Nat.	Log.	Nat.	Log.	
Angle	Cosine		Sine		Cotangent		Tangent		Angle

Trigonometric Functions (Cont.)

Angle	Sine		Cosine		Tangent		Cotangent		Angle
	Nat.	Log.	Nat.	Log.	Nat.	Log.	Nat.	Log.	
30	.3827	5828	.9239	9656	.4142	6172	2.4142	3828	30
40	.3854	5859	.9228	9651	.4176	6208	2.3945	3792	20
50	.3881	5889	.9216	9646	.4210	6243	2.3750	3757	10
23° 00′	.3907	9.5919	.9205	9.9640	4245	9.6279	2.3559	0.3721	67° 00′
10	.3934	5948	.9194	9635	.4279	6314	2.3369	3686	50
20	.3961	5978	.9182	9629	.4314	6348	2.3183	3652	40
30	.3987	6007	.9171	9624	.4348	6383	2.2998	3617	30
40	.4014	6036	.9159	9618	.4383	6417	2.2817	3583	20
50	.4041	6065	.9147	9613	.4417	6452	2.2637	3548	10
24° 00′	.4067	9.6093	.9135	9.9607	.4452	9.6486	2.2460	0.3514	66° 00′
10	.4094	6121	.9124	9602	.4487	6520	2.2286	3480	50
20	.4120	6149	.9112	9596	.4522	6553	2.2113	3447	40
30	.4147	6177	.9100	9590	.4557	6587	2.1943	3413	30
40	.4173	6205	.9088	9584	.4592	6620	2.1775	3380	20
50	.4200	6232	.9075	9579	.4628	6654	2.1609	3346	10
25° 00′	.4226	9.6259	.9063	9.9573	.4663	9.6687	2.1445	0.3313	65° 00′
10	.4253	6286	.9051	9567	.4699	6720	2.1283	3280	50
20	.4279	6313	.9038	9561	.4734	6752	2.1123	3248	40
30	.4305	6340	.9026	9555	.4770	6785	2.0965	3215	30
40	.4331	6366	.9013	9549	.4806	6817	2.0809	3183	20
50	.4358	6392	.9001	9543	.4841	6850	2.0655	3150	10
26° 00′	.4384	9.6418	.8988	9.9537	.4877	9.6882	2.0503	0.3118	64° 00′
10	.4410	6444	.8975	9530	.4913	6914	2.0353	3086	50
20	.4436	6470	.8962	9524	.4950	6946	2.0204	3054	40
30	.4462	6495	.8949	9518	.4986	6977	2.0057	3023	30
40	.4488	6521	.8936	9512	.5022	7009	1.9912	2991	20
50	.4514	6546	.8923	9505	.5059	7040	1.9768	2960	10
27° 00′	.4540	9.6570	.8910	9.9499	.5095	9.7072	1.9626	0.2928	63° 00′
10	.4566	6595	.8897	9492	.5132	7103	1.9486	2897	50
20	.4592	6620	.8884	9486	.5169	7134	1.9347	2866	40
30	.4617	6644	.8870	9479	.5206	7165	1.9210	2835	30
40	.4643	6668	.8857	9473	.5243	7196	1.9074	2804	20
50	.4669	6692	.8843	9466	.5280	7226	1.8940	2774	10
28° 00′	.4695	9.6716	.8829	9.9459	.5317	9.7257	1.8807	0.2743	62° 00′
10	.4720	6740	.8816	9453	.5354	7287	1.8676	2713	50
20	.4746	6763	.8802	9446	.5392	7317	1.8546	2683	40
30	.4772	6787	.8788	9439	.5430	7348	1.8418	2652	30
40	.4797	6810	.8774	9432	.5467	7378	1.8291	2622	20
50	.4823	6833	.8760	9425	.5505	7408	1.8165	2592	10
29° 00′	.4848	9.6856	.8746	9.9418	.5543	9.7438	1.8040	0.2562	61° 00′
10	.4874	6878	.8732	9411	.5581	7467	1.7917	2533	50
20	.4899	6901	.8718	9404	.5619	7497	1.7796	2503	40
30	.4924	6923	.8704	9397	.5658	7526	1.7675	2474	30
40	.4950	6946	.8689	9390	.5696	7556	1.7556	2444	20
50	.4975	6968	.8675	9383	.5735	7585	1.7437	2415	10
	Nat.	Log.	Nat.	Log.	Nat.	Log.	Nat.	Log.	
Angle	Cosine		Sine		Cotangent		Tangent		Angle

Trigonometric Functions (Cont.)

Angle	Sine		Cosine		Tangent		Cotangent		Angle
	Nat.	Log.	Nat.	Log.	Nat.	Log.	Nat.	Log.	
30° 00'	.5000	9.6990	.8660	9.9375	.5774	9.7614	1.7321	0.2386	60° 00'
10	.5025	7012	.8646	9368	.5812	7644	1.7205	2356	50
20	.5050	7033	.8631	9361	.5851	7673	1.7090	2327	40
30	.5075	7055	.8616	9353	.5890	7701	1.6977	2299	30
40	.5100	7076	.8601	9346	.5930	7730	1.6864	2270	20
50	.5125	7097	.8587	9338	.5969	7759	1.6753	2241	10
31° 00'	.5150	9.7118	.8572	9.9331	.6009	9.7788	1.6643	0.2212	59° 00'
10	.5175	7139	.8557	9323	.6048	7816	1.6534	2184	50
20	.5200	7160	.8542	9315	.6088	7845	1.6426	2155	40
30	.5225	7181	.8526	9308	.6128	7873	1.6319	2127	30
40	.5250	7201	.8511	9300	.6168	7902	1.6212	2098	20
50	.5275	7222	.8496	9292	.6208	7930	1.6107	2070	10
32° 00'	.5299	9.7242	.8480	9.9284	.6249	9.7958	1.6003	0.2042	58° 00'
10	.5324	7262	.8465	9276	.6289	7986	1.5900	2014	50
20	.5348	7282	.8450	9268	.6330	8014	1.5798	1986	40
30	.5373	7302	.8434	9260	.6371	8042	1.5697	1958	30
40	.5398	7322	.8418	9252	.6412	8070	1.5597	1930	20
50	.5422	7342	.8403	9244	.6453	8097	1.5497	1903	10
33° 00'	.5446	9.7361	.8387	9.9236	.6494	9.8125	1.5399	0.1875	57° 00'
10	.5471	7380	.8371	9228	.6536	8153	1.5301	1847	50
20	.5495	7400	.8355	9219	.6577	8180	1.5204	1820	40
30	.5519	7419	.8339	9211	.6619	8208	1.5108	1792	30
40	.5544	7438	.8323	9203	.6661	8235	1.5013	1765	20
50	.5568	7457	.8307	9194	.6703	8263	1.4919	1737	10
34° 00'	.5592	9.7476	.8290	9.9186	.6745	9.8290	1.4826	0.1710	56° 00'
10	.5616	7494	.8274	9177	.6787	8317	1.4733	1683	50
20	.5640	7513	.8258	9169	.6830	8344	1.4641	1656	40
30	.5664	7531	.8241	9160	.6873	8371	1.4550	1629	30
40	.5688	7550	.8225	9151	.6916	8398	1.4460	1602	20
50	.5712	7568	.8208	9142	.6959	8425	1.4370	1575	10
35° 00'	.5736	9.7586	.8192	9.9134	.7002	9.8452	1.4281	0.1548	55° 00'
10	.5760	7604	.8175	9125	.7046	8479	1.4193	1521	50
20	.5783	7622	.8158	9116	.7089	8506	1.4106	1494	40
30	.5807	7640	.8141	9107	.7133	8533	1.4019	1467	30
40	.5831	7657	.8124	9098	.7177	8559	1.3934	1441	20
50	.5854	7675	.8107	9089	.7221	8586	1.3848	1414	10
36° 00'	.5878	9.7692	.8090	9.9080	.7265	9.8613	1.3764	0.1387	54° 00'
10	.5901	7710	.8073	9070	.7310	8639	1.3680	1361	50
20	.5925	7727	.8056	9061	.7355	8666	1.3597	1334	40
30	.5948	7744	.8039	9052	.7400	8692	1.3514	1308	30
40	.5972	7761	.8021	9042	.7445	8718	1.3432	1282	20
50	.5995	7778	.8004	9033	.7490	8745	1.3351	1255	10
37° 00'	.6018	9.7795	.7986	9.9023	.7536	9.8771	1.3270	0.1229	53° 00'
10	.6041	7811	.7969	9014	.7581	8797	1.3190	1203	50
20	.6065	7828	.7951	9004	.7627	8824	1.3111	1176	40
	Nat.	Log.	Nat.	Log.	Nat.	Log.	Nat.	Log.	
Angle	Cosine		Sine		Cotangent		Tangent		Angle

Trigonometric Functions (Cont.)

Angle	Sine		Cosine		Tangent		Cotangent		Angle
	Nat.	Log.	Nat.	Log.	Nat.	Log.	Nat.	Log.	
30	.6088	7844	.7934	8995	.7673	8850	1.3032	1150	30
40	.6111	7861	.7916	8985	.7720	8876	1.2954	1124	20
50	.6134	7877	.7898	8975	.7766	8902	1.2876	1098	10
38° 00′	.6157	9.7893	.7880	9.8965	.7813	9.8928	1.2799	0.1072	52° 00′
10	.6180	7910	.7862	8955	.7860	8954	1.2723	1046	50
20	.6202	7926	.7844	8945	.7907	8980	1.2647	1020	40
30	.6225	7941	.7826	8935	.7954	9006	1.2572	0994	30
40	.6248	7957	.7808	8925	.8002	9032	1.2497	0968	20
50	.6271	7973	.7790	8915	.8050	9058	1.2423	0942	10
39° 00′	.6293	9.7989	.7771	9.8905	.8098	9.9084	1.2349	0.0916	51° 00′
10	.6316	8004	.7753	8895	.8146	9110	1.2276	0890	50
20	.6338	8020	.7735	8884	.8195	9135	1.2203	0865	40
30	.6361	8035	.7716	8874	.8243	9161	1.2131	0839	30
40	.6383	8050	.7698	8864	.8292	9187	1.2059	0813	20
50	.6406	8066	.7679	8853	.8342	9212	1.1988	0788	10
40° 00′	.6428	9.8081	.7660	9.8843	.8391	9.9238	1.1918	0.0762	50° 00′
10	.6450	8096	.7642	8832	.8441	9264	1.1847	0736	50
20	.6472	8111	.7623	8821	.8491	9289	1.1778	0711	40
30	.6494	8125	.7604	8810	.8541	9315	1.1708	0685	30
40	.6517	8140	.7585	8800	.8591	9341	1.1640	0659	20
50	.6539	8155	.7566	8789	.8642	9366	1.1571	0634	10
41° 00′	.6561	9.8169	.7547	9.8778	.8693	9.9392	1.1504	0.0608	49° 00′
10	.6583	8184	.7528	8767	.8744	9417	1.1436	0583	50
20	.6604	8198	.7509	8756	.8796	9443	1.1369	0557	40
30	.6626	8213	.7490	8745	.8847	9468	1.1303	0532	30
40	.6648	8227	.7470	8733	.8899	9494	1.1237	0506	20
50	.6670	8241	.7451	8722	.8952	9519	1.1171	0481	10
42° 00′	.6691	9.8255	.7431	9.8711	.9004	9.9544	1.1106	0.0456	48° 00′
10	.6713	8269	.7412	8699	.9057	9570	1.1041	0430	50
20	.6734	8283	.7392	8688	.9110	9595	1.0977	0405	40
30	.6756	8297	.7373	8676	.9163	9621	1.0913	0379	30
40	.6777	8311	.7353	8665	.9217	9646	1.0850	0354	20
50	.6799	8324	.7333	8653	.9271	9671	1.0786	0329	10
43° 00′	.6820	9.8338	.7314	9.8641	.9325	9.9697	1.0724	0.0303	47° 00′
10	.6841	8351	.7294	8629	.9380	9722	1.0661	0278	50
20	.6862	8365	.7274	8618	.9435	9747	1.0599	0253	40
30	.6884	8378	.7254	8606	.9490	9772	1.0538	0228	30
40	.6905	8391	.7234	8594	.9545	9798	1.0477	0202	20
50	.6926	8405	.7214	8582	.9601	9823	1.0416	0177	10
44° 00′	.6947	9.8418	.7193	9.8569	.9657	9.9848	1.0355	0.0152	46° 00′
10	.6967	8431	.7173	8557	.9713	9874	1.0295	0126	50
20	.6988	8444	.7153	8545	.9770	9899	1.0235	0101	40
30	.7009	8457	.7133	8532	.9827	9924	1.0176	0076	30
40	.7030	8469	.7112	8520	.9884	9949	1.0117	0051	20
50	.7050	8482	.7092	8507	.9942	9975	1.0058	0025	10
45° 00′	.7071	9.8495	.7071	9.8495	1.0000	0.0000	1.0000	0.0000	45° 00′
	Nat.	Log.	Nat.	Log.	Nat.	Log.	Nat.	Log	
Angle	Cosine		Sine		Cotangent		Tangent		Angle

Squares, Square Roots, and Reciprocals

no.	square	sq. root	reciprocal	no.	square	sq. root	reciprocal
1	1	1.0000	1.000000000	51	2,601	7.1414	.019607843
2	4	1.4142	.500000000	52	2,704	7.2111	.019230769
3	9	1.7321	.333333333	53	2,809	7.2801	.018867925
4	16	2.0000	.250000000	54	2,916	7.3485	.018518519
5	25	2.2361	.200000000	55	3,025	7.4162	.018181818
6	36	2.4495	.166666667	56	3,136	7.4833	.017857143
7	49	2.6458	.142857143	57	3,249	7.5498	.017543860
8	64	2.8284	.125000000	58	3,364	7.6158	.017241379
9	81	3.0000	.111111111	59	3,481	7.6811	.016949153
10	100	3.1623	.100000000	60	3,600	7.7460	.016666667
11	121	3.3166	.090909091	61	3,721	7.8102	.016393443
12	144	3.4641	.083333333	62	3,844	7.8740	.016129032
13	169	3.6056	.076923077	63	3,969	7.9373	.015873016
14	196	3.7417	.071428571	64	4,096	8.0000	.015625000
15	225	3.8730	.066666667	65	4,225	8.0623	.015384615
16	256	4.0000	.062500000	66	4,356	8.1240	.015151515
17	289	4.1231	.058823529	67	4,489	8.1854	.014925373
18	324	4.2426	.055555556	68	4,624	8.2462	.014705882
19	361	4.3589	.052631579	69	4,761	8.3066	.014492754
20	400	4.4721	.050000000	70	4,900	8.3666	.014285714
21	441	4.5826	.047619048	71	5,041	8.4261	.014084507
22	484	4.6904	.045454545	72	5,184	8.4853	.013888889
23	529	4.7958	.043478261	73	5,329	8.5440	.013698630
24	576	4.8990	.041666667	74	5,476	8.6023	.013513514
25	625	5.0000	.040000000	75	5,625	8.6603	.013333333
26	676	5.0990	.038461538	76	5,776	8.7178	.013157895
27	729	5.1962	.037037037	77	5,929	8.7750	.012987013
28	784	5.2915	.035714286	78	6,084	8.8318	.012820513
29	841	5.3852	.034482759	79	6,241	8.8882	.012658228
30	900	5.4772	.033333333	80	6,400	8.9443	.012500000
31	961	5.5678	.032258065	81	6,561	9.0000	.012345679
32	1,024	5.6569	.031250000	82	6,724	9.0554	.012195122
33	1,089	5.7446	.030303030	83	6,889	9.1104	.012048193
34	1,156	5.8310	.029411765	84	7,056	9.1652	.011904762
35	1,225	5.9161	.028571429	85	7,225	9.2195	.011764706
36	1,296	6.0000	.027777778	86	7,396	9.2736	.011627907
37	1,369	6.0828	.027027027	87	7,569	9.3274	.011494253
38	1,444	6.1644	.026315789	88	7,744	9.3808	.011363636
39	1,521	6.2450	.025641026	89	7,921	9.4340	.011235955
40	1,600	6.3246	.025000000	90	8,100	9.4868	.011111111
41	1,681	6.4031	.024390244	91	8,281	9.5394	.010989011
42	1,764	6.4807	.023809524	92	8,464	9.5917	.010869565
43	1,849	6.5574	.023255814	93	8,649	9.6437	.010752688
44	1,936	6.6332	.022727273	94	8,836	9.6954	.010638298
45	2,025	6.7082	.022222222	95	9,025	9.7468	.010526316
46	2,116	6.7823	.021739130	96	9,216	9.7980	.010416667
47	2,209	6.8557	.021276596	97	9,409	9.8489	.010309278
48	2,304	6.9282	.020833333	98	9,604	9.8995	.010204082
49	2,401	7.0000	.020408163	99	9,801	9.9499	.010101010
50	2,500	7.0711	.020000000	100	10,000	10.0000	.010000000

Squares, Square Roots, and Reciprocals (Cont.)

no.	square	sq. root	reciprocal	no.	square	sq. root	reciprocal
101	10,201	10.0499	.009900990	151	22,801	12.2882	.006622517
102	10,404	10.0995	.009803922	152	23,104	12.3288	.006578947
103	10,609	10.1489	.009708738	153	23,409	12.3693	.006535948
104	10,816	10.1980	.009615385	154	23,716	12.4097	.006493506
105	11,025	10.2470	.009523810	155	24,025	12.4499	.006451613
106	11,236	10.2956	.009433962	156	24,336	12.4900	.006410256
107	11,449	10.3441	.009345794	157	24,649	12.5300	.006369427
108	11,664	10.3923	.009259259	158	24,964	12.5698	.006329114
109	11,881	10.4403	.009174312	159	25,281	12.6095	.006289308
110	12,100	10.4881	.009090909	160	25,600	12.6491	.006250000
111	12,321	10.5357	.009009009	161	25,921	12.6886	.006211180
112	12,544	10.5830	.008928571	162	26,244	12.7279	.006172840
113	12,769	10.6301	.008849558	163	26,569	12.7671	.006134969
114	12,996	10.6771	.008771930	164	26,896	12.8062	.006097561
115	13,225	10.7238	.008695652	165	27,225	12.8452	.006060606
116	13,456	10.7703	.008620690	166	27,556	12.8841	.006024096
117	13,689	10.8167	.008547009	167	27,889	12.9228	.005988024
118	13,924	10.8628	.008474576	168	28,224	12.9615	.005952381
119	14,161	10.9087	.008403361	169	28,561	13.0000	.005917160
120	14,400	10.9545	.008333333	170	28,900	13.0384	.005882353
121	14,641	11.0000	.008264463	171	29,241	13.0767	.005847953
122	14,884	11.0454	.008196721	172	29,584	13.1149	.005813953
123	15,129	11.0905	.008130081	173	29,929	13.1529	.005780347
124	15,376	11.1355	.008064516	174	30,276	13.1909	.005747126
125	15,625	11.1803	.008000000	175	30,625	13.2288	.005714286
126	15,876	11.2250	.007936508	176	30,976	13.2665	.005681818
127	16,129	11.2694	.007874016	177	31,329	13.3041	.005649718
128	16,384	11.3137	.007812500	178	31,684	13.3417	.005617978
129	16,641	11.3578	.007751938	179	32,041	13.3791	.005586592
130	16,900	11.4018	.007692308	180	32,400	13.4164	.005555556
131	17,161	11.4455	.007633588	181	32,761	13.4536	.005524862
132	17,424	11.4891	.007575758	182	33,124	13.4907	.005494505
133	17,689	11.5326	.007518797	183	33,489	13.5277	.005464481
134	17,956	11.5758	.007462687	184	33,856	13.5647	.005434783
135	18,225	11.6190	.007407407	185	34,225	13.6015	.005405405
136	18,496	11.6619	.007352941	186	34,596	13.6382	.005376344
137	18,769	11.7047	.007299270	187	34,969	13.6748	.005347594
138	19,044	11.7473	.007246377	188	35,344	13.7113	.005319149
139	19,321	11.7898	.007194245	189	35,721	13.7477	.005291005
140	19,600	11.8322	.007142857	190	36,100	13.7840	.005263158
141	19,881	11.8743	.007092199	191	36,481	13.8203	.005235602
142	20,164	11.9164	.007042254	192	36,864	13.8564	.005208333
143	20,449	11.9583	.006993007	193	37,249	13.8924	.005181347
144	20,736	12.0000	.006944444	194	37,636	13.9284	.005154639
145	21,025	12.0416	.006896552	195	38,025	13.9642	.005128205
146	21,316	12.0830	.006849315	196	38,416	14.0000	.005102041
147	21,609	12.1244	.006802721	197	38,809	14.0357	.005076142
148	21,904	12.1655	.006756757	198	39,204	14.0712	.005050505
149	22,201	12.2066	.006711409	199	39,601	14.1067	.005025126
150	22,500	12.2474	.006666667	200	40,000	14.1421	.005000000

Squares, Square Roots, and Reciprocals (Cont.)

no.	square	sq. root	reciprocal	no.	square	sq. root	reciprocal
201	40,401	14.1774	.004975124	251	63,001	15.8430	.003984064
202	40,804	14.2127	.004950495	252	63,504	15.8745	.003968254
203	41,209	14.2478	.004926108	253	64,009	15.9060	.003952569
204	41,616	14.2829	.004901961	254	64,516	15.9374	.003937008
205	42,025	14.3178	.004878049	255	65,025	15.9687	.003921569
206	42,436	14.3527	.004854369	256	65,536	16.0000	.003906250
207	42,849	14.3875	.004830918	257	66,049	16.0312	.003891051
208	43,264	14.4222	.004807692	258	66,564	16.0624	.003875969
209	43,681	14.4568	.004784689	259	67,081	16.0935	.003861004
210	44,100	14.4914	.004761905	260	67,600	16.1245	.003846154
211	44,521	14.5258	.004739336	261	68,121	16.1555	.003831418
212	44,944	14.5602	.004716981	262	68,644	16.1864	.003816794
213	45,369	14.5945	.004694836	263	69,169	16.2173	.003802281
214	45,796	14.6287	.004672897	264	69,696	16.2481	.003787879
215	46,225	14.6629	.004651163	265	70,225	16.2788	.003773585
216	46,656	14.6969	.004629630	266	70,756	16.3095	.003759398
217	47,089	14.7309	.004608295	267	71,289	16.3401	.003745318
218	47,524	14.7648	.004587156	268	71,824	16.3707	.003731343
219	47,961	14.7986	.004566210	269	72,361	16.4012	.003717472
220	48,400	14.8324	.004545455	270	72,900	16.4317	.003703704
221	48,841	14.8661	.004524887	271	73,441	16.4621	.003690037
222	49,284	14.8997	.004504505	272	73,984	16.4924	.003676471
223	49,729	14.9332	.004484305	273	74,529	16.5227	.003663004
224	50,176	14.9666	.004464286	274	75,076	16.5529	.003649635
225	50,625	15.0000	.004444444	275	75,625	16.5831	.003636364
226	51,076	15.0333	.004424779	276	76,176	16.6132	.003623188
227	51,529	15.0665	.004405286	277	76,729	16.6433	.003610108
228	51,984	15.0997	.004385965	278	77,284	16.6733	.003597122
229	52,441	15.1327	.004366812	279	77,841	16.7033	.003584229
230	52,900	15.1658	.004347826	280	78,400	16.7332	.003571429
231	53,361	15.1987	.004329004	281	78,961	16.7631	.003558719
232	53,824	15.2315	.004310345	282	79,524	16.7929	.003546099
233	54,289	15.2643	.004291845	283	80,089	16.8226	.003533569
234	54,756	15.2971	.004273504	284	80,656	16.8523	.003521127
235	55,225	15.3297	.004255319	285	81,225	16.8819	.003508772
236	55,696	15.3623	.004237288	286	81,796	16.9115	.003496503
237	56,169	15.3948	.004219409	287	82,369	16.9411	.003484321
238	56,644	15.4272	.004201681	288	82,944	16.9706	.003472222
239	57,121	15.4596	.004184100	289	83,521	17.0000	.003460208
240	57,600	15.4919	.004166667	290	84,100	17.0294	.003448276
241	58,081	15.5242	.004149378	291	84,681	17.0587	.003436426
242	58,564	15.5563	.004132231	292	85,264	17.0880	.003424658
243	59,049	15.5885	.004115226	293	85,849	17.1172	.003412969
244	59,536	15.6205	.004098361	294	86,436	17.1464	.003401361
245	60,025	15.6525	.004081633	295	87,025	17.1756	.003389831
246	60,516	15.6844	.004065041	296	87,616	17.2047	.003378378
247	61,009	15.7162	.004048583	297	88,209	17.2337	.003367003
248	61,504	15.7480	.004032258	298	88,804	17.2627	.003355705
249	62,001	15.7797	.004016064	299	89,401	17.2916	.003344482
250	62,500	15.8114	.004000000	300	90,000	17.3205	.003333333

Squares, Square Roots, and Reciprocals (Cont.)

no.	square	sq. root	reciprocal	no.	square	sq. root	reciprocal
301	90,601	17.3494	.003322259	351	123,201	18.7350	.002849003
302	91,204	17.3781	.003311258	352	123,904	18.7617	.002840909
303	91,809	17.4069	.003300330	353	124,609	18.7883	.002832861
304	92,416	17.4356	.003289474	354	125,316	18.8149	.002824859
305	93,025	17.4642	.003278689	355	126,025	18.8414	.002816901
306	93,636	17.4929	.003267974	356	126,736	18.8680	.002808989
307	94,249	17.5214	.003257329	357	127,449	18.8944	.002801120
308	94,864	17.5499	.003246753	358	128,164	18.9209	.002793296
309	95,481	17.5784	.003236246	359	128,881	18.9473	.002785515
310	96,100	17.6068	.003225806	360	129,600	18.9737	.002777778
311	96,721	17.6352	.003215434	361	130,321	19.0000	.002770083
312	97,344	17.6635	.003205128	362	131,044	19.0263	.002762431
313	97,969	17.6918	.003194888	363	131,769	19.0526	.002754821
314	98,596	17.7200	.003184713	364	132,496	19.0788	.002747253
315	99,225	17.7482	.003174603	365	133,225	19.1050	.002739726
316	99,856	17.7764	.003164557	366	133,956	19.1311	.002732240
317	100,489	17.8045	.003154574	367	134,689	19.1572	.002724796
318	101,124	17.8326	.003144654	368	135,424	19.1833	.002717391
319	101,761	17.8606	.003134796	369	136,161	19.2094	.002710027
320	102,400	17.8885	.003125000	370	136,900	19.2354	.002702703
321	103,041	17.9165	.003115265	371	137,641	19.2614	.002695418
322	103,684	17.9444	.003105590	372	138,384	19.2873	.002688172
323	104,329	17.9722	.003095975	373	139,129	19.3132	.002680965
324	104,976	18.0000	.003086420	374	139,876	19.3391	.002673797
325	105,625	18.0278	.003076923	375	140,625	19.3649	.002666667
326	106,276	18.0555	.003067485	376	141,376	19.3907	.002659574
327	106,929	18.0831	.003058104	377	142,129	19.4165	.002652520
328	107,584	18.1108	.003048780	378	142,884	19.4422	.002645503
329	108,241	18.1384	.003039514	379	143,641	19.4679	.002638522
330	108,900	18.1659	.003030303	380	144,400	19.4936	.002631579
331	109,561	18.1934	.003021148	381	145,161	19.5192	.002624672
332	110,224	18.2209	.003012048	382	145,924	19.5448	.002617801
333	110,889	18.2483	.003003003	383	146,689	19.5704	.002610966
334	111,556	18.2757	.002994012	384	147,456	19.5959	.002604167
335	112,225	18.3030	.002985075	385	148,225	19.6214	.002597403
336	112,896	18.3303	.002976190	386	148,996	19.6469	.002590674
337	113,569	18.3576	.002967359	387	149,769	19.6723	.002583979
338	114,244	18.3848	.002958580	388	150,544	19.6977	.002577320
339	114,921	18.4120	.002949853	389	151,321	19.7231	.002570694
340	115,600	18.4391	.002941176	390	152,100	19.7484	.002564103
341	116,281	18.4662	.002932551	391	152,881	19.7737	.002557545
342	116,964	18.4932	.002923977	392	153,664	19.7990	.002551020
343	117,649	18.5203	.002915452	393	154,449	19.8242	.002544529
344	118,336	18.5472	.002906977	394	155,236	19.8494	.002538071
345	119,025	18.5742	.002898551	395	156,025	19.8746	.002531646
346	119,716	18.6011	.002890173	396	156,816	19.8997	.002525253
347	120,409	18.6279	.002881844	397	157,609	19.9249	.002518892
348	121,104	18.6548	.002873563	398	158,404	19.9499	.002512563
349	121,801	18.6815	.002865330	399	159,201	19.9750	.002506266
350	122,500	18.7083	.002857143	400	160,000	20.0000	.002500000

Squares, Square Roots, and Reciprocals (Cont.)

no.	square	sq. root	reciprocal	no.	square	sq. root	reciprocal
401	160,801	20.0250	.002493766	451	203,401	21.2368	.002217295
402	161,604	20.0499	.002487562	452	204,304	21.2603	.002212389
403	162,409	20.0749	.002481390	453	205,209	21.2838	.002207506
404	163,216	20.0998	.002475248	454	206,116	21.3073	.002202643
405	164,025	20.1246	.002469136	455	207,025	21.3307	.002197802
406	164,836	20.1494	.002463054	456	207,936	21.3542	.002192982
407	165,649	20.1742	.002457002	457	208,849	21.3776	.002188184
408	166,464	20.1990	.002450980	458	209,764	21.4009	.002183406
409	167,281	20.2237	.002444988	459	210,681	21.4243	.002178649
410	168,100	20.2485	.002439024	460	211,600	21.4476	.002173913
411	168,921	20.2731	.002433090	461	212,521	21.4709	.002169197
412	169,744	20.2978	.002427184	462	213,444	21.4942	.002164502
413	170,569	20.3224	.002421308	463	214,369	21.5174	.002159827
414	171,396	20.3470	.002415459	464	215,296	21.5407	.002155172
415	172,225	20.3715	.002409639	465	216,225	21.5639	.002150538
416	173,056	20.3961	.002403846	466	217,156	21.5870	.002145923
417	173,889	20.4206	.002398082	467	218,089	21.6102	.002141328
418	174,724	20.4450	.002392344	468	219,024	21.6333	.002136752
419	175,561	20.4695	.002386635	469	219,961	21.6564	.002132196
420	176,400	20.4939	.002380952	470	220,900	21.6795	002127660
421	177,241	20.5183	.002375297	471	221,841	21.7025	.002123142
422	178,084	20.5426	.002369668	472	222,784	21.7256	.002118644
423	178,929	20.5670	.002364066	473	223,729	21.7486	.002114165
424	179,776	20.5913	.002358491	474	224,676	21.7715	.002109705
425	180,625	20.6155	.002352941	475	225,625	21.7945	.002105263
426	181,476	20.6398	.002347418	476	226,576	21.8174	.002100840
427	182,329	20.6640	.002341920	477	227,529	21.8403	.002096436
428	183,184	20.6882	.002336449	478	228,484	21.8632	.002092050
429	184,041	20.7123	.002331002	479	229,441	21.8861	.002087683
430	184,900	20.7364	.002325581	480	230,400	21.9089	.002083333
431	185,761	20.7605	.002320186	481	231,361	21.9317	.002079002
432	186,624	20.7846	.002314815	482	232,324	21.9545	.002074689
433	187,489	20.8087	.002309469	483	233,289	21.9773	.002070393
434	188,356	20.8327	.002304147	484	234,256	22.0000	.002066116
435	189,225	20.8567	.002298851	485	235,225	22.0227	.002061856
436	190,096	20.8806	.002293578	486	236,196	22.0454	.002057613
437	190,969	20.9045	.002288330	487	237,169	22.0681	.002053388
438	191,844	20.9284	.002283105	488	238,144	22.0907	.002049180
439	192,721	20.9523	.002277904	489	239,121	22.1133	.002044990
440	193,600	20.9762	.002272727	490	240,100	22.1359	.002040816
441	194,481	21.0000	.002267574	491	241,081	22.1585	.002036660
442	195,364	21.0238	.002262443	492	242,064	22.1811	.002032520
443	196,249	21.0476	.002257336	493	243,049	22.2036	.002028398
444	197,136	21.0713	.002252252	494	244,036	22.2261	.002024291
445	198,025	21.0950	.002247191	495	245,025	22.2486	.002020202
446	198,916	21.1187	.002242152	496	246,016	22.2711	.002016129
447	199,809	21.1424	.002237136	497	247,009	22.2935	.002012072
448	200,704	21.1660	.002232143	498	248,004	22.3159	.002008032
449	201,601	21.1896	.002227171	499	249,001	22.3383	.002004008
450	202,500	21.2132	.002222222	500	250,000	22.3607	.002000000

APPENDIX

Plane Figures

Plane Surfaces	
1. Rectangle Area = (base)(altitude) Diagonal = $\sqrt{\text{base}^2 + \text{altitude}^2}$	
2. Right Triangle Area = $\tfrac{1}{2}$(base)(altitude) Hypotenuse = $\sqrt{\text{base}^2 + \text{altitude}^2}$ Angles $A + B = 90°$	
3. Any Triangle Area = $\tfrac{1}{2}$(base)(altitude) NOTE: Altitude h perpendicular to base b Angles $A + B + C = 180°$	
4. Parallelogram Area = (base)(altitude) NOTE: Altitude h perpendicular to base b Sum of angles = $360°$	
5. Trapezoid (Sides a and b parallel) Area = $\tfrac{1}{2}$(sum of parallel sides)(altitude) NOTE: Altitude h perpendicular to a and b	
6. Trapezium (Four sides, none parallel) Area: Draw diagonal BD and get sum of areas of triangles ABD and BCD. Or draw altitudes h and k, then area trapezium = area trapezoid $EBCF$ + triangle ABE − triangle DCF	
7. Regular Polygon NOTE: A regular polygon has equal sides and equal angles and can be inscribed in or circumscribed about a circle. Area = $\dfrac{1}{2}\begin{pmatrix}\text{number}\\\text{of sides}\end{pmatrix}\begin{pmatrix}\text{length of}\\\text{one side}\end{pmatrix}\begin{pmatrix}\text{distance } CP\\\text{to center}\end{pmatrix}$	

SOURCE: Forest C. Dana and Lawrence R. Hillyard, *Engineering Problems Manual*, 5th ed., McGraw-Hill Book Company, Inc., New York, 1958.

Plane Figures (Cont.)

8. Circle

Circumference $= \pi$ (diam)
$\qquad\quad = 2\pi$ (rad)

$\pi = \dfrac{\text{circumference}}{\text{diameter}} = 3.1416$

Area $= \pi(\text{rad})^2 = \dfrac{\pi(\text{diam})^2}{4}$
$\qquad = 0.7854 \,(\text{diam})^2$

1 radian $= \dfrac{180°}{\pi} = 57.2958°$

$\dfrac{\text{Arc } BD}{\text{Circumference}} = \dfrac{\text{angle } BCD°}{360°}$

Also arc $BD =$ (rad) (angle BCD in radians)

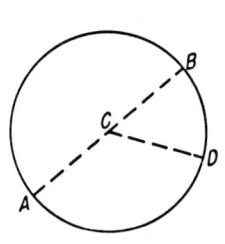

9. Sector of a Circle

Area $= \dfrac{(\text{arc})(\text{rad})}{2}$

$\qquad = \dfrac{\pi(\text{rad})^2 (\text{angle } ACB°)}{360°}$

$\qquad = \dfrac{(\text{rad})^2 (\text{angle } ACB \text{ in radians})}{2}$

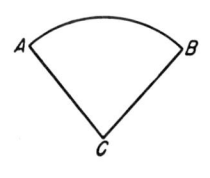

10. Segment of a Circle

Area of segment = area of sector ABC
$\qquad\qquad\qquad\quad$ − area of triangle ABC

Also area $= \dfrac{\text{rad}^2}{2}\left[\dfrac{\pi(\measuredangle ACB°)}{180°} - \sin ACB°\right]$

$\qquad\quad = \dfrac{\text{rad}^2}{2}\left(\measuredangle ACB \text{ in radians} - \sin ACB°\right)$

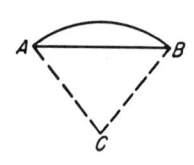

11. Circular Spandrel or Fillet

Area $=$ area of square $- \dfrac{\text{area circle}}{4}$

$\qquad = \text{rad}^2 - \dfrac{\pi(\text{rad})^2}{4}$

$\qquad = 0.2146 \,(\text{rad})^2$

12. Ellipse

Area $= \pi(\text{long rad } AC)(\text{short rad } CE)$

$\qquad = \dfrac{\pi}{4}[\text{long diam } AB][\text{short diam } DE]$

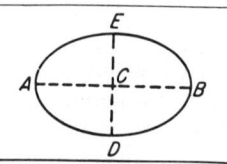

Plane Figures (Cont.)

13. *Parabola*
Rectangular axes, vertex at A
Parabola is tangent to AD at A
Area of segment $ABC = \tfrac{2}{3}bh$
$\qquad\qquad\qquad = \tfrac{2}{3}(AD)(CD)$
Area of spandrel $ADC = \tfrac{1}{3}bh$
$\qquad\qquad\qquad = \tfrac{1}{3}(AD)(DC)$

14. *Parabola*
Rectangular axes, vertex unknown
Parabola is tangent to inclined line AB at A
Area segment $ABC = \tfrac{2}{3}bh$
$\qquad\qquad\qquad = \tfrac{2}{3}(AE)(DC)$
Area spandrel $ADC = \tfrac{1}{3}bh$
$\qquad\qquad D = \tfrac{1}{3}(AE)(DC)$

Solids

15. *Rectangular Prism*
Volume = (area of base) (altitude)

16. *Any Prism*
Axis either inclined or perpendicular to base

Volume = (area of base) $\begin{pmatrix}\text{perpendicular} \\ \text{height}\end{pmatrix}$

or

$\quad = \begin{pmatrix}\text{area of perpen-} \\ \text{dicular cross section}\end{pmatrix}\begin{pmatrix}\text{lateral} \\ \text{length}\end{pmatrix}$

17. *Truncated Prism*
Volume = area of base multiplied by perpendicular distance from base to center of gravity of opposite side

18. *Cylinder*
Axis perpendicular or inclined to base

Volume = (area of base) $\begin{pmatrix}\text{perpendicular} \\ \text{height}\end{pmatrix}$

Also when axis is inclined to base,
Volume = (area of section perpendicular to axis) (length of axis)
Area of cylindrical surface = (perimeter of base) (perpendicular height)

Plane Figures (Cont.)

19. *Pyramid*
Axis either inclined or perpendicular to base
Volume = $\frac{1}{3}$(area of base) $\begin{pmatrix}\text{perpendicular}\\\text{height}\end{pmatrix}$

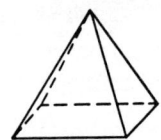

20. *Cone*
Axis either inclined or perpendicular to base
Volume = $\frac{1}{3}$(area of base) $\begin{pmatrix}\text{perpendicular}\\\text{height}\end{pmatrix}$
Right cone, area of conical surface
= $\frac{1}{2}$(circumference of base) (slant height)

21. *Frustum of Pyramid or Cone*
  Ends parallel
Volume = $\frac{1}{3}$(perpendicular height)
  [area of base + area of top
    + $\sqrt{\text{(area of base)(area of top)}}$]

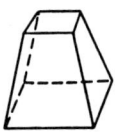

22. *Ungula, slice, of Right Circular Cylinder*
Volume = $\dfrac{\text{height}}{\text{rad} \pm b} [\frac{2}{3}a^3 \pm (b)\text{(area of base)}]$
Use + when base is larger than a half-circle, and − when less.
Area of cylindrical surface
  = $\dfrac{\text{height}}{\text{rad} \pm b} [2a(\text{rad}) \pm (b)\text{ (arc } ABC)]$
Use ± same as in volume

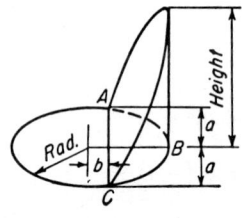

23. *Sphere*
Volume = $\dfrac{4\pi(\text{rad})^3}{3} = \dfrac{\pi(\text{diam})^3}{6}$
Area of surface = $4\pi(\text{rad})^2$
  = $\pi(\text{diam})^2$

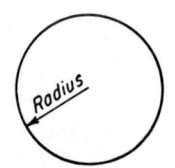

Plane Figures (Cont.)

24. *Spherical Sector*

$$\text{Volume} = \frac{2\pi(\text{rad})^2(h)}{3}$$

Area of surface = area of conical surface + area of spherical segment (see spherical segment)

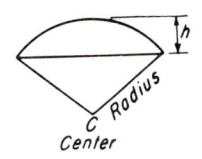

25. *Spherical Segment*

$$\text{Volume} = \pi h^2 \left(\text{rad} - \frac{h}{3}\right)$$
$$= \pi h \left(\frac{\text{chord}^2}{8} + \frac{h^2}{6}\right)$$

Area of spherical surface
$$= 2\pi(\text{rad})h = \pi \left(\frac{\text{chord}^2}{4} + h^2\right)$$

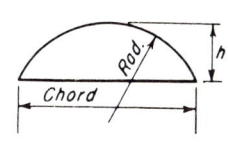

26. *Circular Ring or Torus*
Volume
$$= 2\pi^2 \,(\text{rad of section, } r)^2 \,(\text{mean rad, } R)$$
$$= \frac{\pi^2}{4} \,(\text{diam of section})^2 \,(\text{mean diam})$$

Area of surface
$$= 4\pi^2(\text{mean rad}) \,(\text{rad of section})$$
$$= \pi^2(\text{mean diam}) \,(\text{diam of section})$$

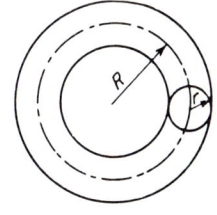

27. *Solids of Revolution* (Theorem of Pappus)

A plane area revolved about an axis in its own plane generates a solid.

Its volume equals the plane area times the length of path followed by the centroid of the area.

Its surface area equals the perimeter of the generating area times the length of path followed by the centroid of the perimeter.

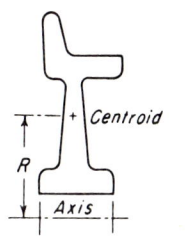

28. *Comparison of Volumes*
Cone, Paraboloid, Sphere, Cylinder having same base diameter and same height
Volumes have following comparative values:

Cone	Paraboloid	Sphere	Cylinder
$\frac{1}{3}$	$\frac{1}{2}$	$\frac{2}{3}$	

Cube would be $\frac{4}{\pi}$ (vol of cyl)

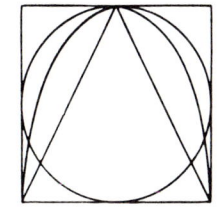

American Standard Graphic Symbols for Piping and Heating*

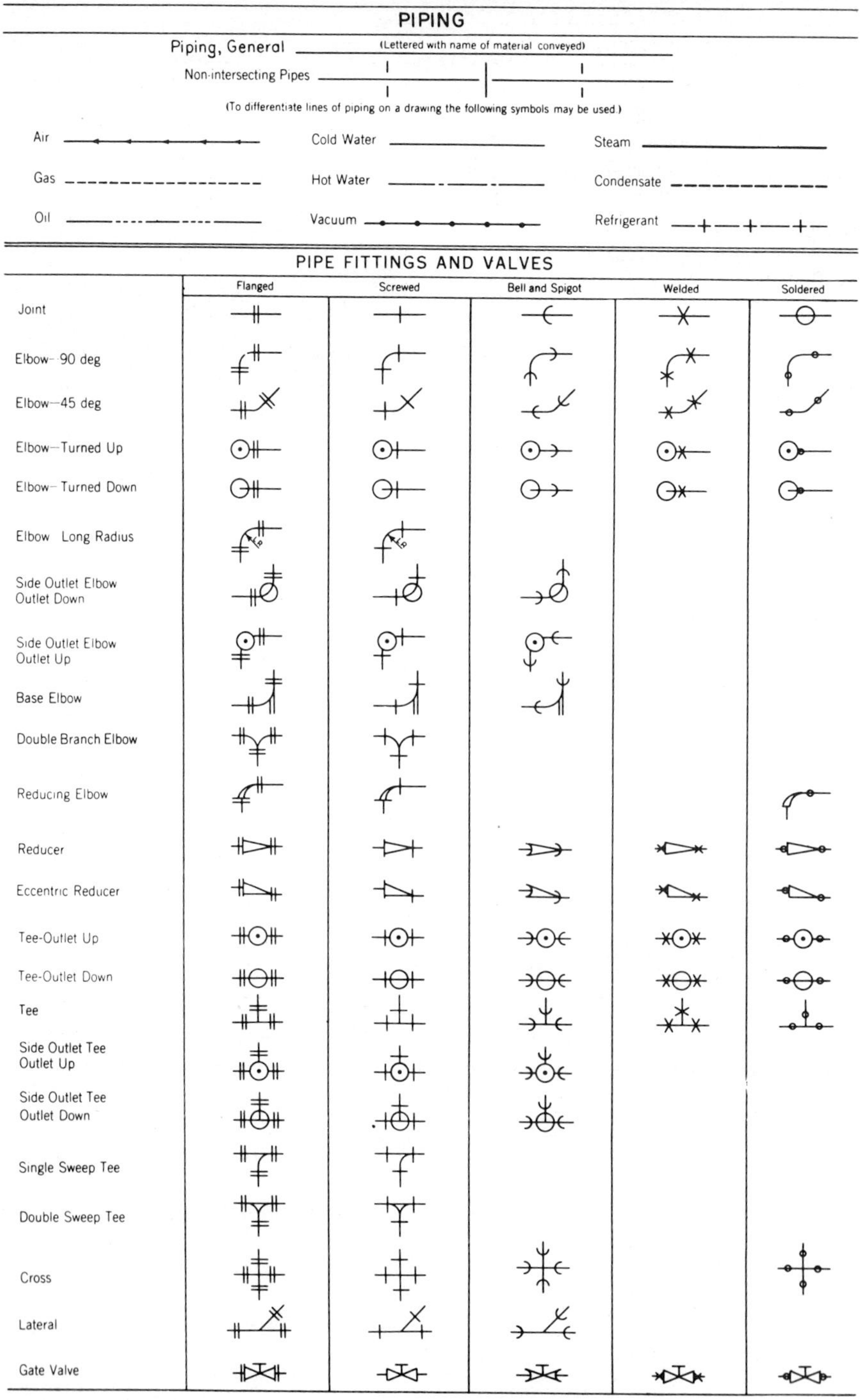

PIPING

Piping, General — (Lettered with name of material conveyed)

Non-intersecting Pipes

(To differentiate lines of piping on a drawing the following symbols may be used.)

- Air
- Gas
- Oil
- Cold Water
- Hot Water
- Vacuum
- Steam
- Condensate
- Refrigerant

PIPE FITTINGS AND VALVES

	Flanged	Screwed	Bell and Spigot	Welded	Soldered
Joint					
Elbow—90 deg					
Elbow—45 deg					
Elbow—Turned Up					
Elbow—Turned Down					
Elbow Long Radius					
Side Outlet Elbow Outlet Down					
Side Outlet Elbow Outlet Up					
Base Elbow					
Double Branch Elbow					
Reducing Elbow					
Reducer					
Eccentric Reducer					
Tee-Outlet Up					
Tee-Outlet Down					
Tee					
Side Outlet Tee Outlet Up					
Side Outlet Tee Outlet Down					
Single Sweep Tee					
Double Sweep Tee					
Cross					
Lateral					
Gate Valve					

* ASA Z14.2—1935.

Symbols for Piping and Heating (Cont.)

PIPING					
	Flanged	Screwed	Bell and Spigot	Welded	Soldered
Globe Valve					
Angle Globe Valve					
Angle Gate Valve					
Check Valve					
Angle Check Valve					
Stop Cock					
Safety Valve					
Quick Opening Valve					
Float Operating Valve					
Motor Operated Gate Valve					
Motor Operated Globe Valve					
Expansion Joint Flanged					
Reducing Flange					
Union	(See Joint)				
Sleeve					
Bushing					

HEATING AND VENTILATING

Lock and Shield Valve		Tube Radiator	(Plan) (Elev.)	Exhaust Duct, Section	
Reducing Valve		Wall Radiator	(Plan) (Elev.)	Butterfly Damper	(Plan or Elev.) (Elev. or Plan)
Diaphragm Valve		Pipe Coil	(Plan) (Elev.)	Deflecting Damper Rectangular Pipe	
Thermostat	(T)			Vanes	
Radiator Trap	(Plan) (Elev.)	Indirect Radiator	(Plan) (Elev.)	Air Supply Outlet	
		Supply Duct, Section		Exhaust Inlet	

INDEX

A

American National Standards Institute, 31, 63
  symbols for piping, heating, and ventilating, 308
Ampersand, 36
Angle(s), definition of, 41
  dimensioning of, 279
  divided into parts, 48
  Greek letters denoting, 46
  between planes, 79
  types of, 41
Arc, 43
  dimensioning of, 280, 284
  length of, 50
Arrowheads, 276
Artgum, 9
  use of, 30, 32
Auxiliary views, 71–83
  functions of, 76
  oblique, 74

B

Bead, 132
Boot or takeoff, 218
Branch fittings, 229–244
  clustered, 239, 240
  design factors, 231–233, 236–238
Breeching, Tees, Ys, 122
Brush, dusting, 18

C

Calculator, 17
Circle, 43, 56
  dimensions for, 287
Circle, dividing into parts, 47
  formulas for, 304
Cloth, glass, 12
  tracing, 12
Compass, 14
  beam, 15, 25
  use of, 23, 25
Cone-to-cone intersections, 165–173, 265
Cone-to-pyramid intersection, 189
Cones, 45, 144
  development of, 143–149
  dimensions for, 281, 284, 287
  formulas for, 306
  frustum, 46, 149–151
    with horizontal takeoff, 254
    with vertical takeoff, 257
  with horizontal duct intersection, 261
  oblique, 154, 229
  truncated, 153
  with vertical duct intersection, 260
  warped, 46, 211, 264–265
Contour principle, 287
Cornice, 131
  square return, 133
    with reduced miter, 134
Cove, 132
Cube, 45
Curves, adjustable, 17
  dimensioning of, 284
  irregular or French, 17
  use of, 28
Cutting-plane method, 125, 128, 165–173, 185–187
Cutting-sphere method, 165, 173
Cyclone collectors, 149, 264
Cylinder, 45, 67
  cut by plane, 125
  development of, 98–101
  dimensions for, 281, 284, 287

Cylinder, formulas for, 305
   and hopper, 257–260
   length of metal needed, 50
Cylindroid, 208

D

Decimal equivalents, 290
Design, 2, 203
Design factors for branch fittings, 231–233, 235–240
   in curved duct elbows, 108
   for offsets, 113
Development, gore method, 173–178
   parallel-line, 97–136
   radial-line, 143–191
   triangulation, 201
   zone method, 175–178
Diagram, true length, 92, 204, 206, 209, 211
Dimensioning, 274–289
   according to contour principle, 287
   aligned and unidirectional, 277
   location, 282
   size, 280
Dividers, 15
   bow, 15
   proportional, 16
   using, 23
Drafting machine, 6
Draftsman, characteristics of, 28
   cleanliness of, 28
   drawing procedure, 26
   function of, 2
   left-handed, 19, 33, 36
   pride in lettering, 40
   true-length diagram, 92, 204, 206, 209, 211
Drawing board, 5
"Ducks," 17

E

Edge-view method, 124, 125n., 129
Elbows, 101

Elbows, conical reducing, 156, 250
   duct, 106–112, 227
   of three or more pieces, 104
   transitional, 107, 218–227
   of two pieces, 104
   warped cone, 264
Elements, 46, 201
Ellipse, 43, 56
   construction of, 54
   formulas for, 304
Equipment, drafting, 4
   cleanliness of, 28
   placement, 19
Erasers, 9
   electric, 30
   use of, 30
Extension lines, 26, 275

F

Fillets, 132
First-angle projection, 63n.
Flute, 132
Formulas, bend allowance, 285
   circle and arc, 50, 285, 304
   cone stretchout, 145–151
   cylinder, 305
   elbow, 105
   offset, 119n., 121n.
   parabola, 305
   pipe cross-section, 232
   polygon, 43, 303
   prism, 305
   pyramid, 306
   sphere, 175, 306

G

Geometric constructions, 41–46
   offsets, 113–122
   plane figures, 303–305
   solids, 44, 305–307
Gore or polycylindric methods, 173–175
Gutters, 131
   ogee eave, 132

H

Hexagon, 53
  formulas for, 303
Hopper and cylinder, 257–260
Hyperbola, 57, 165

I

Instruments, drawing, 13
    bow, 24

L

Leaders for notes in dimensioning, 278
Lettering, 31
  capital, 32
  for left-handers, 36
  lower-case, 38
  numbers and fractions, 36
  pencil for, 31
  position of hand for, 32
  spacing of, 39
Lines, characteristics of, 26
  definition of, 41
  in dimensioning, 275
  dividing into parts, 46
  frontal, 65
  guide, for lettering, 32
  horizontal, 20, 64
  inclined, 22
  of intersection, 124, 161–173, 185–191, 252–265
  oblique, 65
  parallel, construction of, 46
  perpendicular, construction of, 46
  point view of, 78
  position of, in space, 64
  profile, 65
  stretchout, 99, 131
  symbols, 25
  tangent to circle, 49
  tangent to two arcs, 49
  true length of, 76
    by rotation, 91
  types of, 26, 275
  vertical, 21

Location dimensions, 282–284
  according to contour principle, 287

N

Notes in dimensioning, 278

O

Oblong, 58
Offsets, 113
  compound, 120
  curved, 118
  cylindroid, 208
  "run," 114, 119
  tapered, 154
  three-piece, 114
  transition, 215
Ogee, 58, 132
Orthographic projection, 60–67
Oval, 43
Ovolo, 132

P

Paper, drawing, 12
  fastening to board, 19
  tracing, 12
Parabola, 57, 163
  formulas for, 305
Paraboloid, 173
Parallel devices, 5
Patterns, branch fitting, 229–244
  cone: frustum, 252–265
    oblique, 154
    truncated, 153
  cornice, 131–136
  cover: rectangular, 184
    round, 148
  cyclone separator, 149
  cylindroid, 208
  elbow, 218–227, 250
    conical reducing, 156
    rectangular to round, 218, 224
  fishtail, 106
  flop, 101

Patterns, offset, 118
  ogee eave gutter, 132
  paraboloid, 173–178
  pyramid, 178–184
  sphere, 173–178
  tee, 126–129
  tee reducer, 249
  transition: round to elliptical, 211
    round to oblong, 130
    square to hexagonal, 207
    square to round, 211–215
  twist, 204–207
  Y breeching, 129
Pen, ruling, 15
Pencil, 8
  drawing, procedure, 26
  for lettering, 31
  refill, 8, 20
  selection of, 19
  sharpeners, 8
  sharpening of, 20
Pipe, areas, 231
  formulas for, 303–305
  girth, 67, 96, 103
  openings and shapes, 44
Plane, auxiliary, 71
    oblique, 74
  cutting, 125
  definition of, 41
  edge view of, 79
  figures, 41, 303
  H, V, and P, 61
  intersection with cone, 56
  position in space, 65
  of projection, 60
  true size and shape of, 79
    by rotation, 93
Plate, bend allowance for, 285
  definition of, 1
Polyconic method, 175–178
Polycylindric method, 173–175
Polygon, 43, 303
Prisms, 44, 99
  cut by plane, 125
  dimensions for, 280, 282, 287
  formulas for, 305

Profile, dimensioning of, 284
  plane (P), 61
Projections, first-angle, 63n.
  third-angle, 61
Protractor, 17
Pyramid, 45
  compared to cone, 178
  dimensions for, 281, 287
  formulas for, 306
  low-altitude, 184
  oblique, 184
  truncated, 181
  with vertical takeoff, 257
Pyramid-to-cone intersection, 189
Pyramid-to-pyramid intersection, 185–189

Q

Quadrilaterals, definition and types of, 42
  formulas for, 303

R

Reciprocals of numbers, 298–302
Roof collar, 153, 161
Rotation, 91
  true length of line by, 91
  true size and shape of surface by, 93

S

Scales, 9
  architect's, 10
  mechanical engineer's, 11
  SI (metric), 10, 278
Seams, 105, 156
Sheet metal, definition of, 1
  dimensioning of, drawing, 277
  gages, 310
  working drawings, 275
SI measurements, 10, 278, 290
Size description, 274, 280
  according to contour principle, 287
Sketch pad, 18
Soffits, 132
Sphere, 173–178

Spheres, tangent for conical elbow, 159
Spline, 17
Square roots of numbers, 298–302
Squares of numbers, 298–302
Staples, wire, 12
    use of, 19
Surfaces, 173
Symbols, 46
    American National Standards Institute, for piping, heating, and ventilating, 308

T

T square, 5
Take-off or boot, 218
    reducer, 262
Tape, drafting, 13
    use of, 19
Tees, 122
    hopper and cylinder, 257–260
    90-degree, 126
    pitched, 128
    reducer, 249
Third-angle projection, 61
Thumbtacks, 13
    use of, 19
Tracing cloth, 12
Tracing paper, 12
Transitions, elbow, 107, 218–224
    rectangular to round 90-degree, 224
    round to elliptical, 211, 226
    round to oblong, 130, 214
    square of hexagonal, 207
    square to round, 211–215
Triangles, 6, 303
    combining of, 22
    definition of, 41

Triangles, enlargment or reduction with, 53
    equilateral, drawing, 51
    45-degree, 22
    kinds of, 42
    30- to 60-degree, 22
    transfer of, by triangulation, 51, 201
Triangulation, 51, 201
Trigonometric functions, 145
    tables of, 292–297
Twists, 204–206

V

Views, 60
    auxiliary, 71–83
        oblique, 74
    choice of, 66
    essential: for parallel-line development, 97, 101
        for radial line development, 144
    need for special, 67
    principal, 61

W

Working drawings, 275

Y

Ys, 122
    branch, 229
    development of, 129
    three-prong, 240

Z

Zone method, 175–178